THE NINTH TORINO WORKSHOP ON EVOLUTION AND NUCLEOSYNTHESIS IN AGB STARS

AND

THE SECOND PERUGIA WORKSHOP ON NUCLEAR ASTROPHYSICS

THE NINTH TORINO WORKSHOP ON EVOLUTION AND NUCLEOSYNTHESIS IN AGB STARS and THE SECOND PERUGIA WORKSHOP ON NUCLEAR ASTROPHYSICS

Held in Perugia, Italy 21 - 27 October 2007

Conferences organized in the framework of the celebrations for the 7th Centenary of the Perugia University

1308 2008 Centenario Università degli Studi di Perugia

Scientific Advisory Board:
C. Abia, M. Busso (Chair), C. Chiosi, F. D'Antona, R. Gallino, B. Gustafsson, J. C. Lattanzio, O. Straniero, C. Tout, H. Van Winckel, and E. Zinner

Local Organizing Committee:
R. Guandalini, S. Palmerini, M. Busso, and G. Chiocci

Sponsored by: The University of Perugia, INFN (Perugia), INAF (Teramo), and The University of Granada

Edited by: Roald Guandalini, Sara Palmerini, and Maurizio Busso

All papers have been peer reviewed

AMERICAN INSTITUTE OF PHYSICS

AIP CONFERENCE PROCEEDINGS ■ 1001

Editors:

Roald Guandalini
Sara Palmerini
Maurizio Busso

Department of Physics
University of Perugia
via A. Pascoli
I-06123 Perugia
Italy

E-mail: guandalini@fisica.unipg.it
palmerini@fisica.unipg.it
busso@fisica.unipg.it

L.C. Catalog Card No. 2008924313
ISBN 978-0-7354-0520-2
ISSN 0094-243X
Printed in the United States of America

CONTENTS

SESSION IV:
NUCLEOSYNTHESIS COMPUTATIONS

SESSION VII:
SOLAR SYSTEM MATERIALS, PRESOLAR GRAINS, AND NUCLEOSYNTHESIS

SESSION VIII:
PROBLEMS OF STELLAR MIXING

SESSION IX:
CIRCUMSTELLAR ENVELOPES AND MASS LOSS

FOREWORD

This book contains the lectures given at the joint meeting "Nineth Torino Workshop on Evolution and Nucleosynthesis in AGB Stars" and "Second Perugia Workshop in Nuclear Astrophysics", held together in Perugia (Italy) from October 21 to October 26, 2007.

This series of workshops was started some 13 years ago in Torino by R. Gallino and myself and was since then so lucky to succeed in maintaining its name and periodicity, while moving its site around the world. Thus, after three meetings in Torino, we then had one conference in each of the following towns: Rome, Grenoble, Melbourne, Cambridge, Granada, Perugia. We are now waiting for the tenth encounter, to be held in early 2010 in Christchurch, New Zealand. This site in the Antipods further confirms the wide international network of collaboration and research that has formed, in order to understand the many puzzling and stimulating problems still posed by stellar physics in general, and by the evolution and nucleosynthesis of AGB stars in particular.

In the present book the fields covered by the lecturers are quite wide: the joining of the "Torino Workshops", strictly focused on AGB stars, and of the "Perugia Workshops" dedicated to a broader view of Nuclear Astrophysics, resulted in a coordinated, but widely interdisciplinary discussion, where AGB nucleosynthesis could be integrated by complementary issues concerning nuclear processes in massive stars, while the observational sessions usually dominated by spectroscopic results on stellar atmospheres and laboratory measurements on solar and pre-solar system materials could be complemented by the rich information now coming from space-borne infrared photometry.

Among the topics discussed at the meeting in passionate but friendly exchanges, a special place was occupied by the models of non-conventional mixing in evolved stars, a field in which the last three years have seen a wealth of new ideas and proposals presented in the literature. Double diffusive processes like the thermohaline mixing scheme, or faster matter transport possibilities along magnetic flux tubes seem to reopen (with better hopes for finding solutions than in the past) the so far rather frustrating attempts at explaining the origins of subtle abundance anomalies, not accounted for by purely convective motions. We are all very curious to see how much all these ideas will be evolved, by the time of the Christchurch meeting.

A special mention is probably needed for the social activities. It is a tradition of the Torino Workshops to stress that science offers wonderful chances, for people interested in open-minded and unprejudiced exchanges of opinions, to meet and develop a common, collaborative approach to the study of Nature. This is strengthened in the moments free from the official scientific sessions, because we verified that very often it is during the conference tour, at the conference dinner, or strolling around in a relaxed way, that many fruitful collaborations are started and good ideas are born. It is our hope that our efforts to let our guests and friends experience together a few of the appealing features of Central Italy (from both the gastronomic and cultural point of view) may have added something special also to the taste of science, which remains our main interest and commitment.

Maurizio Busso
(on behalf of the organizing committe and of the editors)

SESSION I:

PHYSICAL MODELS FOR THE EVOLUTION OF STARS

Progresses in AGB Modelling

S. Cristallo*, O. Straniero*, R. Gallino†, M.T. Lederer**, L. Piersanti* and I. Domínguez‡

*Osservatorio Astronomico di Teramo (INAF), via M. Maggini 64100 Teramo, Italy
†Dipartimento di Fisica Generale, Universitá di Torino e Sezione INFN di Torino, via P. Giuria 1, 10125 Torino, Italy
**Institut für Astronomie, Türkenschanzstraße 17, A-1180 Wien, Austria
‡Departimento Fisica Teorica y del Cosmos, Universidad de Granada, 18071 Granada, Spain

Abstract. The full understanding of final stellar evolution phases is a fundamental request to properly investigate the Universe at any temporal and spatial scale. While the theoretical scenarios of H- and He-burning have been deeply investigated in the last 30 years, the modelling of stellar evolution beyond the core-helium burning phase and the related nucleosynthesis still present problems related to the physics and to the numerical methods. Low mass AGB Stars ($1<M/M_{\odot}<3$) are among the most important polluters of the Milky Way, because of the strong winds eroding their chemically enriched envelopes. In the AGB phase, the material processed during the alternating series of H and He burnings is recurrently mixed within the convective zones generated by Thermal Pulses and then partially mixed with the surface by the so called Third Dredge Up episodes. As a matter of fact, MS, S, C(N) and some post-AGB stars are enriched in C and s-process elements and the spectroscopic detection of unstable Tc demonstrates that the s process is at work in these stars [1]. Within this context, we started a longstanding program devoted to the study of the physical and chemical evolution of low mass AGB stars by means on the FRANEC stellar evolutionary code [2, 3].
Among the most important results we achieved, we highlight the explanation of the mechanism of neutron production, which controls the s-process nucleosynthesis in low-mass AGB, the coupling of the code with a full nuclear network, from H up to Pb-Bi (at the termination point of the s-process path) and the introduction of C-enhanced low temperature opacities, whose effects are particularly important at low metallicities. In this paper, we describe the current state of the art of our modelling and we present detailed models of low mass AGB ($M=2M_{\odot}$) stars at different metallicities (from $Z=Z_{\odot}$ to $Z=10^{-4}$). The final elemental distributions we find are representative of the ones expected for the intrinsic carbon stars observed in the disk and in the halo of the Milky Way.

Keywords: Nucleosynthesis; AGB stars; C-stars
PACS: 97.10.Cv - 97.10.Tk - 97.30.Hk

1. INTRODUCTION

After the exhaustion of central Helium, low and intermediate mass stars approach the Asymptotic Giant Branch (AGB) phase. The structure of an AGB star consists of a partially degenerate CO core, a thin Helium mantel (He-intershell), separating the H- and He-burning shells, and an H-rich, largely convective, envelope. After a first period, called early-AGB, the He burning dies down and the H burning, whose ashes increase the mass of the He-rich instershell, supplies the total luminosity of the star. Then, when the mass of the intershell approaches a certain critical value, a thermonuclear runaway occurs, driven by the 3-α reactions, the so called thermal pulse (TP). Due to the sudden release of thermonuclear energy, convection arises in the He-intershell, redistributing the large energy excess. The material above the He-burning region is

CP1001, 9th Torino Workshop on Evolution and Nucleosynthesis in AGB Stars and the 2nd Perugia Workshop on Nuclear Astrophysics, edited by R. Guandalini, S. Palmerini, and M. Busso

pushed forward and it is cooled down, so that the H burning switches off. Meanwhile, the He burning becomes quiescent and the convective zone generated by the TP disappears. In this phase, the external convective envelope deeply penetrates in zones that have been previously enriched with the ashes of the H and He burning (Third Dredge Up, TDU). A contraction phase follows, when the residual He buffer becomes so thin that He burning switches off. Then, the temperature at the base of the H-rich envelope increases again and the H burning resumes. This sequence of events is repeated, until the mass loss erodes the (almost) whole envelope and the star leaves the AGB. The duration of the AGB, the strength of the pulses, the efficiency of the TDUs are few examples of quantities affected by the mass loss [4]. The final surface composition is enriched of heavy elements, particularly carbon and s-elements [5, 6, 3].
In Section 1, we describe the most recent features of our code, highlighting in particular how we treat the convection and the adopted opacity treatment. In Section 2, we present new detailed low mass (M=2$M_\odot$) AGB models at different metallicities: Z=1.38×10^{-2} (solar metallicity), Z=6.0×10^{-3}, Z=3.0×10^{-3}, Z=1.0×10^{-3} and Z=1.0×10^{-4}, which almost entirely span the metallicity range of our Galaxy. Finally, we spend a few sentences on very low metallicity models.

2. THE FRANEC CODE

The code used to perform the present calculations is the FRANEC stellar evolutionary code [2, 3]. The borders of the convective regions are defined by the Schwarzschild criterion, while the convective velocities are calculated by means of the mixing length theory. We adopt a revised version of the time-dependent mixing first introduced by [7], in which the degree of mixing is linearly dependent on the ratio between the time step and the mixing time. We derive the value of the mixing length parameter $\alpha = 2.15$ by calculating a Standard Solar Model [8]. We evaluate the AGB mass loss by means of an empirical relation between mass loss and period (see [3] for a description of the adopted mass loss).

2.1. Low Temperature Molecular Opacities

In a stellar model, the determination of the opacity coefficients may results computationally not feasible because of their frequency dependence, so it is often useful to use pre-tabulated mean opacity tables. Tables of radiative opacities have been derived by means of the web facility provided by the OPAL group (http://www-phys.llnl.gov/Research/OPAL/opal.html). These opacity tables extend down to $\log T$ = 3.75: for lower temperatures we refer to [9]. At high density, as in the case of the CO core of an AGB stars, degenerate electrons dominate the energy transfer: in this regime we use the conductive opacities derived by [10]. In principle, a stellar evolutionary code should account for the variations of the envelope chemical composition. In practice, only variations of the main constituents (H and He) are usually considered. Nevertheless, in the envelope of an AGB star, both C dredge-up and the possible conversion of C into N (by CN cycle) substantially affect the atomic and the molecular contribution to the radiative opacity. For temperatures lower than 4000 K, well below the lower limit for the OPAL tables, this contribution becomes very important and affects the estimated stellar radius, temperature and, in turn, the mass loss rate. Note that at low metallicity, the C enhancement may be as large as a factor of 1000, while the iron content remains practically untouched. Solar-scaled low temperatures opacities are therefore inadequate to properly calculate AGB models. In order to overcome this limitation, we calculate new low temperature C- and N-enhanced opacity tables, suitable for AGB model calculations [11]. In the following, we simply report the C-enhancement factor used in the models (F_i in Table 1). Note that, changing the metallicity, different C- and N-enhancements have been used, because of the different C/O ratios attained in the envelope.

3. THE MODELS

In this Section we compare evolution and nucleosynthesis of low mass AGB models with different metallicity.

As an example, the three lines in in Fig.1 illustrate the evolution of the positions, in mass coordinates, of (top to bottom): the inner border of the convective envelope, the maximum nuclear energy production within the H-burning shell and the maximum nuclear energy production within the He-intershell, for the solar metallicity model. The squared spikes in the latter line marks the activation of the $^{13}C(\alpha,n)^{16}O$ reaction. The calculations have been stopped when the dredge up ceases, because the mass-loss has already consumed most of the envelope. In the central column of Table 2 we report the amount of material cumulatively dredged up during the AGB at different metallicities. A maximum is attained at $Z=1.0\times10^{-3}$. It is the result of the complex interplay between the efficiency of the H burning and the opacity of the envelope. A less efficient H burning, as indeed occurs reducing the metallicity, favors a deeper dredge up [12]. On the contrary, a lower metallicity implies a reduced envelope opacity and, in turn, a smaller dredge up. This scenario is, however, complicated by the enhancement of carbon caused by the dredge up itself. At the lowest metallicity, this primary carbon dominates over the original composition on both the efficiency of the H burning and the envelope opacity. The differences in the dredge up of the models at different metallicity is even more evident in Fig.2. For each evolutionary sequence, the amount of mass that is dredged up initially increases, because the core mass increases, and then decreases, as a consequence

TABLE 1. C- and N-enhancement for low temperature molecular opacities at various metallicities.

Z	F_1	F_2	F_3	F_4	F_5
1.38×10^{-2}	1	1.5	1.8	2.2	4
6.0×10^{-3}	1	2	5	10	50
3.0×10^{-3}	1	2	5	10	50
1.0×10^{-3}	1	5	10	50	200
1.0×10^{-4}	1	10	100	500	2000

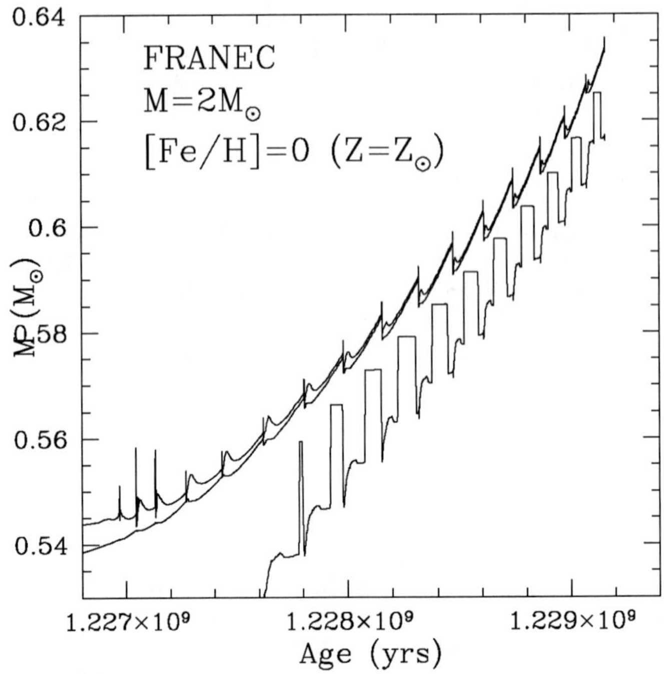

FIGURE 1. Evolution of the positions in mass of the inner border of (top to bottom): convective envelope, H-burning shell and most energetic mesh of the He-burning shell, during the thermally pulsing AGB phase of the solar metallicity model. See text for details.

of the significant reduction of the envelope mass caused by the huge mass loss occurring when the star approaches the AGB tip.

It is also interesting to note that thermal pulses are hotter at low metallicity. As shown in the third column of Table 2, the maximum temperature attained at the base of the

TABLE 2. Cumulative dredged up material and maximum temperature attained at the base of the TPs during the AGB phase of a star with M=2M$_\odot$ at different metallicities.

Z	$\mathbf{M}_{TDU}^{tot}$	$\mathbf{T}_{MAX}^{TP}$
1.38×10^{-2}	3.62×10^{-2}	2.92×10^{8}
6.0×10^{-3}	7.27×10^{-2}	3.00×10^{8}
3.0×10^{-3}	8.95×10^{-2}	3.06×10^{8}
1.0×10^{-3}	1.31×10^{-1}	3.19×10^{8}
1.0×10^{-4}	9.54×10^{-2}	3.24×10^{8}

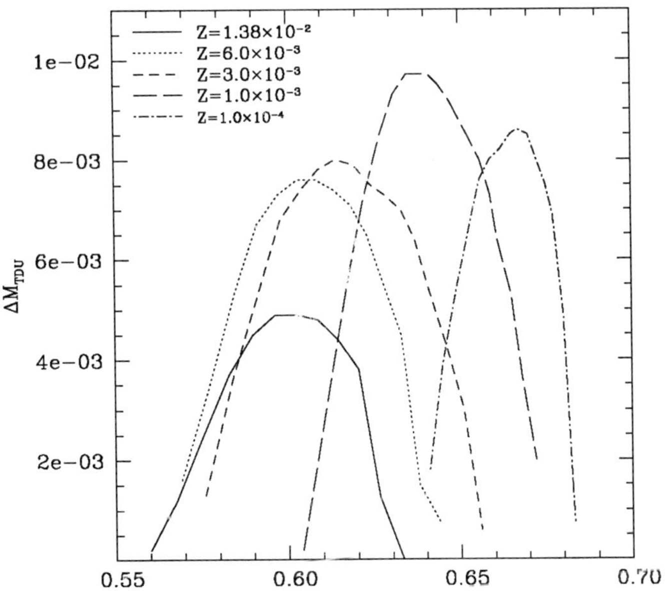

FIGURE 2. Per pulse dredged up material as a function of the core mass at different metallicities.

convective zones generated by TPs monotonically grows with decreasing the metallicity, making the nucleosynthesis induced by the activation of the $^{22}Ne(\alpha,n)^{25}Mg$ neutron source more important. A detailed analysis of these problems will be presented in a forthcoming paper (Cristallo et al, *in preparation*).

In Fig.3 we report the surface enrichments, in the usual spectroscopic notation[1]: for each models we plot the [El/Fe] after the last pulse with TDU. A strong evidence resulting from the models is that the s-process distribution is changing depending on the metallicity, as predicted by [13]. We find that, at solar metallicity, the abundance of light s elements (*ls*) at the first s peak (Sr, Y and Zr) is comparable with that of the heavy s elements (*hs*) at the second s peak (Ba, La, Ce, Pr, and Nd). Lead is under-produced with respect to barium. When C/O=1 in the envelope, the [hs/ls] index (representative of the mean neutron exposure that characterizes the s-process distribution) is in very good agreement with those measured in galactic C(N) stars [14]. On the other hand, at lower metallicities the higher neutrons to seeds ratio favors the production of heavier elements. Therefore, lowering the metallicity below the solar value, the chemical pattern results more peaked around the *hs* elements (for Z$\sim$ 3÷6 $\times 10^{-3}$) and then, further decreasing

[1] [El/Fe]=log(N(El)/N(Fe))−log(N(El)/N(Fe))$_\odot$; the label El stands for the generic element.

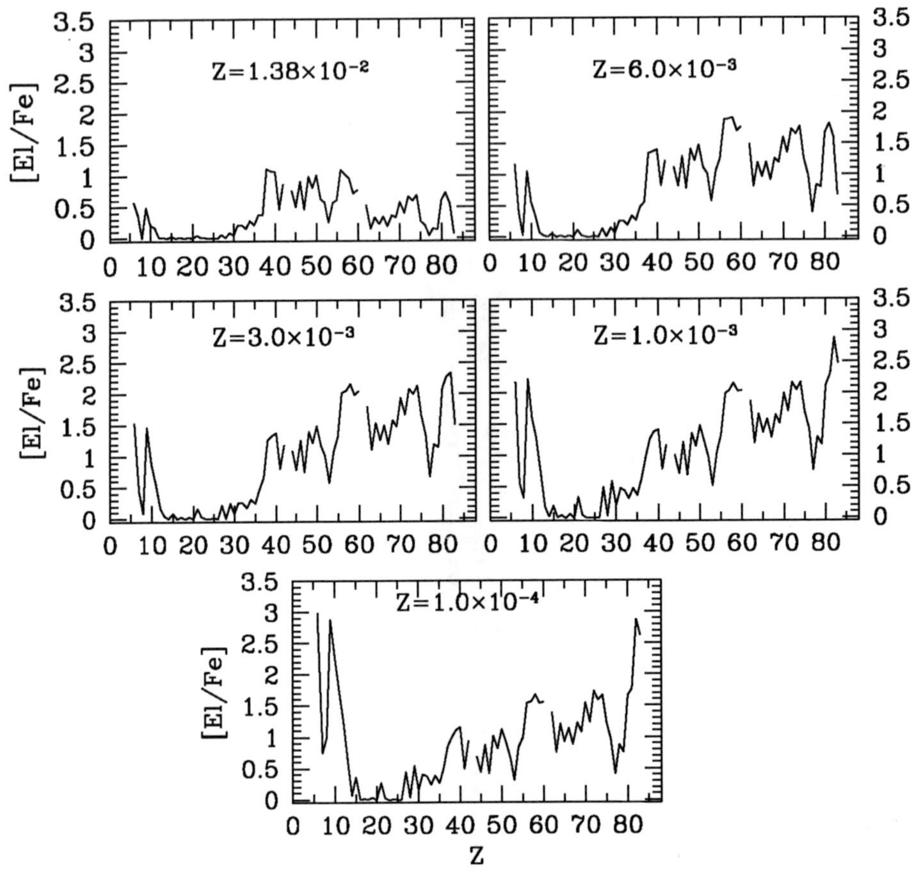

FIGURE 3. Final surface distribution at different metallicities.

the metallicity, the s process path favors the production of ^{208}Pb (for Z< 1 $\times 10^{-3}$). It is worth to note that, for the first time, it is possible to simultaneously compare physical properties (like luminosities, surface temperatures, etc) and chemical properties (light and heavy element abundances) of the models to the corresponding stellar properties.

Let us finally stress an important limit of low metallicity models. Our knowledge of the AGB mass loss comes from observation of Galactic and Magellanic Clouds stars For this reason, extant observational studies can not help in the parameterizations of this phenomenon at low metallicity, even if some hints could come from observations in dwarf Spheroidal galaxies [15, 16]. It emerges that the gas to dust ratio, which is needed to properly estimate the total mass loss, can not be derived from first principles.

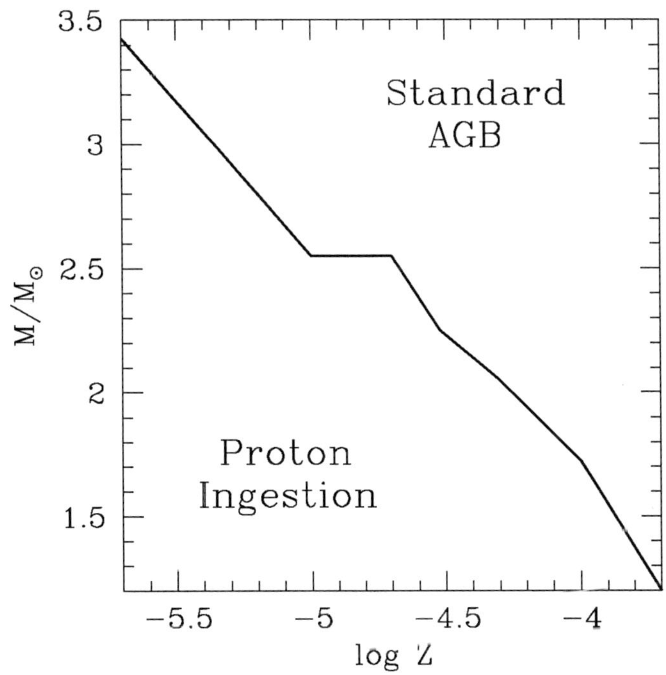

FIGURE 4. In a metallicity-mass diagram, we identify the two regions where AGB models follow a standard evolution (upper region) or experiment proton ingestions at the beginning of the AGB phase (lower region) (Figure from [11]). See text for details.

Due to this lack of information, we adopt the same mass loss prescriptions as for the solar metallicity. We have estimated that at $Z=10^{-4}$ this assumption could result in an uncertainty of a factor of 2 in the predicted yields (Cristallo et al., *in preparation*).

3.1. Very Low Metallicity Models

Finally, when modelling very metal poor AGBs, we found that, for a given metallicity, there exists a lower mass limit (see Fig. 4) for which the "normal" AGB evolution breaks down [11]. Stars with smaller initial mass experience a protons engulfment during the first fully developed TP (see, e.g., [17]), which leads to a peculiar s-process nucleosynthesis, a low $^{12}C/^{13}C$ and a significant synthesis of primary N. Owing to the low entropy barrier, protons are engulfed within the He-intershell, where the high temperature and the large C abundance induce a violent H-burning flash. The consequent synthesis of ^{13}C allows the activation of the $^{13}C(\alpha,n)^{16}O$ reactions and a very large neutron density is attained. Thus, the production of neutron-rich nuclei, normally by-passed by a standard s-process, can occur in these stars. Later on, when the intershell cools down, a deep TDU occurs. Since in the envelope the C/O becomes greater than 1, the large N abundance favors the formation of CN molecules, thus inducing a further increase of the radiative opacities. This new set of very low metallicity models, which represents the current frontier of AGB modelling, is still under analysis.

ACKNOWLEDGMENTS

This work has been supported by the PRIN-INAF program 2006.

REFERENCES

1. P.W. Merrill, *ApJ* **116**, 21 (1952)
2. A. Chieffi, M. Limongi, and O. Straniero, *ApJ* **502**, 737 (1998)
3. O. Straniero, R. Gallino, and S. Cristallo, *Nucl. Phys. A* **777**, 311 (2006)
4. O. Straniero, I. Domínguez, S. Cristallo, and R. Gallino, *PASA* **20**, 389 (2003)
5. O. Straniero, R. Gallino, M. Busso, A. Chieffi, M. Limongi, and M. Salaris, *ApJL* **440**, 85 (1995)
6. R. Gallino, C. Arlandini, M. Busso, M., Lugaro, C. Travaglio, O. Straniero, A. Chieffi, and M. Limongi, *ApJ* **497**, 388 (1998)
7. W. M. Sparks and A. S. Endal, *ApJ* **237**, 130 (1980)
8. L. Piersanti, O. Straniero, and S. Cristallo, *A&A* **462**, 1051 (2007)
9. D. R. Alexander and J. W. Ferguson, *ApJ* **437**, 879 (1994)
10. A. Y. Potekhin, *A&A* **351**, 787 (1999)
11. S. Cristallo, O. Straniero, M.T. Lederer, and B. Aringer, *ApJ* **667**, 489 (2007)
12. O. Straniero, M. Limongi, A. Chieffi, I. Domínguez, M. Busso, and R. Gallino, *Mem.S.A.It.* **71**, 719 (2000)
13. M. Busso, R. Gallino, D.L. Lambert, C. Travaglio, and V.V. Smith, *ApJ* **557**, 802 (2001)
14. C. Abia, and 8 coauthors, *ApJ* **579**, 817 (2002)
15. E. Lagadec, A.A. Zijlstra, M. Matsuura, J.W. Menzies, J.Th. van Loon, and P.A. Whitelock, *MNRAS*, in press (2007)
16. M. Matsura, and 12 coauthors, *MNRAS*, in press (2007)
17. N. Iwamoto, T. Kajino, G.J. Mathews, M.Y. Fujimoto, and W. Aoki, *ApJ* **602**, 377 (2004)

Low Temperature Mean Opacities for the Carbon-Rich Regime

Michael T. Lederer* and Bernhard Aringer*,†

* *Department of Astronomy, University of Vienna, Türkenschanzstraße 17, A-1180 Wien, Austria*
† *Dipartimento di Astronomia, Università di Padova, Vicolo dell'Osservatorio 3, 35122 Padova, Italy*

Abstract. Asymptotic Giant Branch (AGB) stars undergo a change in their chemical composition during their evolution. This in turn leads to an alteration of the radiative opacities, especially in the cool layers of the envelope and the atmosphere, where molecules are the dominant opacity sources. A key parameter in this respect is the number ratio of carbon to oxygen atoms (C/O). In terms of low temperature mean opacities, a variation of this parameter usually cannot be followed in stellar evolution models, because up to now tabulated values were only available for scaled solar metal mixtures (with C/O $\simeq$ 0.5). We thus present a set of newly generated tables containing Rosseland mean opacity coefficients covering both the oxygen-rich (C/O $<$ 1) and the carbon-rich (C/O $>$ 1) regime. We compare our values to existing tabular data and investigate the relevant molecular opacity contributors.

Keywords: Molecular Opacities, Stellar Evolution
PACS: 95.30.Ky - 97.10.Cv - 97.10.Ex

1. INTRODUCTION

Low and intermediate mass stars, depending on their actual mass and metallicity, change their surface chemical composition in the AGB phase of their evolution due to a phenomenon that is usually dubbed "third dredge-up" (see [1] for a detailed review). This process denominates a series of so-called thermal pulses (TP), where at the end of each pulse the convective envelope penetrates the sites of nucleosynthesis and the freshly produced elements are brought to the surface of the star. Of particular concern are the ashes of He burning, which is mainly ^{12}C. From the several reasons of why the dredge-up of carbon is of great importance for the AGB evolution, the one being particularly considered here is the change in opacity due to the carbon enhancement. Radiative opacities are provided by the OPAL collaboration [2] as well as the Opacity Project [3]. In both cases the effects of single element abundance variations are covered by the data. Unfortunately they do not include the most important molecular opacity sources but are basically limited to the atomic contributions. In the cool layers (i.e., temperatures lower than about 4000 K) of the convective envelope and the atmosphere of late-type stars, however, molecules become the dominant opacity source.

For the temperature range below 10^4 K, a common choice are the tabulated Rosseland opacities from [4][F05]. This compilation contains only tables for a scaled solar metal composition (plus tables with varied abundances of the alpha elements). As outlined above, the chemical composition of an AGB star changes during its evolution. The amount of molecules formed for certain species depends, beside the local thermody-

CP1001, *9th Torino Workshop on Evolution and Nucleosynthesis in AGB Stars and the 2nd Perugia Workshop on Nuclear Astrophysics,*
edited by R. Guandalini, S. Palmerini, and M. Busso

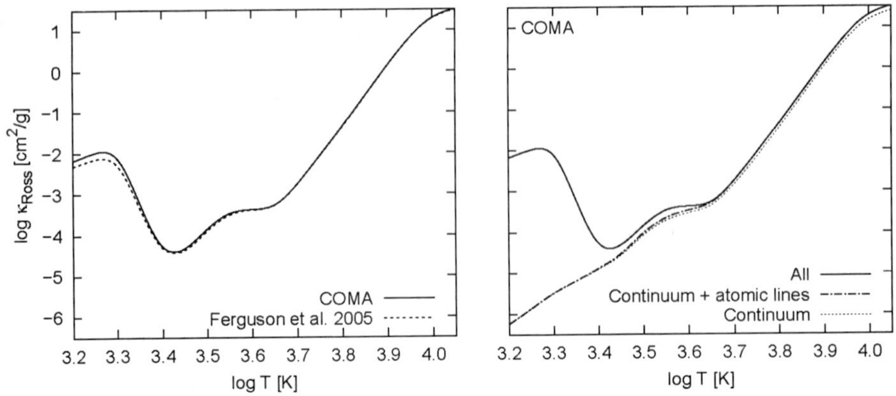

FIGURE 1. Left panel: Comparison of the Rosseland mean opacity calculated with COMA to the values of F05 for solar metallicity ($X = 0.7, Z = 0.02$) at $\log R = -3.0$. Metal abundances are taken from [5]. The data match for high temperatures but do differ at lower temperatures which is mostly due to the different set of molecular line lists used in the particular calculations.
Right panel: Contribution of continuum, atomic (evaluated by omitting all line data and only molecular line data in the calculations, respectively) and molecular sources to the total Rosseland mean opacity (abundances and parameters as given above). Molecules are the dominating opacity source at low temperatures, whereas at higher temperatures the opacity is almost all due to continuum sources. The contribution of atomic line opacity is relatively minor but non-negligible.

namic conditions, crucially on the number ratio of carbon to oxygen atoms (C/O) which is about 0.5 for solar abundances (the exact value depends on the adopted abundance prescription). Despite the fact that the use of the F05 tables substantially underestimates the radiative opacity in the cool atmosphere of an evolved AGB star eventually enriched in C and N, they are commonly adopted by stellar modelers. [6] made a first step toward a correct description of the abundance changes in the calculation of opacity coefficients by estimating molecular concentrations through dissociation equilibrium calculations. These were combined with approximation formulae for the contribution of each molecular species to the Rosseland mean opacity. Although the results of this work definitively demonstrate the importance of a correct opacity treatment, its simplified approach suffers from some drawbacks, for instance the limited number of molecular species included.

In the following we present a set of new opacity tables that appropriately take into account possible overabundances of carbon and nitrogen with respect to a scaled solar metal mixture. In this way one can in terms of radiative opacity correctly follow the evolution of stars undergoing third dredge-up. Additionally, the consequences of a slow deep mixing process that is generally considered responsible for the partial conversion of carbon into nitrogen in low-mass AGB stars can be investigated.

2. OPACITY TABLES

The new opacity tables have been calculated by means of the COMA code from [7]. Beside continuous opacity sources and atomic line opacities, the latter ones being derived from VALD [8], we included data for a total number of 20 molecules in our calculations. The list of molecules given in [9] including references to corresponding thermodynamic data has recently been extended: we added data for the molecules ZrO (Bertrand Plez, priv. comm.; this list was used for the generation of a model atmosphere grid for S stars by [10]), YO (John Littleton, priv. comm.), FeH [11] and CrH [12]. Grain opacities have not been included in our calculations, however, the supposed scope of the produced data are stellar evolution models that usually do not attain such low temperatures.

In order to ease the integration of our tables into existing stellar evolutions codes we adopted the format of the [13] tables. In more detail, we tabulated values for the logarithm of the Rosseland mean opacity (in $\mathrm{cm^2\,g^{-1}}$) as a function of $\log T$ and $\log R$[1]. We cover a range in temperature from $\log T = 3.2$ to $\log T = 4.05$ with steps of $\Delta \log T = 0.05$, and in the $\log R$ dimension from -7 to $\log R = 1$ with steps of $\Delta \log R = 0.5$.

Up to now, tables have been calculated for metallicities of $Z = 1 \times 10^{-4}$, 1×10^{-3}, 3×10^{-3}, 6×10^{-3} and $Z = Z_{\odot}$. Starting from a given scaled solar metal composition — the element abundances and the isotopic composition were taken from [5] — with the overall metallicity as given above, we increased the mass fractions of ^{12}C and ^{14}N by factors depending on the initial metallicity. All other isotopes were left untouched. This results in a general increase of the metallicity, thus we reduced the mass fraction of ^{4}He in turn (in consistency with the OPAL approach). For both ^{12}C and ^{14}N we applied 5 enhancement factors for three different values of X (0.5, 0.7, 0.8), whereby we get 75 tables per initial metallicity. Variations in the mass fraction of ^{16}O were also investigated. Compared to the effects of the carbon enhancement, this contribution was found to be negligible. For this reason, alterations of the O abundance were not included in our current set of tables. However, we plan to assess the implications of varying the alpha element abundances as a whole.

Before the generation of our set of tables we compared the results to data from [4]. In Fig. 1 we show an example for the solar metallicity case ($X = 0.7$, $Z = 0.02$) at $\log R = -3$. The agreement for temperatures higher than $\log T \gtrsim 3.6$ is very good, whereas the results differ somewhat for the lower temperatures. The reason for the discrepancy is partly due to the different values adopted for the microturbulent velocity (COMA: $\xi = 2.5\,\mathrm{km\,s^{-1}}$, F05: $\xi = 2.0\,\mathrm{km\,s^{-1}}$) in the evaluation of the line opacities, but the major part can be ascribed to the differing molecular line dataset. In the same figure we indicate the contributions of continuous, atomic and molecular opacity to the mean opacity coefficient. Molecules are the dominant opacity source at low temperatures, continuous opacities are relevant at high temperatures. Atomic line opacities play a comparably minor role but have to be taken into account to arrive at accurate results.

---

[1] $R \equiv \rho / T_6^3$ where $[\rho] = \mathrm{g\,cm^{-3}}$ and T_6 is the temperature in millions of Kelvin.

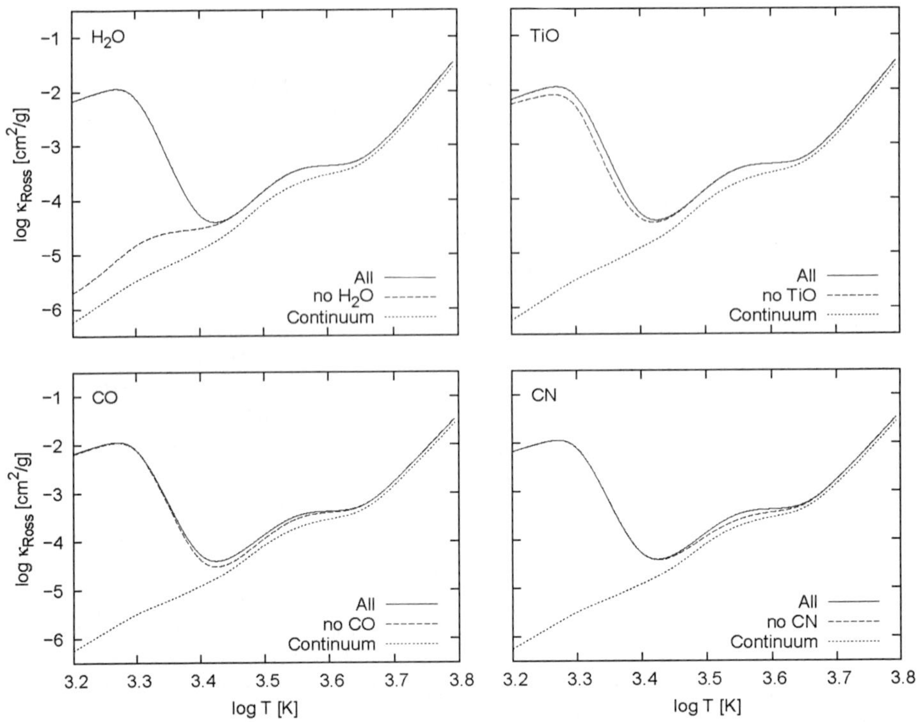

FIGURE 2. Molecules that contribute most to the Rosseland mean opacity in the oxygen-rich case (C/O = 0.50). At the lowest temperatures κ_{Ross} is completely dominated by the water molecule (H_2O). In this region, titanium oxide (TiO) also contributes significantly. At somewhat higher temperatures, CO and CN are important opacity sources. This is true in both the oxygen-rich and the carbon-rich regime. The solar metallicity case ($X = 0.7$, $Z = 0.02$) is shown at a value of $\log R = -3.0$. "All" denotes the calculation with the full molecular dataset, "no ... " means that the indicated molecule was omitted in the calculation.

2.1. Molecular Contribution

For the evaluation of the molecular contribution to the Rosseland mean opacity we calculated tables omitting the respective molecule under consideration and compared them to calculations with the full set of molecular data. Due to the non-linear character of the Rosseland mean one can only estimate the contribution of a single molecule in this indirect way.

In Fig. 2 we identify H_2O, TiO, CN and CO to be the main constituents of κ_{Ross} in the oxygen-rich case. Other molecules that should be taken into account are VO, SiO and OH. For the value of $\log R = -3$ shown in Fig. 2, VO contributes at $\log T \simeq 3.3$, SiO and OH are relevant around $\log T = 3.4$. A calculation based on these seven molecules results in opacity coefficients that only deviate by a few percent from a calculation using the full dataset.

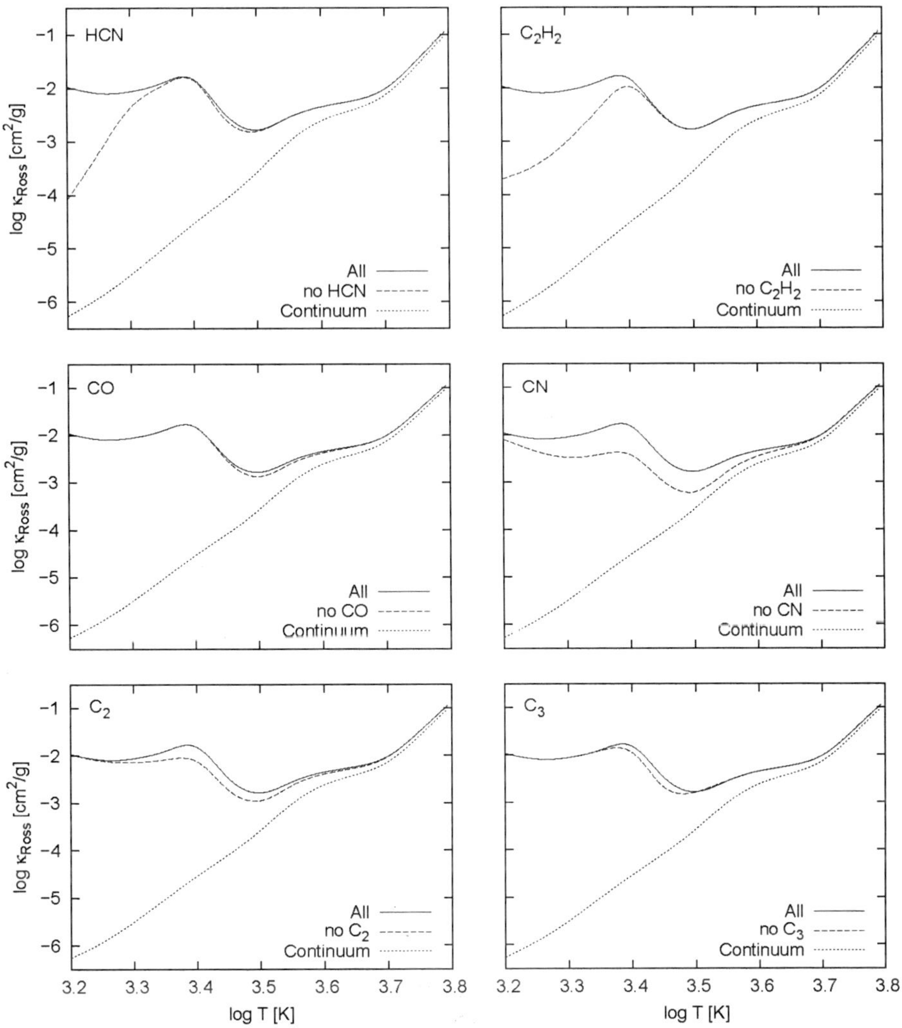

FIGURE 3. Molecules that contribute most to the Rosseland mean opacity in the carbon-rich case (C/O = 2.00). C_2H_2 and HCN deliver the bulk of the opacity at low temperatures. The contribution of CO is a little bit lower than in the oxygen-rich case, whereas CN contributes considerably more to κ_{Ross} over a wider temperature range. C_2 and C_3 become important especially for higher C/O ratios and cause the peaked structure around $\log T = 3.4$ in the plots shown here. Cf. also Fig. 2.

In the carbon rich case, again CO and CN are important opacity sources. The largest component of κ_{Ross} at lower temperatures is however generated by the polyatomic molecules C_2H_2 and HCN. C_2 and C_3 are relevant at slightly higher temperatures and give rise to a peak in the opacity around $\log T \simeq 3.4$ for increasing C/O ratios.

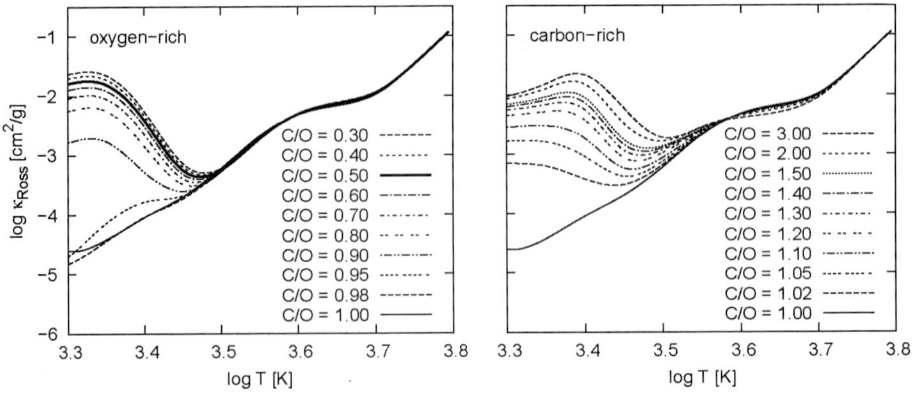

FIGURE 4. Changes of the mean opacity coefficient for an altered carbon-to-oxygen ratio. The solar value of $C/O = 0.50$ is indicated by the thick solid line. In the oxygen-rich regime ($C/O < 1$, left panel), an increasing C/O ratio causes the opacity at low temperatures to drop as the number of oxygen atoms that are free to form other molecules than CO decreases. After the transition to the carbon-rich regime, $C/O > 1$, right panel) the opacity rises again as more and more carbon-bearing molecules are formed. The parameters for the plots are $X = 0.7, Z = 0.02$ and $\log R = -1.5$. Note that for this plot the mass fractions of all the metals have been rescaled after enhancing the amount of ^{12}C to arrive at $Z = 0.02$ again.

2.2. C/O Variations

The reason why the number ratio of carbon to oxygen atoms C/O is the decisive quantity for molecular opacities is the large dissociation energy of the CO molecule. For $C/O < 1$ almost all carbon atoms are bound in CO while the oxygen atoms in excess are free to form other molecules as for example H_2O and TiO. As C/O increases, less oxygen-bearing molecules (apart from CO) are formed which causes the mean opacity to drop (see Fig. 4, left panel). Around $C/O = 1$ the Rosseland mean opacity reaches a minimum at low temperatures. A further increase in C/O beyond 1 (when the star becomes carbon-rich) induces the formation of carbon-bearing molecules such as C_2, C_3, C_2H_2, HCN etc. and the opacity rises again (right panel of Fig. 4). It is thus crucial to follow the chemical changes in the envelopes of AGB stars as the dredge-up of carbon results in significantly different opacities not only when transiting from the oxygen-rich to the carbon-rich regime, but even within both regimes.

In Fig. 5 we show an example for a set of opacity tables at a metallicity of $Z = 1 \times 10^{-3}$. As described above (and in contrast to the data shown in Fig. 4) no rescaling of the metals is done but the ^{4}He mass fraction is reduced instead. When the star becomes carbon-rich, molecules do form that are more opaque than the ones in an oxygen-rich regime, which causes a steeper temperature gradient. A consequence is for instance a decrease of the effective temperature in stellar evolution models. The stellar radius increases, and the average mass-loss rate increases, thus eroding at a faster rate the envelope mass. As shown by [14], changes of the core and the envelope masses affect all the fundamental properties of AGB stars, such as the TP strength or the total amount of mass that is dredged up.

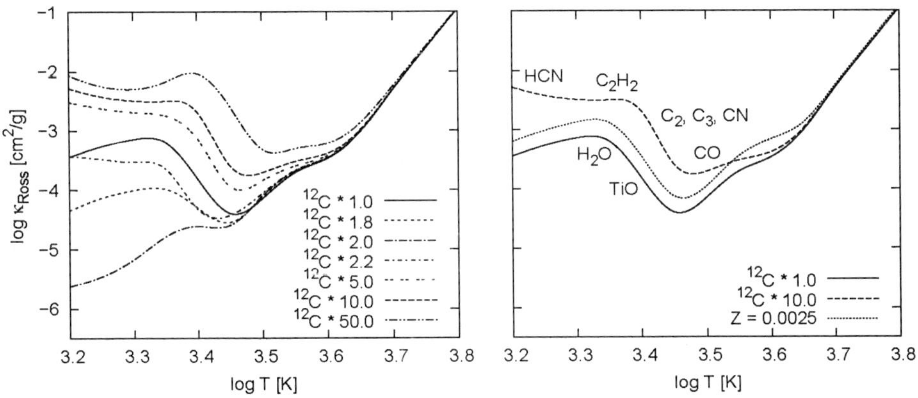

FIGURE 5. Left panel: Example of a set of opacity tables for $Z = 1 \times 10^{-3}$ (metal composition from to [5]). The mass fraction of ^{12}C was enhanced according to the factors indicated. This results, beside a higher C/O ratio, in an overall increase of Z. Thus the ^{4}He mass fraction was reduced instead to fulfil $X+Y+Z=1$. Parameters for both plots in this figure are $X = 0.7$ and $\log R = -1.5$
Right panel: Comparison of data obtained by only enhancing the ^{12}C (dashed line) abundance with a table based on a scaled solar metal mixture with the same overall metallicity (dotted line). The mean opacity for a metal mixture with C/O > 1 cannot be approximated by a table for the oxygen-rich case with a higher metallicity. Both the shape of the curve (relevant molecules are indicated for both cases) as well as the order of magnitude differ significantly.

Moreover, the right panel of the figure underlines the fact that one cannot mimic the increase in the mass fraction of carbon in a metal mixture by simply scaling up all the metals starting from solar values. As different molecules contribute to the mean opacity it is qualitatively and quantitatively different in each case.

3. APPLICATIONS

The new opacities were first applied in a stellar evolution model with a metallicity of $Z = 1 \times 10^{-4}$ calculated with the FRANEC code (the release described in [15]). To summarize the results briefly, the larger opacity coefficients in the carbon-rich regime imply cooler envelopes, larger mass-loss rates and thus shorter AGB lifetimes. Moreover, the variation of the surface composition and the global yields of low-mass AGB stars are affected. It turns out that the models computed were able to reproduce the physical properties of their observational counterparts. For an in-depth discussion we refer to the paper by [9]. For the implications of the new opacity coefficient at other metallicities see the contribution of Cristallo et al., this volume. Details will be given in a series of forthcoming papers.

4. CONCLUSIONS

The new low temperature Rosseland mean opacities introduced here represent a step ahead in the modelling of low and intermediate mass stars. We want to point out that whenever temperatures lower than approximately $\log T = 3.7$ are attained in a non-solar metal mixture (i. e. for C/O ratios other than 0.5), the usually adopted scaled solar values for the mean opacity coefficients are inappropriate to describe the actual physics. The implications for stellar evolution could already be anticipated from the analysis of the data shown here, and the application of the tables in detailed model calculations confirmed our expectations. It appears that the inconsistency between surface temperatures and s-process enhancements arising from a comparison of the models to observations of carbon-enhanced metal-poor stars could be solved by using appropriate opacity coefficients (cf. the contribution of Cristallo et al., this volume). An interesting question would also be the interdependency of the new opacity coefficients with different mass loss prescriptions, as these issues are physically closely coupled.

The next step will be to establish a large grid of opacity tables and to collect them in a database. With the help of a complete dataset one will be able assess the impact of varying opacity coefficients on stellar evolution over a wide metallicity range.

ACKNOWLEDGMENTS

MTL and BA acknowledge funding by the Austrian Research Fund FWF (projects P-18171 and P-19503). MTL has been supported by the Austrian Academy of Sciences (DOC programme).

REFERENCES

1. M. Busso, R. Gallino, and G. J. Wasserburg, *ARA&A* **37**, 239 (1999)
2. F. J. Rogers and C. A. Iglesias, *in J. M. Nemec, J. M. Matthews, eds, IAU Colloq.* **139**, 221 (1993)
3. M. J. Seaton, *MNRAS* **362**, L1 (2005)
4. J. W. Ferguson, D. R. Alexander, F. Allard, T.Barman, J. G. Bodnarik, P. H. Hauschildt, A. Heffner-Wong, and A. Tamanai, *ApJ* **623**, 585 (2005)
5. K. Lodders, *ApJ* **591**, 1220 (2003)
6. P. Marigo, *A&A* **387**, 507 (2002)
7. B. Aringer, *PhD thesis, University of Vienna*, (2000)
8. F. G. Kupka, T. A. Ryabchikova, N. E. Piskunov, H. C. Stempels, and W. W. Weiss, *BaltA* **9**, 590 (2000)
9. S. Cristallo, O. Straniero, M. T. Lederer, and B. Aringer, *ApJ* **667**, 489 (2007)
10. B. Plez, S. van Eck, A. Jorissen, B. Edvardsson, K. Eriksson, and B. Gustafsson, in *N. Piskunov, W. W. Weiss, D. F. Gray, eds, Modelling of Stellar Atmospheres, IAU Symposium* **210**, 2P (2003)
11. M. Dulick, C. W. J. Bauschlicher, A. Burrows, C. M. Sharp, R. S. Ram, and P. Bernath, *ApJ* **594**, 651 (2003)
12. C. W. J. Bauschlicher, R. S. Ram, P. F. Bernath, C. G. Parsons, and D. Galehouse, *J.Chem.Phys.* **115**, 1312 (2001)
13. D. R. Alexander and J. W. Ferguson, *ApJ* **437**, 879 (1994)
14. O. Straniero, I. Domínguez, S. Cristallo, and R. Gallino *PASA* **20**, 389 (2003)
15. A. Chieffi, M. Limongi, and O. Straniero, *ApJ* **502**, 737 (1998)

The Influence of Mass Loss on the Evolution and Nucleosynthesis of Massive Stars

A. Chieffi* and M. Limongi†

*Istituto Nazionale di Astrofisica - Istituto di Astrofisica Spaziale e Fisica Cosmica, Via Fosso del Cavaliere, I-00133, Roma, Italy

†Istituto Nazionale di Astrofisica - Osservatorio Astronomico di Roma, Via Frascati 33, I-00040, Monteporzio Catone, Italy

Abstract. We briefly discuss the influence of mass loss on the internal evolutionary properties of massive stars. In particular we show that the C abundance left by the central He burning depends significantly on the adopted mass loss rate after the H rich mantle has been completely ejected.

Keywords: Stellar Evolution; Massive Stars; Mass Loss; Supernovae
PACS: 97.10.Cv;97.10.Me;97.10.Nf;97.20Ec;97.60Bw

1. INTRODUCTION

We have recently recomputed a full set of solar metallicity models in the mass range 11-120 $M_\odot$ taking into account mass loss. The yields produced by this new set of models have already been used to discuss the galactic abundances of ^{26}Al and ^{60}Fe [1]. The main difference between this set of models and our previous ones is the inclusion of mass loss all along the evolution, and hence the possibility to extend the grid of masses well above the 35 $M_\odot$ where mass loss plays a crucial role. In the following we will briefly rediscuss the H and He burning phases paying particular attention to the influence of mass loss on the interior evolution of these stars. The paper is organized as follows: Section 2 is devoted to the central H burning while the central He burning is addressed in Section 3. A final brief conclusion follows.

2. THE MODELS

The models discussed in this paper are the ones presented by [1] and computed with the latest release of the stellar evolutionary code FRANEC (5.050218). Let us briefly remind that the interaction between convection and local nuclear burning has been taken into account by fully coupling together and solving simultaneously the set of equations controlling the local nuclear burning and those describing the convective mixing. Convective mixing is treated by means of a diffusion equation where the diffusion coefficient is computed by means of the mixing-length theory. The nuclear network is the same adopted in [2] while the nuclear cross sections have been updated whenever possible (see Table 1 in [1]). A moderate amount of overshooting of 0.2 H_p has been included into the calculation just in central H burning. Mass loss has been taken into account following the prescriptions of [3] for the blue supergiant phase ($T_{eff} > 12000$ K), [4] for the

CP1001, 9th Torino Workshop on Evolution and Nucleosynthesis in AGB Stars and the 2nd Perugia Workshop on Nuclear Astrophysics, edited by R. Guandalini, S. Palmerini, and M. Busso

red supergiant phase ($T_{eff} < 12000$ K), [5] for the WNL Wolf-Rayet phase and [6] during the WNE/WCO Wolf-Rayet phases. We assume that the star enters the WR phase when $Log(T_{eff}) > 4$ and $H_{surf} < 0.4$ and we adopt the following usual definitions for the various WR phases: WNL ($10^{-5} < H_{surf} < 0.4$), WNE ($H_{surf} < 10^{-5}$ and $(C/N)_{surf} < 0.1$), WNC ($0.1 < (C/N)_{surf} < 10$) and WCO ($(C/N)_{surf} > 10$).

3. THE CENTRAL H BURNING PHASE

Current models for mass loss from OB stars predict that its efficiency mainly depends on the amount of momentum deposition of the radiative flux on to the outgoing material [3]. As the surface global properties of a star change because of its continuous evolution, the structure of the atmosphere modifies as well as the mass loss rate. In addition to that, there are also sharp changes in the atmospheric structure, as the one due to a shift in the ionization balance of Fe in correspondence of an effective temperature of the order of 25000 K, that lead to strong discontinuities in the mass loss rates (bistability jumps, [3]). As a consequence of these phenomena, a massive star may change its mass loss rate even considerably during just its central H burning phase. Figure 1 shows, as a blue solid line the mass loss rate experienced by an 80 $M_\odot$ during the central H burning. The one order of magnitude jump in the mass loss rate (marked by a red dotted line) occurs in this case when the central H abundance drops down to 0.108 by mass fraction. Obviously the timescale over which the higher mass loss rate may operate will depend on the amount of H left in the convective core at the time at which the critical temperature is crossed. Any change in the computation of a stellar model, including e.g. the adopted size of the convective core, may therefore lead to substantial differences in the total amount of mass lost by a massive star in central H burning.

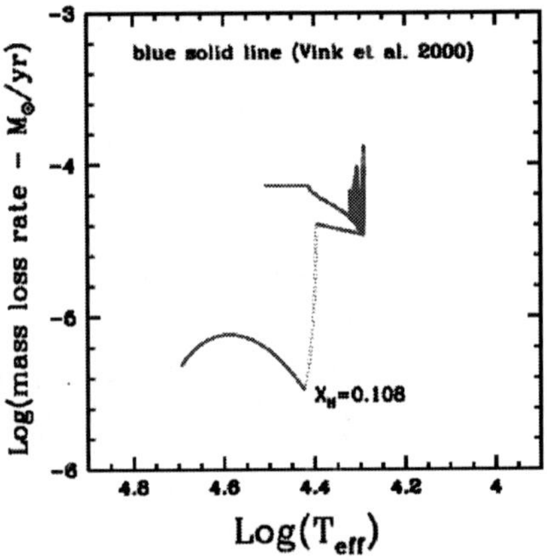

FIGURE 1. Blue solid line: mass loss rate in central H burning for an 80 $M_\odot$ of solar metallicity. Red dashed line: bistability predicted by the theoretical mass loss rate proposed by [3].

The existence of such a variety of phenomena that may drastically change the mass loss rate in central H burning poses a quite simple but important question: does the structure of a star at the end of the central H burning depend on the detailed mass loss rate adopted or not? In other words, it is important to understand in detail how the mass loss rate changes in time or it is just possible to adopt some "average" mass loss rate all along the H burning. To answer to such a question we have recomputed the evolution of the 80 $M_{\odot}$ by adopting a constant average mass loss rate equal to 10^{-5} $M_{\odot}/yr$. This value corresponds to the average mass loss rate experienced by the reference model (that loses roughly 32.4 $M_{\odot}$ on a timescale of 3.16 10^6 yr)

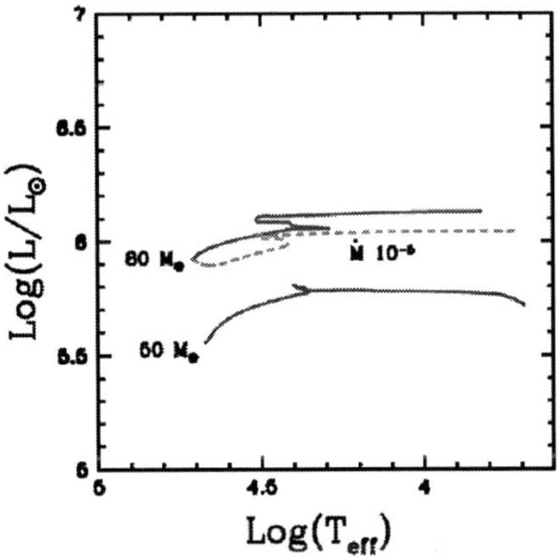

FIGURE 2. Blue solid lines: standard evolutions of an 80 $M_{\odot}$ and a 50 $M_{\odot}$. Red dashed line: test evolution in which the mass loss rate is kept constant and equal to the average value obtained for the 80 $M_{\odot}$.

The path followed by this test model in the HR diagram is shown in Fig. 2 as a red dashed line. The two blue solid lines refer to two standard models of 80 and 50 $M_{\odot}$. The run computed by adopting the constant mass loss rate evolves initially towards lower luminosities and only when the central H abundance is almost halved it recovers towards a more "classical" path. Though the final mass at the end of the central H burning is very similar in the test and the standard case, the size of the He core is substantially reduced. While the standard model leaves the H burning with a He core mass of 33.3 $M_{\odot}$ the test model forms a significantly smaller He core mass of 28.3 $M_{\odot}$. The behavior of the test model may be understood by reminding that a star quickly readjusts on a physical structure corresponding to its current total mass only if it is chemically homogeneous. Such a direct correspondence between center and surface lowers progressively as a mean molecular weight gradient forms. This means that the center becomes progressively insensible to a variation of the total mass: the limiting case in which the H is fully exhausted in a central region of the star corresponds to the situation in which the total mass – luminosity relation is fully replaced by the well known He core mass – luminosity relation. In the present case, the star begins initially to lower its luminosity because it still "feels" the reduction of the total mass. However, when roughly half of the H is converted in He, the core and the envelope begin to discouple progressively and the luminosity bends upward again and the star recovers a more "classical" pattern. The mass size of the He core plays a pivotal role in the further evolution of a star and hence we must conclude that a proper computation of the evolution of a massive star requires a detailed knowledge of the mass loss rate as a function of the physical and chemical parameters since its feedback on the stellar interior depends on the evolutionary stage at which it occurs.

It is worth mentioning that these results obviously hold also if the mass is lost not through the wind but by interactions in a binary system. Hence they must be properly taken into account in the study of these systems.

4. THE CENTRAL HE BURNING PHASE

The evolution of a star beyond the H burning depends on the mass size of the He core and not on the total mass: this is a well known property of the stellar models that immediately tells us in which mass interval mass loss influences the advanced evolutionary phases of a massive star (and consequently the collapse, explosion and yields). First of all mass loss globally affects the full mass interval 30-120 $M_{\odot}$ because it is efficient enough to determine a reduction of the mass size of the H convective core (see above) that in turn is the main responsible for the initial size of the He core mass. But there is an additional phenomenon that occurs in all masses that completely lose their H rich mantle together to part of their He core mass.

Stars initially less massive than the minimum mass that completely loses its H rich mantle develop a central He burning within a He core that never recedes in mass, it either advances or at least remains constant in mass. In this case the He convective core never recedes and the reason is that the conversion of He in C and O increases the opacity in all the convective core so that its outer border is continuously pushed outward rather than inward. Hence a strong chemical discontinuity forms at the border of the He convective core where the He abundance changes from roughly zero to roughly one at the central He exhaustion. Hence the He convective shell will form outside the maximum extension of the convective core in a region whose chemical composition is still the one left by the H burning.

Viceversa stars whose initial mass exceeds a threshold value (that in our set of models is of the order of 35-40 $M_{\odot}$) not only lose their whole H rich envelope but also a fraction of their He core mass. In this mass interval, therefore, the central He burning occurs, partly or even completely, in a He core mass that progressively reduces: such an occurrence has a profound impact on the central He burning itself and therefore on the whole advanced burnings. As the He core mass reduces because of the mass loss, the central He burnings readjusts on a physical structure, and hence on a convective core size, that would compete to the current He core mass. Since the size of the convective core scales directly with that of the He core, the convective core recedes progressively in mass leaving behind a He profile. In this case the He convective shell will form in a region with a He profile whose steepness reflects the speed at which the He core mass has been eroded in He burning. In addition to that, also the amount of carbon left in the He exhausted core is affected by mass loss because the central physical conditions readjust continuously on the ones proper of a star of progressively smaller He core mass. Since the C abundance left by the He burning scales inversely with the He core mass, the smaller the final He core mass the larger the abundance of C. Figure 3 shows as a blue line the C abundance for the full set of models we have recently computed [1].

By the way, let us remind that the amount of carbon left by the He burning is the result of the delicate balance among the 3α process, the $^{12}C(\alpha,\gamma)^{16}O$, the run of the central density and temperature with time and eventually the behavior of the convective core. Generally speaking, keeping fixed the cross sections of the processes involved, the larger the He core mass the lower the C abundance at the central He exhaustion: the reason is that the central temperature increases and the central density decreases with the growth of the He core mass and both these behaviours favor the O production with respect to the C one. Hence this explain why the stronger the mass loss rate (once the H rich envelope has been lost) the larger the amount of C at the end of the central He burning. We refer the reader to the paper by [7] for a comprehensive discussion of the dependence of the C abundance on the initial mass and its interplay with the convective mixing.

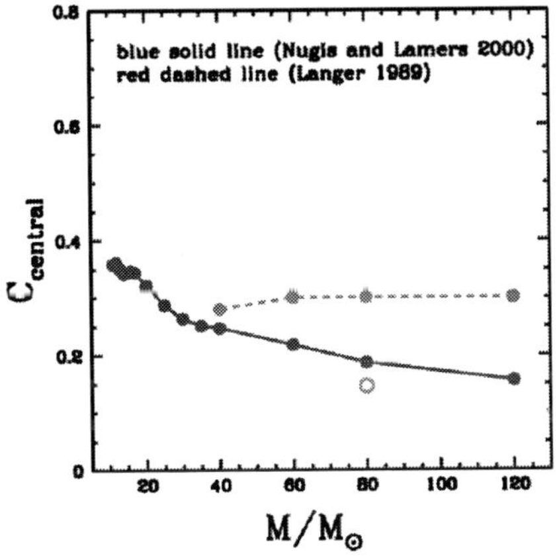

FIGURE 3. Blue solid line: C abundance left by the central He burning for our standard set of models [1]. Magenta open circle: C abundance at the central He exhaustion for an 80 $M_\odot$ evolved at constant mass. Red dashed line: C abundances obtained by adopting the mass loss rate proposed by [6].

Coming back to Fig. 3, only stars more massive than 35 $M_\odot$ experience a reduction of the He core. In this phase we have adopted the mass loss rate suggested by [5]. If we would have switched off the mass loss in this phase, the 80 $M_\odot$, for example, would have ended the central He burning with a C abundance of 0.146 instead of 0.187 (magenta open dot in Fig. 3). If we would have adopted the mass loss rate proposed for this phase by [6] (that is much stronger than the [5] one) the final C abundance would have been the one shown as a red dashed line in Fig. 3.

5. CONCLUSIONS

We have seen that the influence of the mass loss in H burning depends on the specific central H abundance at which it occurs. The mass loss that occurs at the very beginning of the evolution pushes the star towards a configuration typical of a structure of smaller mass. As the H is converted in He, the interior of a star becomes progressively less sensible to a variation of the total mass. Since the mass loss rates available in the literature scale directly with the luminosity of a star and inversely with its effective temperature, and since during the central H burning massive stars evolve towards higher luminosities and cooler temperatures, most of the mass is actually lost once the central H has been substantially converted in He. This is the reason why even the adoption of quantitatively very different mass loss rates does not affect the general trend of the evolution that always occurs towards luminosities higher than the Zero Age Main Sequence one.

The influence of the mass loss during the central He burning occurs in two ways. A first indirect one is due to the fact that mass loss in H burning may affect the size of the H convective core and therefore the size of the He core mass. A second direct one occurs if mass loss reduces the He core mass while the star is burning He in the center. This occurs for masses larger than, say, 35 $M_{\odot}$, that lose their whole H rich envelope plus a fraction of the He core mass. In this case the He convective core shrinks progressively in mass leaving behind a He profile. Also the C abundance changes accordingly and its abundance increases as the mass loss rate increases in this phase.

Before closing let us stress here that the reactivity of the interior of a star to the mass loss is the same either if the mass loss is due to the physical behavior of the atmosphere or to the environment, for example to the evolution in a binary (multiple) system. This means that in principle it would be possible to have a *variety* of C abundances at the central He exhaustion (and hence of final explosions and yields) for the same initial mass depending on the environment in which each specific star lives.

ACKNOWLEDGMENTS

I (A.C.) warmly thank the organizers for the kind hospitality and the very nice time all of us had in Perugia.

REFERENCES

1. M. Limongi and A. Chieffi, *ApJ* **647**, 483 (2006)
2. M. Limongi and A. Chieffi, *ApJ* **592**, 404 (2003)
3. J.S. Vink, A. de Koter, and H.J.G.L.M. Lamers, *A&A* **362**, 295 (2000)
4. C. de Jager, H. Nieuwenhuijzen, and K.A. van der Hucht, *A&A Suppl. Ser.* **72**, 259 (1988)
5. T. Nugis and H.J.G.L.M. Lamers, *A&A* **360**, 227 (2000)
6. N. Langer, *A&A* **220**, 135 (1989)
7. G. Imbriani, M. Limongi, L. Gialanella, F. Terrasi, O. Straniero, and A. Chieffi, *ApJ* **558**, 903 (2001)

Single Star Progenitors for Type Ia Supernovae

Christopher A. Tout*, Dayal T. Wickramasinghe† and John C. Lattanzio**

*Institute of Astronomy, The Observatories, Madingley Road, Cambridge CB3 0HA, England
†Department of Mathematics, The Australian National University, Canberra, ACT 0200, Australia
**Centre for Stellar and Planetary Astrophysics, School of Mathematics, Monash University, Clayton, Victoria 3800, Australia

Abstract. Type Ia supernovae are identified as exploding degenerate stars. Their luminosity is due to the radioactive decay of about a solar mass of ^{56}Ni through ^{56}Co to ^{56}Fe. Although it is generally accepted that a degenerate carbon/oxygen white dwarf explodes as it accretes material from a binary companion, the progenitors of type Ia supernovae have not been categorically identified. The cores of intermediate mass asymptotic giant branch stars are composed of carbon and oxygen in a degenerate state. We discuss conditions under which carbon can ignite at the centre of such a core and when the ensuing explosion would appear as a type Ia supernova.

Keywords: Stars: AGB and Post-AGB, Stars: Mass-Loss, Supernovae: General, White Dwarfs
PACS: 97.60.Bw

1. INTRODUCTION

Luminous type Ia supernovae (hereinafter SNe Ia) are amongst the brightest objects in the Universe. Observations indicate that their absolute magnitudes lie in a narrow range $M_{\rm B} = -18.5 \pm 0.5$, or $\pm 60\%$ in luminosity (from Table 1 of [1]). Furthermore observations reveal a correlation between maximum absolute brightness and light curve shape [2] that facilitates an effective reduction in the standard deviation of absolute luminosities to $\pm 15\%$. This small spread coupled with the fact that they can be seen to great distances make SNe Ia excellent standard candles for study of the cosmology of the Universe as first envisaged by [3]. Observations have, quite precisely, determined the rate of expansion, immortalised in the Hubble constant, [4] and have further determined that this rate is accelerating with time, a measurement that has led to the invocation of a cosmological constant contributing about 70% of the critical density of the Universe to be added to matter's contribution of 30% [5, 6].

Supernovae types are differentiated by their spectra and characteristic light curves. SNe II show hydrogen in varying amounts while SNe I generally show no evidence of hydrogen. SNe I are further subdivided into SNe Ia which show prominent silicon lines and generally no helium, SNe Ib which show no silicon but possibly helium and SNe Ic which show neither silicon nor helium lines. The SNe II generally are thought to be the deaths of massive stars (initial mass $M_{\rm i} > 10\,{\rm M}_\odot$) whose cores reach the Chandrasekhar mass and collapse. SNe Ib/c, which tend to be associated with regions of star formation, are similarly thought to be caused by core collapse of initially very massive stars ($M_{\rm i} > 40\,{\rm M}_\odot$) that have blown off their envelopes. SNe Ia, on the other hand, are almost certainly powered by the radioactive decay of ^{56}Ni, via ^{56}Co, to ^{56}Fe with the ^{56}Ni created in a thermonuclear runaway in a degenerate white dwarf. If $1\,{\rm M}_\odot$

CP1001, 9th Torino Workshop on Evolution and Nucleosynthesis in AGB Stars and the 2nd Perugia Workshop on Nuclear Astrophysics, edited by R. Guandalini, S. Palmerini, and M. Busso

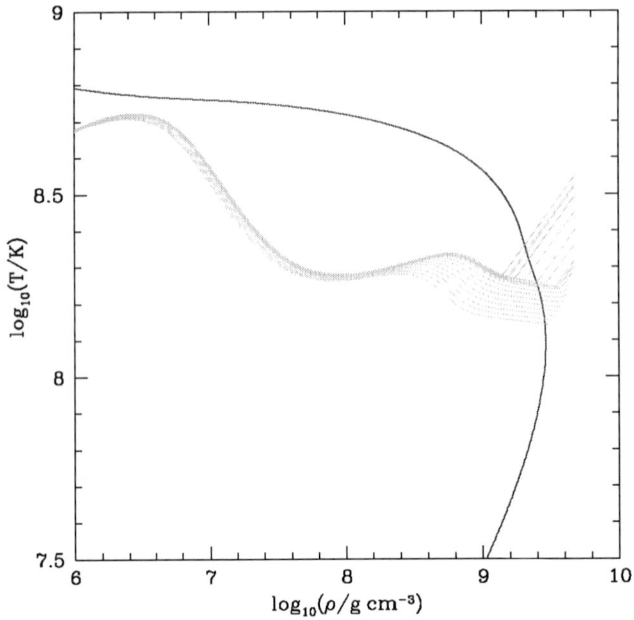

FIGURE 1. The line crossing both axes is the carbon ignition curve on which energy generation by carbon burning equals the energy loss in neutrinos. The other lines are the run of temperature and density through the white dwarf, at late times, increasingly later from bottom to top. In this case these cross the ignition curve at the centre and a thermonuclear runaway can begin there under degenerate conditions. The accretion rate is $2 \times 10^{-6}\,M_\odot yr^{-1}$.

of material composed of one fifth carbon and four-fifths oxygen is converted to ^{56}Fe the energy released amounts to 1.8×10^{44} J. Of this, 2×10^{43} J are liberated in the radioactive decay and this is enough to power the supernova at $5 \times 10^9\,L_\odot$ for 80d. The silicon results from incomplete burning in the outer parts of the core. Degenerate cores form from the ashes of nuclear burning at the centre of red giants. When the envelope of the giant is removed the core cools to a white dwarf.

2. CARBON IGNITION

If matter is added to the surface of the white dwarf it shrinks because the electrons must be squashed closer together into higher momentum states as more pressure support is required. The compression heats the degenerate material which cools mainly by neutrino losses through as electrons and positrons are created and destroyed in the dense material. As the temperature and density rises carbon burning begins. In the centre it is mainly driven by the density rise while near the surface it is the accretion driven heating. When the energy generation rate by carbon burning exceeds the neutrino loss rate convection

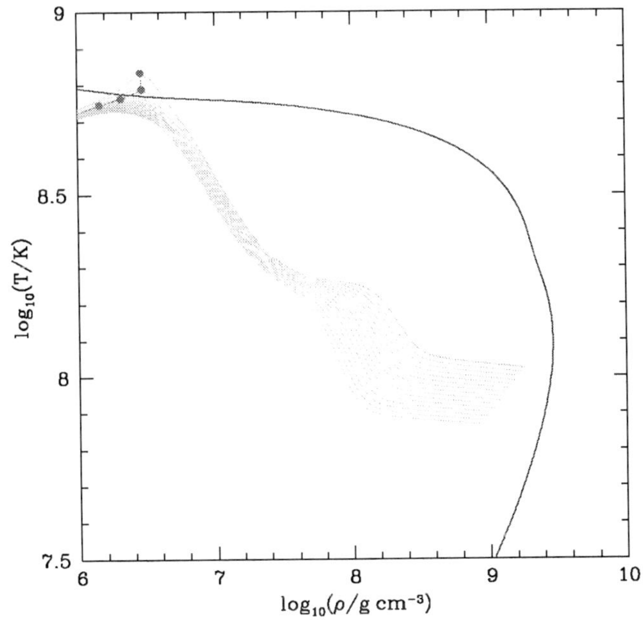

FIGURE 2. Similar to Fig. 1 but with a higher accretion rate of $3 \times 10^{-6}\,M_\odot yr^{-1}$. Ignition occurs close to the surface and carbon ignites under much less degenerate conditions. Grey filled circles indicate the path of the mass shell where carbon first ignites at mass $1.354\,M_\odot$ just $0.008\,M_\odot$ below the surface.

sets in and attempts to carry off the excess energy assisted by further neutrino losses in URCA pairs but a thermonuclear runaway is inevitable [7, 8]. If the accretion rate is low ignition begins in the cooler centre. If it is high it begins in the hottest region near the surface.

Such a runaway it manifests itself by a sudden rise in carbon burning luminosity. Figures 1 and 2 show how the internal run of temperature and density changes during accretion at rates just above and just below the critical rate at which ignition moves from the centre to the surface. We determine a carbon ignition curve as a function of temperature and density by equating the carbon burning energy production with that lost in neutrinos. Above and to the right of this line conditions are right for carbon ignition to run away. Below it, any carbon burning ceases if the accretion is turned off. If a line from our internal structure crosses the ignition curve then we know that the carbon in the star has ignited at that particular point. There is then a clear distinction between those cases that ignite at the centre (Fig. 1) and those which ignite in the outer layers just below the surface (Fig. 2). There is not a smooth transition of the ignition point from the centre to the surface because, in both cases, the intervening parts of the white dwarf lie well away from the ignition curve. The transition is sudden and so the critical rate at which it occurs is well determined for each of the cases we consider.

Ignition at the degenerate centre is expected to runaway through most of the white

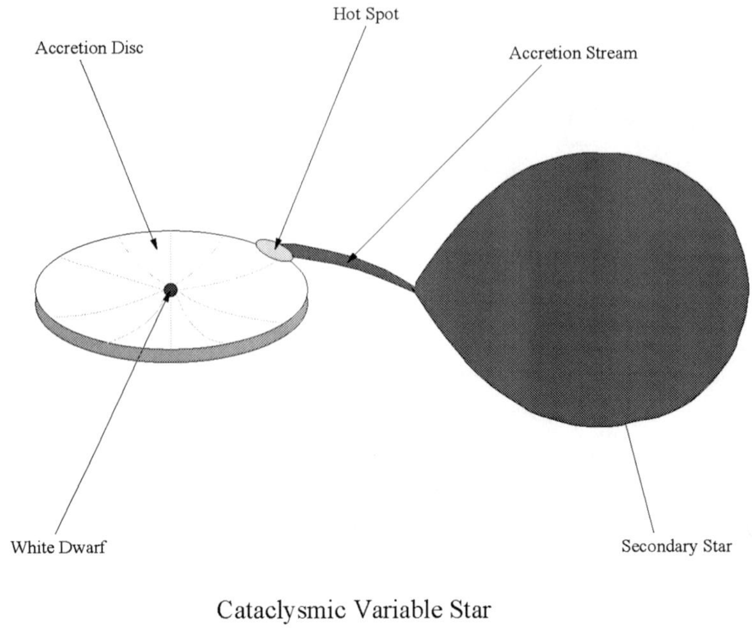

FIGURE 3. A schematic diagram of a cataclysmic variable showing the Roche-lobe-filling secondary star overflowing on to an accretion disc from which material falls on to the white dwarf primary. Where the stream impacts with the disc shocked gas produces a hot spot.

dwarf burning everything except the outer layers to nuclear statistical equilibrium around the iron peak at ^{56}Ni. If it occurs near the outside thermal expansion stops the burning at ^{20}Ne. Subsequently a slightly deeper layer reaches the temperature for carbon ignition and it too burns to neon. Gradually the entire star's carbon burns to neon and it cannot again reach the conditions for thermonuclear runaway before collapsing beyond the point of no return to a neutron star.

The process of ignition requires addition of mass to the white dwarf and it has usually been imagined that this occurs through mass transfer in a binary system. But precisely the effects can if the growing core of an AGB star itself approaches the Chadrasekhar mass. Though this was considered in the past the slow core growth during thermal pulses and the large mass loss rates expected in superwinds have led us to believe that AGB cores never explode.

3. BINARY STAR MODELS

In a close binary system, mass transfer can increase the mass of a white dwarf. Accreting white dwarfs have been known for some time as the engines of cataclysmic variables (Fig. 1), the source of novae and dwarf novae [9] and so are the first candidate to be considered. However if the accreting material is hydrogen-rich, accumulation of a layer of only $10^{-5} - 10^{-3}\,M_\odot$ leads to ignition of hydrogen burning sufficiently violent to eject most, if not all of or more than, the accreted layer in a nova eruption. The white dwarf mass does not significantly increase and ignition of its interior is avoided. However if the accretion rate is high $\dot{M} > 10^{-7}\,M_\odot\,yr^{-1}$ hydrogen can burn as it is accreted, bypassing novae explosions [10], and allowing the white dwarf mass to grow. Though, if it is not much larger than this, $\dot{M} > 3 \times 10^{-7}\,M_\odot\,yr^{-1}$, hydrogen cannot burn fast enough so that accreted material builds up a giant-like envelope around the core and burning shell which eventually leads to more drastic interaction with the companion and the end of the mass transfer episode. Rates in the narrow range for steady burning are found only when the companion is in the short-lived phase of thermal-timescale expansion as it evolves from the end of the main sequence to the base of the giant branch. Super-soft X-ray sources [11] are probably in such a state but cannot be expected to remain in it for very long and white dwarf masses almost never increase sufficiently to explode as SNe Ia.

The standard model overcomes this problem by postulating that, when the mass-transfer rate exceeds that allowed for steady burning, only just the right fraction of the mass transferred is actually accreted by the white dwarf. The standard mechanism currently invoked is a strong wind from the accretion disc that expels the material from the system before it reaches the white dwarf [12]. An alternative might be that the white dwarf does indeed swell up to giant dimensions but that the resulting common-envelope evolution is very efficient so that the small amount of excess material can be ejected without the cores spiralling in. This is quite consistent with the findings of [13] that such efficiency is necessary for at least one phase of common-envelope evolution in the formation of close double white dwarf systems.

A more promising scenario is mass transfer from one white dwarf to another [14]. In a very close binary orbit gravitational radiation can drive two white dwarfs together until the less massive fills its Roche lobe. If both white dwarfs are CO and their combined mass exceeds $1.38\,M_\odot$ enough mass could be transferred to set off a SN Ia. However if the mass ratio $M_{\rm donor}/M_{\rm accretor}$ exceeds 0.628 mass transfer is dynamically unstable because a white dwarf expands as it loses mass. Based on calculations at somewhat lower, steady accretion rates, [15] have shown that the ensuing rapid accretion of material allows carbon to burn in mild shell flashes converting the white dwarf to ONe and ultimately leading to AIC and not a SN Ia. [16] have reanalysed this result and find that the critical mass could be reached in some cases if the accretion rate can be limited to about half the Eddington rate.

4. SINGLE STAR MODELS

In the absence of mass loss any star less massive than about $7\,M_\odot$ develops a degenerate CO core as the star ascends the asymptotic giant branch (AGB). If this core can grow to $1.38\,M_\odot$ or so it ignites and explodes. The core behaves as in Fig. 1 with the accretion rate controlled by the rate of burning. However it is believed that in practice mass loss from red giants and AGB stars is such that the entire stellar envelope is stripped off before this mass is reached to leave a lower mass CO white dwarf.

This is certainly the case when the usual empirical mass-loss formulae such as that

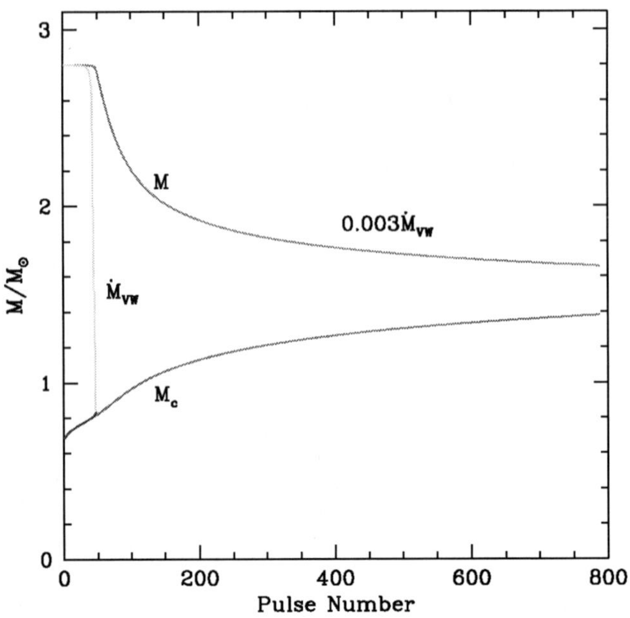

FIGURE 4. Evolution of the core mass and total mass of a $2.8\,M_{\odot}$ star of solar metallicity under normal and substantially reduced mass loss.

of [17] or [18] are applied to population-I, $Z = 0.02$, stars. However [19] demonstrate that, even with what would usually be considered extreme overestimates of the mass-loss rates, population-III, $Z = 0$, stars of $7\,M_{\odot}$ do not grow cores to the critical mass before they have lost their envelopes. Given that, even with no mass loss, a star of $1.39\,M_{\odot}$ cannot reach a CO core mass of $1.38\,M_{\odot}$ with more than $0.01\,M_{\odot}$ of hydrogen in its envelope, there must be a range of zero-metallicity single stars that would end their lives as SNe Ia. As metallicity, and with it presumably mass loss, increases so this range of masses moves to higher and higher mass and eventually moves out of the range for degenerate ignition.

Mass-loss rates at metallicities other than solar are very poorly constrained. Even at $Z = 0.02$ the rates from giants are based on observations of a small sample of stars for which they can be measured together with models of the initial/final mass for stars near the turn off in globular clusters. It is not unlikely that some fraction of giants lose mass at somewhat lower rates.

We can investigate this more quantitatively with a synthetic AGB evolution algorithm [Tout, Wickramasinghe & Lattanzio in preparation]. Figure 4 shows the evolution of a $2.8\,M_{\odot}$ star of solar metallicity. When mass loss is applied at the standard rate proposed by [18], $\dot{M}_{\rm VW}$, a superwind sets in when the core mass, $M_{\rm c}$ is just over $0.8\,M_{\odot}$ and the envelope is stripped off before it can grow much at all leaving a white dwarf. In an extreme case of mass loss at $0.003\dot{M}_{\rm VW}$ the superwind is not strong enough to prevent

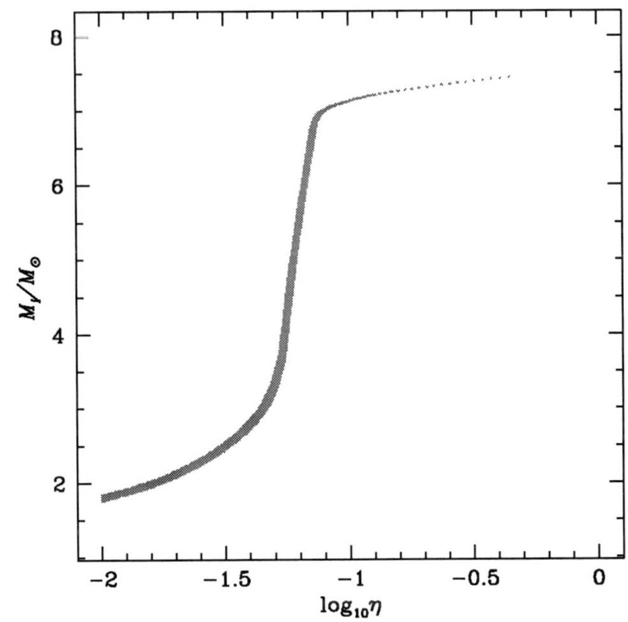

FIGURE 5. The effect of reducing the normal [18] stellar wind mass-loss rate by a factor η. The mass of the progenitor star is M_i. In the bottom right region the star's envelope is lost before it's core reaches its ignition mass. In the upper left region the core reaches the ignition mass while the star still has a substantial envelope. If the core explodes these would still appear as SNe II. In the strip in between the core reaches the ignition mass when less than $0.1\,M_\odot$ of the envelope (equivalent to $0.07\,M_\odot$ of hydrogen) remains and could appear as a SN Ia.

core growth and M_c reaches $1.38\,M_\odot$ while there is still about $0.3\,M_\odot$ of hydrogen-rich envelope left. The core of this star would explode in a thermonuclear runaway but the hydrogen would make it look like a SN II rather than an SN Ia. Somewhere in between these rates the mass loss is such that the envelope is almost entirely removed at the time the core explodes. It is not certain how much hydrogen can remain in the envelope if the explosion is to appear as an SN Ia but an upper limit seems to be about $0.1\,M_\odot$ [Schmidt, private communication].

To assess the potential of unusually low rates of mass loss giving rise to SNe IA from single star progenitors we have performed a set of calculations with the mass-loss rate reduced by factors of up to 100 systematically throughout the AGB lifetime of stars of various masses (Fig. 5). For any given rate less than about one-twentieth of what is considered normal for AGB stars, there is always a small range of masses in which the hydrogen envelope is very thin and yet the core mass reaches the ignition mass.

5. CONCLUSIONS

Type Ia SNe are thermonuclear explosions of degenerate carbon/oxygen cores at just below the Chandrasekhar mass. The progenitors are otherwise unidentified but are nearly always assumed to be accreting white dwarfs. However we have demonstrated that single stars with reduced mass loss are not to be ignored. We have not however presented a mechanism for reducing the mass-loss rate but note that at lower metallicities the mass loss is likely to be slower and the core can grow somewhat faster.

We would expect several consequences of these ideas. Any single star progenitors are likely to be over $2M_\odot$ (Fig. 5) and so such SNe Ia should be associated with younger galaxies with recent star formation. There is some indication that there is such a subset of progenitors [20]. The incidence of these SNe Ia should increase with red shift as metallicity falls. There should be a set of SNe II that have very little hydrogen and evolve into SNe Ia as they decay. Again there is some evidence that this is the case [Eldridge private communication].

ACKNOWLEDGMENTS

CAT thanks Churchill College for his Fellowship and Monash University and the Australian National University for hospitality. We thank Brian Schmidt for interesting conversations and Rebecca Martin for the preparation of Fig. 1 and 2.

REFERENCES

1. M. Rowan-Robinson, *MNRAS* **332**, 352 (2002)
2. M.M. Phillips, *ApJ* **413**, L105 (1993)
3. W. Baade, *ApJ* **88**, 285 (1938)
4. D. Branch, *ARA&A* **36**, 17 (1998)
5. S. Perlmutter, et al., *ApJ* **517**, 565 (1999)
6. A.G. Riess, et al., *AJ* **116**, 1009 (1998)
7. D. Arnett and E. Livne, *ApJ* **427**, 315 (1994)
8. D. Arnett and E. Livne, *ApJ* **427**, 330 (1994)
9. B. Warner, *Cataclysmic Variables, CUP, Cambridge*, (1995)
10. B. Paczyński and A.N. Zytkow, *ApJ* **222**, 604 (1978)
11. P. Kahabka and E. P. J. van den Heuvel, *ARA&A* **35**, 69 (1997)
12. I. Hachisu, M. Kato, and K. Nomoto, *ApJ* **470**, L9 (1994)
13. G. Nelemans and C.A. Tout, *MNRAS* **356**, 753 (2005)
14. R.F. Webbink, *ApJ* **277**, 355 (1984)
15. K. Nomoto and I. Iben Jr., *ApJ* **297**, 531 (1985)
16. R.G. Martin and C.A. Tout, *MNRAS* **373**, 263 (2006)
17. D. Reimers, *Mem.Soc.R.Sci.Liège 6e Ser.* **8**, 369 (1975)
18. E. Vassiliadis and P.R. Wood, *ApJ* **413**, 641 (1993)
19. H.B. Lau, R. Stancliffe, and C.A. Tout, *MNRAS, submitted*, (2007)
20. F. Mannucci, M. Della Valle, and N. Panagia, *MNRAS* **370**, 773 (2006)

The Mysterious R Stars

Robert G. Izzard[*,†], C. Simon Jeffery[**] and John Lattanzio[†]

[*]Sterrekundig Institute, Universiteit Utrecht, Postbus 80000, 3508 TA, Utrecht, The Netherlands.
[†]School of Mathematical Sciences, PO Box 28M, Monash University, Victoria 3800, Australia.
[**]Armagh Observatory, College Hill, Armagh, BT61 9DG, Northern Ireland.

Abstract. The R stars are a rare class of K-type giant carbon stars. Canonical stellar evolutionary theory cannot explain their existence, yet they have been observed for more than a century. The early-R stars, the warmest in the R class, are enhanced in ^{12}C, ^{13}C and ^{14}N relative to the Sun, but not in *s*-processes elements or oxygen, and are all single stars. We test the idea that binary mergers lead to the formation of the early-R stars by a comparison of binary population synthesis model results with observations.

Keywords: R Stars, Binary Stars, Stellar Mergers, Carbon Stars, Nucleosynthesis, Red Clump
PACS: 97.10.Cv, 97.10.Kc, 97.10.Yp, 97.20.Li, 97.30.Hk, 97.80.Fk

1. INTRODUCTION

First observed by Secchi [1] and later classified by Fleming and Pickering [2], the R stars show molecular absorption bands which are typical of carbon stars, but are bluer ($T_{\rm eff} \gtrsim 4000$K) than normal N-type carbon stars ($T_{\rm eff} \lesssim 3500$K). Of the two R sub-classes, cool/late-R stars are probably misclassified N-type giants (i.e. thermally pulsing asymptotic giant branch, or TP-AGB, stars). The warm/early-R stars are chemically peculiar: enhancements are seen in ^{12}C, ^{13}C and ^{14}N but not oxygen, iron and *s*-process abundances [3]. Also, they are *all* single stars [4] which is very odd compared to a typical field binary fraction of 20% (among giants). Their luminosity is similar to the core helium burning *red clump* stars, of which they number 0.1% [5].

The early-R stars are Galactic thick disk objects of solar, or slightly sub-solar, metallicity, so are old and hence low-mass stars. They are not enhanced in *s*-process elements so have not been enriched by a companion TP-AGB star. They are too warm to be TP-AGB stars.

In canonical stellar evolution models the surface abundance of carbon *drops* during the first dredge up as a star evolves up the giant branch. After this the surface abundance does not change again until the AGB. In the case of the R stars something odd happens which makes them carbon rich throughout their core helium burning (red clump) phase, i.e. after they reach the red giant branch tip, and before they ascend the AGB.

We test the idea that the early-R stars form from the merger of a low-mass helium white dwarf and a red giant during a common envelope phase. The merger leads to the formation of a rapidly rotating core inside the giant star, which ignites helium and mixes the resulting carbon into the stellar envelope. A carbon-enhanced core-helium burning star results, i.e. an early R star.

For a more detailed review please see [6].

CP1001, 9th Torino Workshop on Evolution and Nucleosynthesis in AGB Stars and the 2nd Perugia Workshop on Nuclear Astrophysics,
edited by R. Guandalini, S. Palmerini, and M. Busso

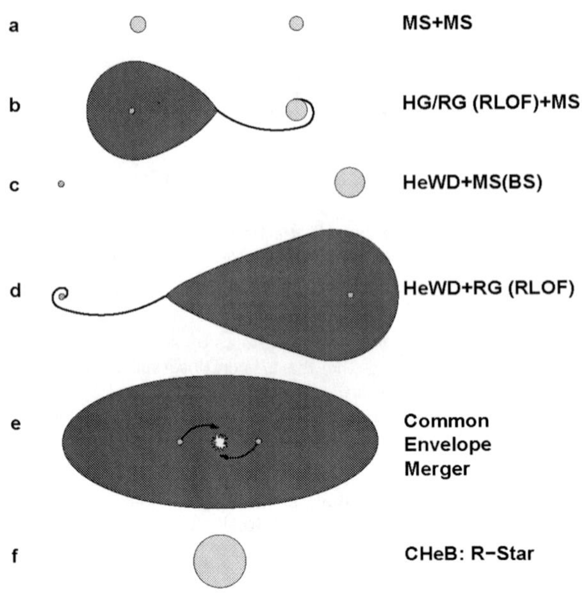

a The R-star begins as a zero-age binary with masses $M_1 \sim 1-1.7\,\mathrm{M}_\odot$ (left) and $M_2 \sim 0.3-1\,\mathrm{M}_\odot$ (right) in a short-period orbit of $P \sim 1-2\,$days (separation $\sim 5-8\,\mathrm{R}_\odot$).

b The primary evolves and expands, transferring material by stable Roche-lobe overflow to the companion. As a result, the orbit widens.

c The primary becomes a low-mass ($0.15-0.2\,\mathrm{M}_\odot$) helium white dwarf. The secondary is now a blue straggler: it is brighter and bluer than a star of its age should be.

d The secondary evolves and expands as it ascends the giant branch. The stars are more separated than at birth, so the secondary core grows to be more massive ($0.16-0.24\,\mathrm{M}_\odot$) than that of the primary. Roche-lobe overflow begins again.

e Mass transfer is unstable and both stars are engulfed in a common envelope. The cores spiral together and merge. Helium ignites in the rapidly rotating merged core and carbon is mixed into the envelope.

f The star, now single, settles into its core helium burning phase. Its envelope is carbon rich, so it is observed as an R-type star.

FIGURE 1. The binary evolutionary path to an early-R star.

2. MODEL

A binary merger model is a natural solution to the early R-star formation problem because it both leads to exclusively single stars and is a good method for spinning up the core, which may lead to unusual helium ignition and mixing of carbon.

We applied the population nucleosynthesis code of [7], based on the codes of [8] and [9], to simulate early R-star formation. Binary systems were tagged as early-R stars when:

- They were old, i.e. thick disk objects older than about 5 Gyr. This excludes TP-AGB stars which undergo third dredge up, as they are more massive ($\gtrsim 1.25\,M_{\odot}$) and hence evolved too quickly to be early-R stars.
- They pass through a common envelope phase and the two cores, both made of helium, merge. The most common channel is a red giant merging with a low-mass helium white dwarf.

We assumed a constant star formation rate, the [10] initial mass function and assumed all binaries are circular. All stars in the core-helium burning phase were classed as *red clump* stars. We changed the free parameters in our common envelope prescription within reasonable limits and found there was little impact on our results.

Note that at solar metallicity ($Z = 0.02$) the early-R stars *cannot* be formed by mass accretion from a companion TP-AGB star because there is simply too much oxygen in the star for $C/O > 1$ to be possible: accreted carbon is locked up in carbon monoxide.

The main evolutionary path which makes the early-R stars is that of a helium white dwarf merging with a red giant, as shown in Fig. 1. After helium ignition, we assume enough carbon is mixed to the stellar surface to convert the star into a carbon star. This material must mix through the hydrogen-burning shell, where some of it can be converted into ^{13}C and ^{14}N. If this mixing is fast enough, and not too hot, the ON cycle is not activated and oxygen is not depleted.

3. RESULTS

If we assume that *all* red giant-helium white dwarf mergers make R stars we find, from our $Z = 0.02$ models, an early-R to red clump (core helium burning) number ratio of about 1%. This should be compared to about 0.1% from Hipparcos [5]. Our results are a factor of ten too high but this should be considered as a positive result: it is very likely that we have neglected some physics of the merger which prevents mixing of carbon from the core to the envelope. Similarly, common envelope evolution is poorly understood and our simple α-prescription may be inadequate.

We have ignored other merger channels, such as red giant-Hertzsprung gap and Hertzsprung gap-helium white dwarf mergers, which are less numerous by an order of magnitude or more in our simulations ([6] give a detailed discussion). These cannot be ruled out – they may well make some early-R stars.

In a population with a metallicity less than $Z \sim 0.002$ the (extrinsic) CH stars dominate the helium-burning carbon star population. These are made by accretion of carbon from a companion TP-AGB star and hence are both *s*-process rich and binaries. This metallicity corresponds to the Galactic halo, which is indeed observed to be rich in CH stars. These stars are *not* the early-R stars but in surveys detailed abundance and/or binarity data are often not available, so there may be some confusion between R and CH stars at intermediate metallicities in old populations.

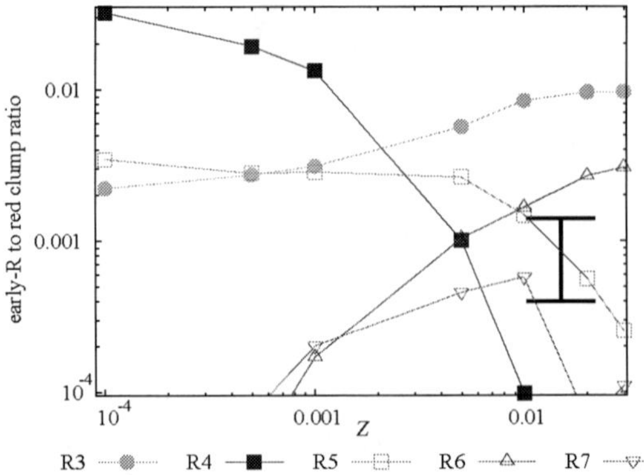

FIGURE 2. Our simulated early-R to core helium burning/red clump number ratios. The R*n* types are: **R3** Red giant-helium white dwarf merger, **R4** CH star (accretion from a TP-AGB primary, now a carbon-oxygen white dwarf), **R5** Hertzsprung gap-helium white dwarf merger, **R6** Hertzsprung gap-red giant merger, **R7** Hertzsprung gap-Hertzsprung gap merger. The Hipparcos ratio [5] is shown by the black error bar. Our simulations make too many early-R stars but this is a positive result because we assumed that *all* helium-core mergers make early-R stars.

4. CONCLUSIONS AND OUTSTANDING ISSUES

Our simulations lead us to conclude that the red giant-helium white dwarf merger channel is the likely route by which the early-R stars are formed. Although our modelled early-R to red clump ratio is too large by an order of magnitude compared to Hipparcos observations, we consider this a positive result. We briefly describe some issues we did not address.

If we assume that angular momentum is conserved in the helium cores during the spiral-in process and resulting merger, and if we assume a typical (non-rotating, spherical!) helium white dwarf radius for the merged core, then the merged core rotates faster than its breakup velocity. A true merged core will be deformed by its own rotation, so may remain stable to breakup, but may also lose some mass into and exchange angular momentum with the common envelope e.g. by interacting magnetic fields.

Detailed simulations of the above mechanisms are some way off, but we can speculate that if interaction between the merged core and surrounding material is strong enough, the core may slow its rotation sufficiently *before* helium ignition so that a helium flash with mixing does not occur (i.e. the helium flash follows the canonical model). This may reduce our model early-R star formation rate to better fit the observations. On the other hand, if there is a region of strong shear outside the spinning core any magnetic field present will be wound up and there may be little transfer of angular momentum between the core and the envelope, at least before helium ignition.

If the post-merger objects are rapidly rotating they may resemble the FK Com stars. However, the early-R stars are not observed to be rapidly rotating. Mass loss and/or magnetic braking during the evolution from the tip of the giant branch to the red clump may be responsible for slowing the R-star spin.

After the first phase of mass transfer but prior to the merger the pre-R star binary consists of a helium white dwarf and a blue straggler, i.e. a main sequence star which is bluer and more massive than stars of the same age. In a globular cluster such stars would be spotted quite easily (even if their white dwarf companion would not).

To conclude, in the hundred years since the discovery of the early-R stars we have made some progress toward understanding their origin. We do not understand binary mergers very well but perhaps we have some idea of the binaries which are progenitors of the early-R stars. In the future we require detailed simulations of both the core-merger process and subsequent stellar evolution. Only then will we start to truly understand these strange and fascinating objects.

ACKNOWLEDGMENTS

We began work on the early-R stars at the *Nucleosynthesis in Binary Stars* workshop at the Lorentz Centre, Leiden, The Netherlands. RGI thanks the Nederlandse Organisatie voor Wetenschappelijk Onderzoek (NWO), Armagh Observatory and Monash University for funding and allowing him to visit for extended periods. RGI also thanks JL's mother for the wonderful food and JL for much fine wine. CSJ acknowledges financial support from the Northern Ireland Dept. of Culture, Arts and Leisure (DCALNI), the UK Particle Physics and Astronomy Research Council (PPARC, now STFC) and the Netherlands Research School for Astronomy (NOVA). This research has made use of the SIMBAD database, operated at CDS, Strasbourg, France. Other aspects of this research were supported by the DEST and Australian Research Council.

REFERENCES

1. A. Secchi, *MNRAS* **28**, 196 (1868)
2. W. Fleming and E. C. Pickering, *Harvard College Observatory Circular* **145**, 1 (1908)
3. J. F. Dominy, *ApJS* **55**, 27 (1984)
4. R. D. McClure , *PASP* **109**, 256 (1997)
5. G. Knapp, D. Pourbaix, and A. Jorissen, *A&A* **371**, 222 (2001)
6. R. G. Izzard, C. S. Jeffery, and J. Lattanzio, *A&A* **470**, 661 (2007)
7. R. G. Izzard, L. M. Dray, A. I. Karakas, M. Lugaro, and C. A. Tout, *A&A* **460**, 565 (2006)
8. J. R. Hurley, C. A. Tout, and O. R. Pols, *MNRAS* **329**, 897 (2002)
9. R. G. Izzard, C. A. Tout, A. I. Karakas, and O. R. Pols, *MNRAS* **350**, 407 (2004)
10. P. Kroupa, C. Tout, and G. Gilmore, *MNRAS* **262**, 545 (1993)

SESSION II:

AGB STARS IN CLUSTERS AND THE CHEMICAL EVOLUTION OF GALAXIES

Lithium, AGB and the Formation of Globular Clusters

Francesca D'Antona and Paolo Ventura

INAF-OAR, via di Frascati 33, I-00040 Monteporzio (Italy)

Abstract. The possible sites of hot CNO processing for the gas which allowed the formation of "second generation", chemically anomalous, stars in Globular Clusters are either the envelopes of massive rotating stars or the envelopes of massive asymptotic giant branch (AGB) stars. As the massive stars destroy Lithium, while the massive AGBs can also produce it, the Lithium abundance of unevolved stars of GCs could be a powerful discriminant between the two scenarios. We show the status of art of the computation of Lithium production in massive AGBs for different population II metallicities, provide explanations for the trends with metallicity and mass loss rate of the models, and compare the results with observations. Unfortunately, today's determinations of Lithium abundance in GC stars are not sufficient, and not enough precise, to allow a choice between self–enrichment by AGB or massive star models.

Keywords: Globular Clusters; Formation; Chemical Evolution; Lithium
PACS: 97.10.Cv, 97.10.Me, 97.10.Vm, 97.20.Tr, 98.20.Gm

1. INTRODUCTION

In the last decade, spectroscopic and photometric observations have clarified that most of Globular Clusters (GCs) in our Galaxy are not simple stellar populations. Together with "normal" stars, having abundances of light elements very similar to those found among the halo stars, they contain an important fraction (at least 30-50% in most clusters) of objects showing signatures of hot CNO – NeNa – MgAl processing, also among the scarcely evolved turnoff and subgiant stars. It is now commonly believed that, at the beginning of the GC formation, there has been a first epoch of star formation that gave origin to the "normal" (first generation) stars, with CNO and other abundances similar to the population II field stars of the same metallicity. Afterwards —within a few million years, or a few tens of million years— there has been some other epoch of star formation, including material heavily processed through the CNO cycle. This material either was entirely ejected by stars belonging to the first stellar generation, or, more probably, it is a mixture of ejected and pristine matter of the initial star forming cloud. This tentative scheme of evolution relies on the evidence that there is no appreciable difference in the metallicity between the "normal" and the chemically anomalous stars belonging to the same GC. The matter must have been processed through the hot CNO cycle, and not, or only marginally, through the helium burning phases, as the sum of CNO elements is the same in the "normal" and anomalous stars [1], or at most it is a factor $\sim$2 larger in the anomalous stars [2]. The progenitors then may be either massive Asymptotic Giant Branch (AGB) stars [3, 4] or fast rotating massive stars [5]. In the first case, the hot

CP1001, *9th Torino Workshop on Evolution and Nucleosynthesis in AGB Stars and the 2nd Perugia Workshop on Nuclear Astrophysics*, edited by R. Guandalini, S. Palmerini, and M. Busso

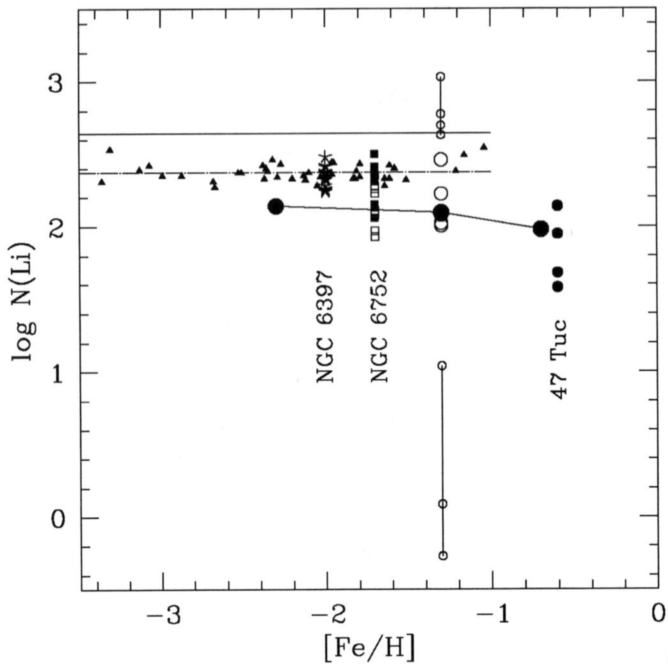

FIGURE 1. The Lithium abundances in halo dwarfs, as a function of the metallicity, from Meléndez and Ramírez [6] are plotted as small full triangles. Their average at $\log \varepsilon(Li) = 2.37$ is also drawn dash dotted, while the full line at $\log \varepsilon(Li) = 2.64$ represents the Big Bang abundance based on the WMAP baryon density from [7]. The points at [Fe/H]=-2, -1.7, and -0.6 are the data available for dwarfs in NGC 6397 [8], NGC 6752 [9], and 47 Tuc [10]. At [Fe/H]=-1.3, we have plotted as large open circles the Lithium yields from massive AGBs according to the models by Ventura and D'Antona [11] with standard mass loss rate (see text). The big full dots are the yields from $5M_\odot$ models of different metallicities (Ventura and D'Antona, in preparation). At [Fe/H]=-1.3, the small open dots at the bottom represent the Lithium yields from models adopting Vassiliadis and Wood [12] mass loss formulation, while the upper small open dots represent yields of models having mass loss rate a factor five larger than in the standard models.

CNO processing occurs at the hot bottom of the deep convective envelopes during the thermal pulse (TP) phase, and a few dredge up episodes may contribute to alter the elements abundance. The envelopes are lost in low velocity winds and planetary nebula ejection, and easily remain into the cluster and may form stars, possibly mixing with remnant pristine material. In the case of the envelopes of fast rotating massive stars, mixing occurs through meridional circulation and, if the stars are rotating at break up velocity, the matter enriched by the CNO burning products is easily lost with small momentum, mixes with pristine material, and may contribute to further star formation.

A key element to discriminate between these two scenarios is certainly the light and fragile element Lithium, which is only destroyed in massive stars, but can also be manifactured in the massive AGBs, through Hot Bottom Burning (HBB). In the following we

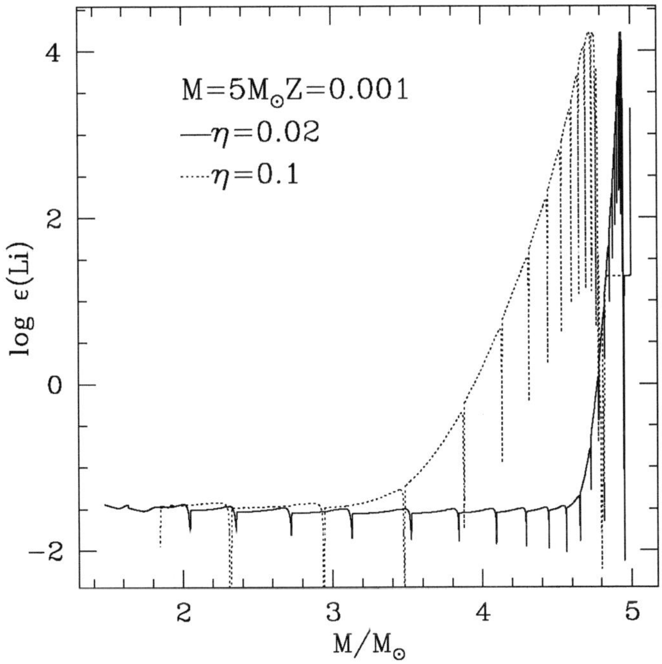

FIGURE 2. Lithium abundance in the envelope of a $5M_{\odot}$ star along its AGB evolution. The abscissa is the current mass of the star, which is reduced by mass loss. The dashed line refers to a mass loss 5 times larger than the mass loss in the full line.

review the importance of Lithium in the cosmological context, and consider the differences between the measured Lithium abundances in halo stars and in GCs. New results of AGB modelling of Lithium production are discussed. Unfortunately, there are still too few measurements of Lithium in the non evolved GC stars. In one case (NGC 6752) the resulting abundance may be compatible with formation of second generation stars in primordial matter mixed with the matter processed in the envelopes of fast rotating massive stars, while other cases are more ambiguous, and preliminary new results on NGC 6397 are better accomodated in the AGB scenario.

2. THE IMPORTANCE OF LITHIUM IN COSMOLOGY

Until a few years ago, measurements of the primordial light element abundances were regarded as a relevant test of the standard Big Bang Nucleosynthesis (BBN), and as a way to constrain the mean baryon density of the Universe. After this latter has been more precisely determined through the cosmic microwave background fluctuations, and especially through the WMAP [7], now the attention has shifted to understand the "disagreements" of the primordial abundances with those inferred by BBN. The most prominent problem is with the ^{7}Li abundance measured at the surface of population II dwarfs, that, in a simple minded interpretation, should be representative of the Big Bang abundance. The WMAP Lithium, in fact, has a logarithmic abundance $\log\varepsilon(Li) = 2.64 \pm 0.03$[1] a value about a factor three larger than that for the metal poor halo stars by [13] ($\log\varepsilon(Li) = 2.24$) and in any case larger than the recent value given by [6] ($\log\varepsilon(Li) = 2.37 \pm 0.05$). An accepted physical interpretation of this discrepancy is that slow diffusion processes at the surface of population II dwarfs deplete Lithium (as well as iron). Korn et al. [14], on a dataset of 18 turnoff stars in the GC NGC 6397 have shown that both the iron and Lithium abundances vary and may be consistently explained in the framework of the models by [15] that include consideration of diffusion. The predictions of models including atomic diffusion are tested [14, 16] from an analysis of stars at different evolutionary stages in NGC 6397, finding a very good agreement for a precise choice for parameterizing the turbulent mixing below the formal convection zone. The analysis by Korn et al. is aimed at finding differences due to the evolution, so the authors average the values of Lithium abundance of the turnoff stars —these, however, are very similar for the 5 stars examined. On the other hand, Lithium abundances in GC turnoff stars exhibit much larger variations than in field stars of similar metallicity. This is shown in Fig. 1, where we collect the halo data from [6] and recent data for three GCs at different metallicities.

Therefore, the abundances of Population II stars seem to be affected by atomic diffusion, but, in addition, the abundances in GCs show variations which are probably connected to the other chemical anomalies.

3. THE LITHIUM YIELDS FROM AGBS

Together with the observational data, we show in Fig. 1 the results of model computation for Lithium in the ejecta of AGB stars. Lithium is temporarily produced in the envelopes of massive AGBs, those suffering the "Hot Bottom Burning" (HBB), through the Cameron–Fowler [17] mechanism: ^{3}He left in the envelope by previous incomplete p–p burning can nuclearly react with ^{4}He producing the unstable nucleus of ^{7}Be. Convection brings the Beryllium into the outer cooler parts of the envelope, where it decays into Lithium. Of course, Lithium goes back to the hot bottom where it burns, but the whole operation timespan is long enough that Lithium in the envelope reaches even very high abundances before all the ^{3}He is exhausted. For this reason the most luminous AGBs in the Magellanic Clouds show large overabundances of Lithium [18], and many of the most massive AGBs of the Galaxy —objects which are heavily obscured by their circumstellar envelope, and have been first discovered as OH–masers— have high Lithium too [19][2]. Thus observations show that the Cameron–Fowler mechanism works in the massive AGBs we see today. It must have been working also in the massive AGBs of the GCs, those which could be responsible for self–enrichment. Notice however that the Lithium yields from the models are *very much dependent on the adopted mass loss*

[1] the Lithium abundance is expressed as a logarithmic abundance, $\log\varepsilon(Li) = 12 + \log_{10}(Li/H)$.
[2] in addition to extremely high Rubidium [20], but that is another story.

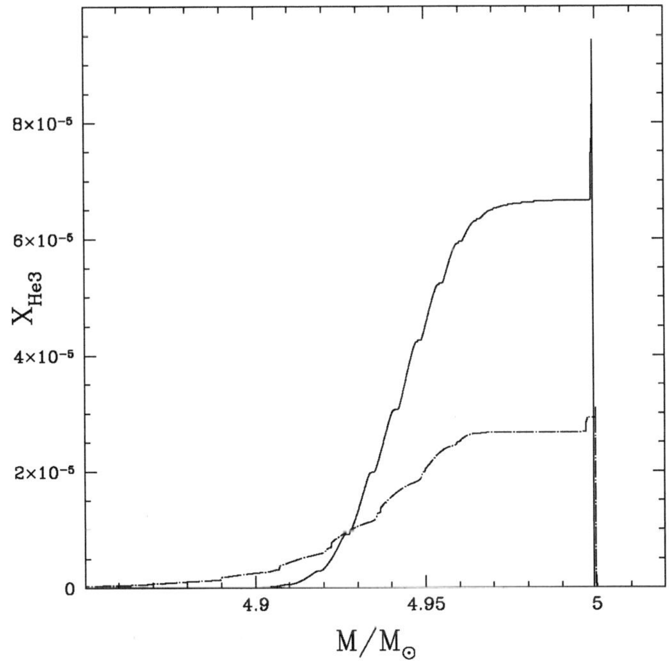

FIGURE 3. ^{3}He abundance in the envelope of a 5M$_\odot$star along its AGB evolution. The abscissa is the current mass of the star, which is reduced by mass loss. The dashed line refers to a metallicity Z=0.004, and the full line to Z=0.001.

rate. The mass loss rate we adopt in our models (see, e.g. [21]) follows the formulation by Blöcker [22], and has been calibrated in order to be consistent with the fact that *all* the most luminous AGBs in the Magellanic Clouds are Lithium rich [23]. For such a choice, we see that the yields provide an abundance close to the abundances found in GC stars. If we take into account that diffusion will act to reduce the atmospheric abundance, we probably need a somewhat larger efficiency of mass loss. In Fig. 1 we see that an increase of the rate by a factor 5 produces Lithium abundance much larger than needed. Viceversa, using the Vassiliadis and Wood [12] rates, the Lithium production in our models is insignificant. The behaviour of Lithium yield with mass loss is easily understood if we take into account that Lithium survives only for a short time in the HBB envelope: the faster is mass loss, the more Lithium will be ejected before it burns at the bottom of the envelope. This is shown in Fig. 2, where the Lithium abundance in a 5M$_\odot$ models is shown as function of the total stellar mass (due to mass loss, this is a kind of temporal abscissa, and provides a simple idea of the global yield in the ejecta).

The GC data show that the largest metallicity cluster examined, 47 Tuc, has on average smaller Lithium abundances. There may be several different reasons for this fact, but,

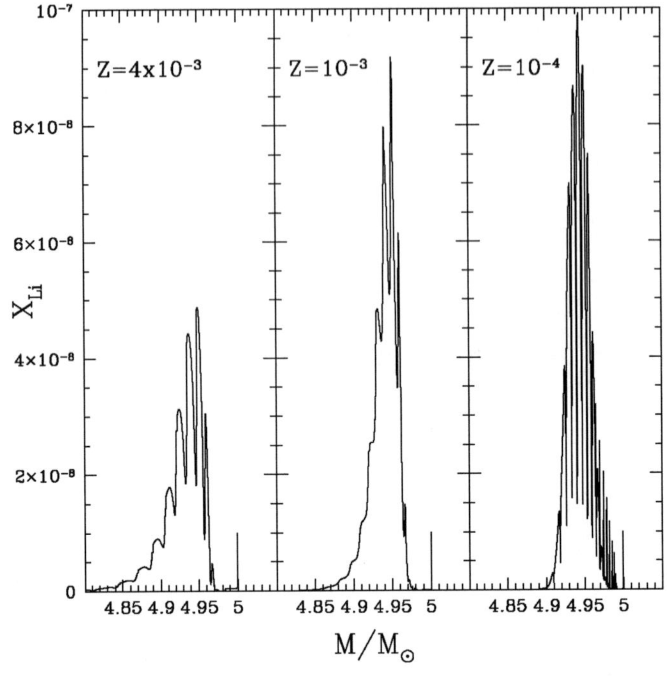

FIGURE 4. Lithium abundance along the evolution of a $5M_\odot$ of three different metallicities. The maximum production of Lithium is obtained at the lowest $Z=10^{-4}$.

interestingly, our new yields predict a decrease of the Lithium yield with Z[3], as shown by the big dots in Fig. 1. Figure 3 shows the reason for this behaviour: the ^{3}He reservoir in the star is larger for smaller Z, so that the production of Lithium (Fig. 4) decreases with Z.

4. DILUTION MODELS

A way to understand whether AGBs or rotating massive stars are the cause of the chemical anomalies is to compare the Lithium variations versus the variations in the abundance some other elements taking part in the hot CNO burning. For instance, the data for NGC 6752 [9] show an anticorrelation Lithium vs. sodium which can be interpreted as the result of mixing of high Lithium – low sodium primordial matter with matter processed in massive rotating stars and having no Lithium – high sodium, finding a good agreement [5]. On the other hand, also the self–enrichment by AGBs needs some dilution with pristine matter to reproduce the Na–O anticorrelation [24, 11] and may be

[3] actually, our old yields [4] predicted an increase of Lithium with Z

able to reproduce the same data, although the dilution factors for the AGB matter can be smaller than in the case of massive stars, thanks to the fact that some Lithium is present in the AGB ejecta (Ventura and D'Antona, in preparation). Concerning the other GCs, the data are scarce and do not form a coherent pattern. We wish to mention an intriguing unpublished result by Pasquini et al. (2007, in preparation) concerning two stars in NGC 6397: these have quite close Lithium abundance ($\log\varepsilon$(Li)$\sim$2.3), similar Sodium and Nitrogen abundances, but their Oxygen abundance differ by $\sim$0.6dex! If one star has "primordial" Lithium, the one depleted in Oxygen by a factor 4 can not have primordial Lithium only. Why not? The first one can have been born in the first generation, with primordial Li and α–enhanced O_{in}, and that the second one is a mixture of primordial matter and matter in which Oxygen is depleted to the level O_{depl}, depending on the fraction β of pristine gas so that:

$$\frac{O_{in}}{4} = \beta O_{in} + (1-\beta) O_{depl} \tag{1}$$

we have simply that

$$O_{depl} = O_{in} \frac{1/4-\beta}{1-\beta} \tag{2}$$

So the fraction β depends on O_{depl}. As it is difficult that this value is smaller than $\sim$–0.8 dex in models either of AGB or of massive stars, the maximum fraction of pristine matter in the star with depleted oxygen should be $\beta \sim 0.1$. Starting from primordial Lithium 2.64, if Lithium is only destroyed in the processed matter, the maximum abundance in the second star should be $\sim$ 1.5 and not 2.3 as observed. These observations, therefore, are more in favour of AGB pollutors than of massive stars pollutors: obviously, we need many more observations before we can solve the conundrum.

ACKNOWLEDGMENTS

We thank the organizer for this very successful and stimulating meeting. F.D. dedicates this work to the colleague and friend Mario Perinotto.

REFERENCES

1. I. G. Cohen and J. Meléndez, *AJ* **129**, 303(2005)
2. E. Carretta, R. G. Gratton, S. Lucatello, A. Bragaglia, and P. Bonifacio, *A&A* **433**, 597 (2005)
3. P. Ventura, F. D'Antona, I. Mazzitelli, and R. Gratton, *ApJ Letters* **550**, L65 (2001)
4. P. Ventura, F. D'Antona, and I. Mazzitelli, *A&A* **393**, 215 (2002)
5. T. Decressin, G. Meynet, C. Charbonnel, N. Prantzos, and S. Ekström, *A&A* **464**, 1029 (2007)
6. J. Meléndez and I. Ramírez, *ApJ Letters* **615**, L33 (2004)
7. D. N. Spergel, et al., *ApJ Suppl.* **170**, 377 (2007)
8. P. Bonifacio, et al., *A&A* **390**, 91 (2002)
9. L. Pasquini, P. Bonifacio, P. Molaro, P. Francois, F. Spite, R. G. Gratton, E. Carretta, and B. Wolff, *A&A* **441**, 549 (2005)
10. P. Bonifacio, et al., *A&A* **470**, 153 (2007)
11. P. Ventura and F. D'Antona, *A&A*, in press (2008)
12. E. Vassiliadis and P. R. Wood, *ApJ* **413**, 641 (1993)

13. P. Bonifacio and P. Molaro, *MNRAS* **285**, 847 (1997)
14. A. J. Korn, F. Grundahl, O. Richard, P. S. Barklem, L. Mashonkina, R. Collet, N. Piskunov, and B. Gustafsson, *Nature* **442**, 657 (2006)
15. O. Richard, G. Michaud, and J. Richer, *ApJ* **580**, 1100 (2002)
16. A. J. Korn, F. Grundahl, O. Richard, L. Mashonkina, P. S. Barklem, R. Collet, B. Gustafsson, and N. Piskunov, *ApJ* **671**, 402 (2007)
17. A.G.W. Cameron and W.A. Fowler, *ApJ* **164**, 111 (1971)
18. B. Plez, V. V. Smith, and D. L. Lambert, *ApJ*, **418**, 812 (1993)
19. D. A. García-Hernández, P. García-Lario, B. Plez, A. Manchado, F. D'Antona, J. Lub, and H. Habing, *A&A* **462**, 711 (2007)
20. D. A. García-Hernández, P. García-Lario, B. Plez, F. D'Antona, A. Manchado, and J. M. Trigo-Rodríguez, *Science* **314**, 1751 (2006)
21. P. Ventura and F. D'Antona, *A&A* **439**, 1075 (2005)
22. T. Blöcker, *A&A* **297**, 727 (1995)
23. P. Ventura, F. D'Antona, and I. Mazzitelli, *A&A* **363**, 605 (2000)
24. F. D'Antona and P. Ventura, *MNRAS* **379**, 1431 (2007)

The Self-Enrichment Scenario in Intermediate Metallicity Globular Clusters

P. Ventura and F. D'Antona

INAF - Observatory of Rome. Via Frascati 33 - 00040 MontePorzio Catone - Italy

Abstract. We test the reliability of the self-enrichment scenario for Globular Clusters of intermediate metallicity, i.e. [Fe/H]~ -1.3. To this scope, we calculated new models of intermediate mass stars evolved during the Asymptotic giant branch phase, with the latest updates of the input physics and with an $\alpha-$enhanced mixture $[\alpha/\mathrm{Fe}]$=+0.4. We compare the theoretical yields with the observed patterns of stars in Globular Clusters, and confirm that when convection is modelled efficiently, the chemical composition of the ejecta of the most massive models are in agreement with the observed abundance patterns.

Keywords: Stars: Abundances – Stars: AGB and Post-AGB – Globular Clusters: General
PACS: 43.35.Ei, 78.60.Mq

1. INTRODUCTION

Globular Clusters (GC) stars are far from representing a chemically homogeneous sample: deep spectroscopic analysis evidentiate clear abundance patterns among the light elements [1, 2]. Oxygen is anticorrelated with sodium and alluminum, and is correlated with fluorine and, at a smaller extent, with magnesium. The total C+N+O content is observed to be approximately constant within a factor 2 [3]. Since these patterns suggest nuclear processing, in situ mechanisms related to some non canonical extra mixing during the RGB phase were invoked to account for these chemical anomalies [4]; yet, the discovery of such patterns also in TO and SGB stars [5], and the absence of such patterns in field stars, point in favour of a self-enrichment scenario working within GCs. An appealing explanation might be that an early generation of intermediate mass stars (4-6$M_\odot$) evolved rapidly ($100-200$Myr), and during their Asymptotic Giant Branch (AGB) phase polluted the interstellar medium with matter which, due to Hot Bottom Burning (HBB) within their massive convective envelope, should show the signature of advanced CNO nucleosynthesis [6].

In this contribution we concentrate on GCs of intermediate metallicity ([Fe/H]~ -1.3, i.e. Z=0.001), and present the latest generation of AGB models calculated by means of the ATON code for stellar evolution, with the most updated equation of state and conductive opacities. The initial mixture is assumed to be $\alpha-$enhanced, to allow a better investigation of our working hypothesis.

The results obtained are compared with the observed abundance patterns of GCs stars. Our points is clear: do exist some assumptions concerning the chemical and physical inputs that render the chemical composition of the ejecta of massive AGBs compatible with the patterns evidentiated by the spectroscopic investigations of GCs? In particular, we will focus on the O-Al and O-Na anticorrelations, which are the most analyzed in the

CP1001, 9th *Torino Workshop on Evolution and Nucleosynthesis in AGB Stars and the* 2nd *Perugia Workshop on Nuclear Astrophysics*, edited by R. Guandalini, S. Palmerini, and M. Busso literature, and neglect the Mg-Al anticorrelation, as it is still under debate [7], and the quantitative reproduction of the fluorine depletion (which we find in our models), since ^{19}F has been measured so far only in bright giants [8], so that a possible role played by some non canonical extra-mixing during the RGB phase cannot be ruled out.

2. THE ATON CODE

The evolutionary computations discussed here have been calculated with the ATON code for stellar evolution, a full description of which can be found in [9]. The most recent updates are discussed and commented in [10].

The nuclear network includes 30 species, from hydrogen to silicum; all the reactions relevant for this study, involving the light elements mostly investigated in the spectroscopic analysis of GC stars, are properly considered.

Mass loss can be modelled according to the traditional Reimers' recipe, or with the two prescriptions more appropriate to describe the stellar winds during the AGB phase, i.e. [11] and [12].

Convection can be modelled either with the traditional Mixing Length Theory (MLT) [13], or by the Full Spectrum of Turbulence (FST) model [14].

Nuclear burning is coupled with mixing inside convective regions; the approach is diffusive-like, and the appropriate equation by Cloutman and Eoll [15] are solved; this, though more time consuming, is mandatory to allow a consistent description of the outer convective zone of AGBs, since a considerable fraction of the nuclear energy is generated within the envelope [16]. The overshoot of the convective eddies within radiative regions is represented by an exponential decay of convective velocities, with an e-folding distance given by ζH_p, where ζ is a free parameter.

The latest updates, specifically used for the present investigation, are the followings:

- We use the latest release of the conductive opacities by Potechkin [17]
- The new OPAL (2005) equation of state was implemented into the code
- The mixture can be assumed to be solar scaled, or $\alpha-$enhanced, with $[\alpha/\mathrm{Fe}]$[1]$=$ $+0.2, +0.4$.

3. RESULTS

We consider GCs of intermediate metallicity, with [Fe/H]~ -1.3. We followed the evolutions of stars with initial mass $3M_\odot \leq M \leq 6.3M_\odot$ with chemical composition Z=0.001 and Y=0.24. We chose the lower and upper limits for the mass, respectively, because of the absence of any HBB in masses with $M < 3M_\odot$, and of the off center ignition of carbon in models whose mass exceeds $6.3M_\odot$. The evolutions are followed until the mass of the envelope becomes so small to prevent HBB, i.e. $\sim 0.5 - 0.7M_\odot$. We modelled convection according to the FST treatment. The parameter giving the e-folding

---

[1] We use the standard notation: given the mass fraction X of a given element, and the iron abundance Fe, we define [X/Fe]=log(X/Fe)-log(X/Fe)$_\odot$

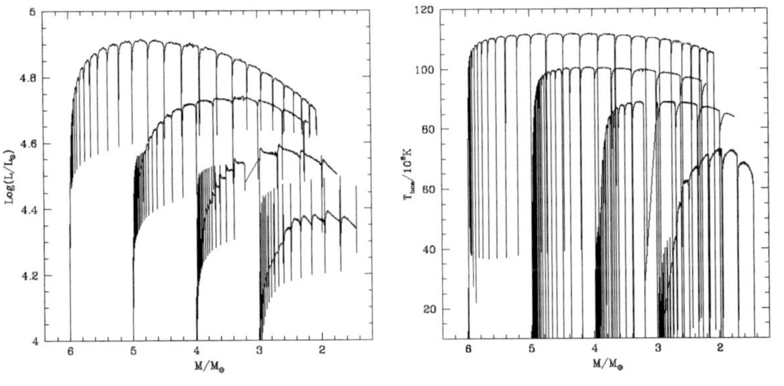

FIGURE 1. Variation with the total mass of the Luminosity (Left) and Temperature at the bottom of the convective envelope (Right) of some intermediate mass models during the AGB phase. For clarity reasons, only models corresponding to initial masses 3,4,5,6$M_\odot$ are shown.

distance of the exponential decay of velocities inside radiative regions is $\zeta = 0.02$, in agreement with a calibration based on the observed widths of the main sequences of open clusters given in [9]; no extra-mixing was assumed during the TDU. For mass loss we used the Blöcker recipe, with the free parameter η_R entering the formula chosen as $\eta_R = 0.02$, in agreement with a previous calibration based on the luminosity function of lithium rich stars in the Magellanic Clouds given in [18].

The evolution of the luminosity and the temperature T_{bce} at the bottom of the envelope of the models is shown in Fig.1. The rapid increase of the luminosity when T_{bce} exceeds $\sim 40 \times 10^6$K, which is the temperature at which HBB occurs, is a typical consequence of the AGB modelling based on the FST treatment of convection. The right panel of Fig.1 indicates a correspondence between initial mass and a typical T_{bce}, which remains approximately constant during the almost totality of the AGB evolution: this fixes the degree of the nucleosynthesis which is achieved at the bottom of the envelope. In the present investigation, the maximum temperature at the bottom of the envelope, found in the most massive models, is $T_{bce} = 110$MK.

Table 1 gives the chemical content of the ejecta of the models presented here. The high values of the helium mass fraction are almost entirely due to the second dredge-up: this provides robustness to this prediction, since it is independent of the choice of the free parameters entering the following AGB phase. The strong HBB achieved at the bottom of the envelope of these stars is also confirmed by the large depletion of fluorine, shown in col.5. Col.8 gives the ratio between the overall CNO content of the ejecta and the initial value: the most massive models, characterized by a strong mass loss and thus a small number of TPs, are expected to show only a modest increase of the CNO, due to the very few TDU episodes experienced.

The variation of the surface abundances of oxygen (left) and aluminum (right) during the AGB phase are shown in Fig.2; we choose the mass as abscissa, to have a direct idea of the chemistry of the ejecta. In the left panel we note that only masses $M \geq 4.5M_\odot$ destroy surface oxygen, whereas the less massive models, due to the lower T_{bce} and the

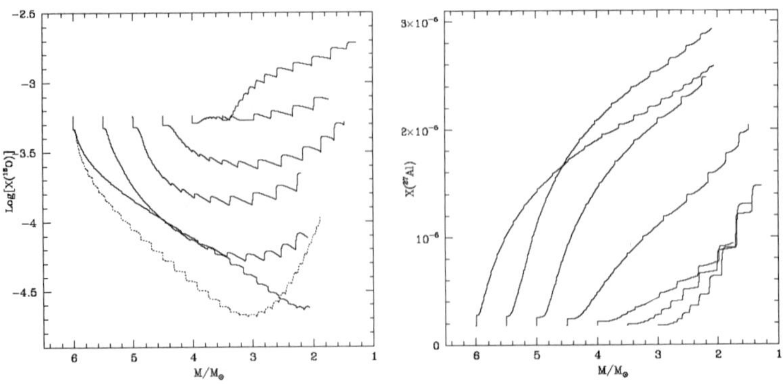

FIGURE 2. Variation with the total mass of the surface abundances of oxygen (Left) and alluminum (Right) of some intermediate mass models during the AGB phase. For clarity reasons, models corresponding to initial masses 3.5 and $6.3M_{\odot}$ were omitted. The dashed track corresponds to a $6M_{\odot}$ model calculated with a lower mass loss ($\eta_R = 0.01$).

many TDUs, produce it. In the upper range of masses investigated here the oxygen yield is approximately independent of mass, because increasing the mass determines a larger luminosity, hence a stronger mass loss, with a consequent higher mass loss in the early AGB phase, when the surface oxygen abundance is still high (note the different slopes of the oxygen-mass trend corresponding to the 5 and $6M_{\odot}$ cases in the left panel of Fig.2): this compensates the larger extent of the oxygen depletion expected on the basis of the larger T_{bce}. The dashed track, which represents a $6M_{\odot}$ model calculated with a mass loss rate artificially decreased by a factor 2, shows that the oxygen yield is also approximately independent of mass loss: when $\dot{M}$ is reduced, the star first ejects into the interstellar medium more oxygen poor material, but this is compensated by the larger number of TPs experienced, which acts to increase the oxygen abundance.

The oxygen content of the ejecta of the most massive models, which is found to be [O/Fe]=-0.4, turns out to be practically independent of the uncertainties associated to mass loss, cross sections, and the exact upper limit of the initial mass of stars undergoing the AGB phase, and is thus a very robust prediction of the present analysis. Within the context of the self-enrichment scenario, we identify this abundance of oxygen with the oxygen content of the most oxygen poor stars observed in the GCs: if we limits this comparison to scarcely evolved stars, to rule out any possible additional contribution of some kind of extra-mixing acting during the RGB phase, we see a very good agreement between the theoretical predictions and the observational evidence, as can be seen by looking at the oxygen abundances of the stars shown in Fig.4.

The right panel of Fig.2 evidentiates that Alluminum is produced in all AGBs, due to proton capture by the heavy magnesium isotopes, which are synthesized by HBB via ^{24}Mg burning, and also dredged-up after each TP. In this case, however, the reliability of the results is undermined by the uncertainties associated to the relevant reactions rates [19], which introduce an error on the theoretical predictions on [Al/Fe] of 0.3-0.4 dex; we can anyway confirm that massive AGBs produce great amounts of alluminum within

TABLE 1. Chemical content of the ejects of the AGB models.

$M/M_\odot$	Y	A(Li)	[^{16}O/Fe]	[^{19}F/Fe]	[^{23}Na/Fe]	[^{27}Al/Fe]	R(CNO)
3.0	.248	2.77	0.92	0.10	1.16	0.65	9.6
3.5	.265	2.43	0.77	-0.26	1.30	0.66	7.9
4.0	.281	2.20	0.44	-0.61	1.18	0.55	4.9
4.5	.310	2.00	0.19	-0.90	0.97	0.85	3.1
5.0	.324	1.98	-0.06	-1.16	0.60	1.02	2.1
5.5	.334	1.93	-0.35	-1.39	0.37	1.10	1.3
6.0	.343	2.02	-0.40	-1.36	0.31	1.04	0.97
6.3	.348	2.06	-0.37	-1.28	0.30	0.99	0.94

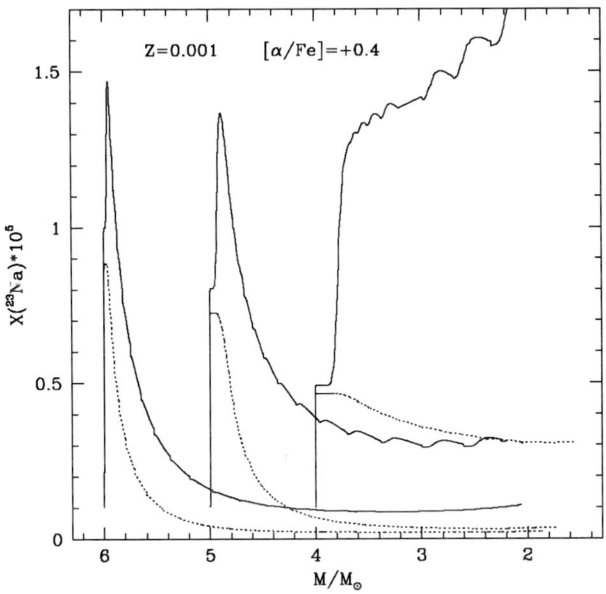

FIGURE 3. The evolution with mass of the surface abundance of sodium for models of 4,5,6$M_\odot$. The dashed tracks indicate the results obtained when the lower limits of the cross sections of the reaction ^{22}Ne(p,γ)^{23}Na are adopted.

their envelope (see col.7 of Table 1); the most massive models of our sample achieve an Al enhancement of [Al/Fe]$\sim$ 1 dex. These results are in good agreement with the largest alluminum abundances detected in GCs stars, which in fact show up an increase by a factor 10 compared to the solar scaled abundance [20]. We are left with sodium, which has been extensively measured in GCs stars. Figure 4 evidentiates that a clear oxygen-sodium anticorrelation is present in all the clusters analyzed, the stars most oxygen poor showing a sodium enrichment of the order of $\sim$ 0.5dex. The amount of sodium which may be produced within this class of objects is hard to predict, for the following reasons:

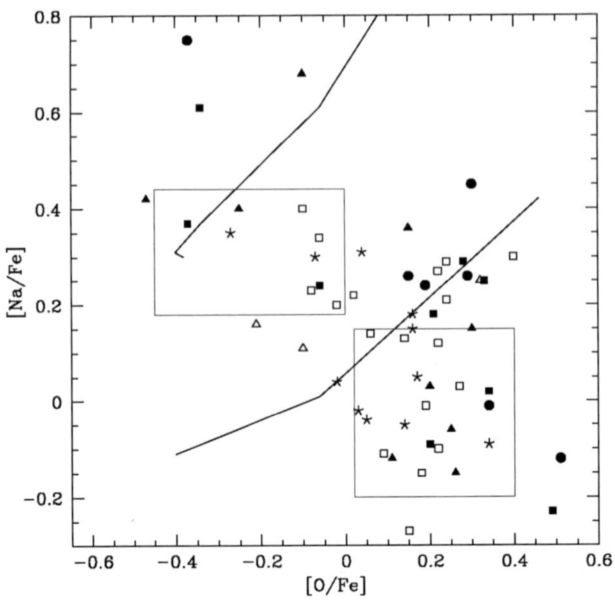

FIGURE 4. The oxygen and sodium abundances observed in scarcely evolved GC stars of intermediate metallicity. The meaning of symbols is as follows: Full squares: NGC 6752 [5]; Open squares: M5 stars with $V > 16$ [24]; Stars: M13 stars with $V > 15$ [7]; Open triangles: M3 stars with $V > 15$ [7]; Full points: NGC 6218 stars [23]; Open squares: High gravity M3 giants [20]; Full triangles: High gravity M13 giants [20]. The upper and lower solid tracks indicate the theoretical predictions of our models when the upper and lower limits of the cross sections of the proton capture reactions by ^{22}Ne nuclei are adopted.

a) the behaviour of sodium is not linear, because it is first produced at the bottom of the envelope due to proton capture by ^{22}Ne nuclei, and later destroyed when the destruction rate exceeds the production (see Fig.3); b) the uncertainties associated to the reaction rates are huge, of the order of 3 dex [21, 22]. Figure 3 compares the results in terms of the surface sodium abundance when the upper (solid) and lower (dotted) limits for the ^{22}Ne(p,γ)^{23}Na are adopted: these results, completely different between each other, give an idea of the modest predictive power of our models on this issue.

Comparing our models with the observed abundances (see Fig.4), we may only conclude that when the larger cross-sections for proton capture by ^{22}Ne nuclei are adopted, the yields of the most massive models are compatible with the observed abundances.

4. CONCLUSIONS

We presented new AGB models of intermediate mass with the latest updates regarding the input physics used to calculate the evolutionary sequences. In order the test the reliability of the self-enrichment idea for intermediate metallicity Globular Clusters, we assumed an $\alpha-$enhanced mixture, with [α/Fe]=+0.4.

We confirm previous results published by our group, i.e. that when convection is modelled efficiently, the bottom of the convective envelope of the most massive AGBs undergoes a strong nuclosynthesis, with the consequent change of the surface chemistry; physically, this leads to a larger luminosity, a higher mass loss rate, and reduced the number of thermal pulses and of third dredge-up episodes experienced by these stars.

All the models achieving Hot Bottom Burning destroy the surface oxygen: the oxygen content of the ejecta of models whose mass exceeds $5M_{\odot}$ is [O/Fe]=-0.4. This prediction is very robust, because it is independent of the free parameter entering the mass loss prescription, and of the uncertainties associated to the relevant cross sections.

The same models also produce sodium and alluminum, thus confirming a qualitative agreement with the observed anticorrelations of GCs stars. A confirm of this scenario on quantitative basis requires a better knowledge of the proton capture reactions by ^{22}Ne, ^{25}Mg and ^{26}Mg, which at the moment undermine the predictive power of the results obtained.

REFERENCES

1. R.P. Kraft, *PASP* **106**, 553 (1994)
2. E. Carretta, *AJ* **131**, 1766 (2006)
3. I.I. Ivans, C. Sneden, R.P. Kraft, et al., *AJ* **118**, 1273 (1999)
4. P. Denissenkov and A. Weiss, *ApJ* **559**, L115 (2001)
5. R. Gratton, P. Bonifacio, A. Bragaglia, et al., *A&A* **369**, 87 (2001)
6. P. Ventura, F. D'Antona, I. Mazzitelli, and R. Gratton, *ApJ* **550**, L65 (2001)
7. J.G. Cohen and J. Meléndez, *AJ* **129**, 303 (2005)
8. V.V. Smith, K. Cunha, I.I. Ivans, J.C. Lattanzio, S. Campbell, and K.H. Hinkle, *ApJ* **633**, 392 (2005)
9. P. Ventura, A. Zeppieri, F. D'Antona, and I. Mazzitelli, *A&A* **334**, 953 (1998)
10. P. Ventura and F. D'Antona, *A&A* **431**, 279 (2005)
11. T. Blöcker, *A&A* **297**, 727 (1995)
12. E. Vassiliadis and P.R. Wood, *ApJ* **413**, 641 (1993)
13. E. Vitense, *Zs.Ap.* **32**, 135 (1953)
14. V.M.C.Canuto and I. Mazzitelli, *ApJ* **370**, 295 (1991)
15. L. Cloutman and J.G. Eoll, *ApJ* **206**, 548 (1976)
16. I. Mazzitelli, F. D'Antona, and P. Ventura, *A&A* **348**, 846 (1999)
17. S. Cassisi, A.Y. Potekhin, A. Pietrinferni, M. Catelan, and M. Salaris, *ApJ* **661**, 1094 (2007)
18. P. Ventura, F. D'Antona, and I. Mazzitelli, *A&A* **363**, 605 (2000)
19. R.G. Izzard, M. Lugaro, A.I. Karakas, C. Iliadis, and M. van Raai, *A&A* **466**, 641 (2007)
20. C. Sneden, R.P. Kraft, P. Guhathakurta, R.C. Peterson, and J.P. Fulbright, *AJ* **127**, 2162 (2004)
21. S.E. Hale, A.E. Champagne, C. Iliadis, V.Y. Hansper, D.C. Powell, and J.C. Blackmon, *PhysRevC* **65**, 5801 (2002)
22. S.E. Hale, A.E. Champagne, C. Iliadis, V.Y. Hansper, D.C. Powell, and J.C. Blackmon, *PhysRevC* **70**, 5802 (2004)
23. E. Carretta, A. Bragaglia, R.G. Gratton, F. Leone, A. Recio-Blanco, and S. Lucatello, *A&A* **450**, 523 (2006)
24. S.V. Ramirez and J.G. Cohen, *AJ* **125**, 224 (2003)

Observing Third Dredge Up in NGC 1846

T. Lebzelter*, K.H. Hinkle†, M.T. Lederer*, S. Cristallo**, O. Straniero**, W. Nowotny* and P. Wood‡

*Institute f. Astronomy, Tuerkenschanzstrasse 17, A1180 Vienna, Austria
†NOAO, P.O.Box 26732, Tucson, AZ 85726, USA
**INAF, Osservatorio Astronomico di Collurania, 64100 Teramo, Italy
‡RSAA, Australian National University, Weston Creek, ACT 2611, Australia

Abstract. In this contribution we present the first measurements of the C/O ratio for a sample of AGB stars in the LMC globular cluster NGC 1846. From a variability study we set strong constraints on the mass of the cluster stars, which is an important quantity for comparison with evolutionary models. Variations of the surface abundance both with luminosity and pulsation properties were found and the various relevant factors are discussed. We find a good agreement with our synthetic models if we assume [O/Fe]=+0.2 dex. The F abundance seems to increase with the C/O ratio. As variability plays a key role on the AGB we give some preliminary results from our study of abundance determination in the dynamical case at the end of this paper.

Keywords: Stellar Evolution, Nucleosynthesis; Globular Clusters in External Galaxies
PACS: 97.10.Cv, 98.20.Jp

1. INTRODUCTION

During the short but decisive Asymptotic Giant Branch (AGB) phase a series of third dredge-up events is responsible for mixing nucleosynthesis products to the surface from where it can be ejected into the interstellar medium by stellar mass loss (see e.g. [1] for a review). Changes in the surface abundance of various key elements can be observed and allow to set constraints both on the nucleosynthesis going on in the stellar interior and the role of these stars in the cosmic matter cycle. Among the best indicators of past or ongoing third dredge-up is the abundance of ^{12}C. This product of He burning can be measured in the form of the C/O and $^{12}C/^{13}C$ ratios, respectively. ^{12}C has also a strong influence on the atmospheric structure and the overall appearance of the spectrum as its dredge up can slowly change a star from oxygen-rich to carbon-rich.

Current stellar evolution models (e.g. [2, 3] and various authors in this volume) provide quite detailed predictions for the surface abundance changes during the AGB phase. Different models agree qualitatively, quantitative checks of the model predictions for third dredge-up efficiency so far are restricted to AGB stars in the solar neighborhood (e.g.[3, 4]) or to indirect methods like the study of the progeny of AGB stars or synthetic stellar evolution (e.g. [5, 6]).

Clusters of stars offer an excellent possibility to investigate several questions of stellar astronomy. They provide samples of stars that are homogeneous in age and in general also in metallicity located at the same distance. Thus it is possible to determine the evolutionary status of cluster stars and their mass more accurately than for field stars. This provides an important advantage for comparison with models of stellar atmospheres

CP1001, 9th Torino Workshop on Evolution and Nucleosynthesis in AGB Stars and the 2nd Perugia Workshop on Nuclear Astrophysics, edited by R. Guandalini, S. Palmerini, and M. Busso

and evolution.

To study the third dredge-up the globular clusters of the Milky Way are too old. The Magellanic Clouds, however, contain a population of intermediate age clusters (e.g. [7]) with AGB stars in the mass range 1.5 to $2\,M_{\odot}$. In this paper we discuss three aspects in the context of investigating the properties of the third dredge-up observationally. First, we will summarize a study on the mass determination of AGB stars in clusters [8] – stellar variability offers here some new possibilities. Second, we will present results on the C/O ratio in AGB stars in one LMC cluster, NGC 1846, illustrating the change of abundances along the AGB (Lebzelter et al., submitted). Finally, we will briefly discuss the problems in abundance determination that can arise from the large amplitude and long period variability found among AGB stars (Lebzelter et al., in preparation).

NGC 1846 is an intermediate age cluster belonging to the LMC. It has a metallicity of [Fe/H]$= -0.49$ [9]. A number of AGB stars have been identified by Lloyd Evans [10] and we will use his naming convention throughout this paper. Near infrared measurements and estimates of the bolometric magnitudes of the AGB stars have been presented by [11] (and references therein). No stars with a high mid-infrared excess, an indicator of a very high mass loss rate, have been found [12, 8]. A few luminous C-stars are known in this cluster, indicating the occurrence of third dredge-up.

2. THE AGE AND MASS OF THE AGB STARS IN NGC 1846

Determining the mass of the AGB stars in this cluster more accurately is a challenging but necessary task to allow to set tight constraints for stellar evolution and dredge-up models. Two characteristics of cluster AGB stars turn out to be very helpful in this context, namely that these stars are typically long period variables and that, as cluster members, they all should have a very similar mass. Thus we can compare the observed period-luminosity relations of the cluster stars with P-L-relations from pulsation models of varying mass. The best-fit set of model relations would then constrain the mass of the AGB stars. This task has been undertaken by [8] for NGC 1846. As the visual amplitudes of long period variables typically exceed 0.1 mag the variability can be easily detected. However, their long periods require long term monitoring. In the case of NGC 1846 this was achieved by combining data from the MACHO archive with more recent data from CTIO obtained by the authors. The static and linear non-adiabatic pulsation models were created with updated versions of the pulsation codes described in [13]. The metallicity was chosen to be close to the literature value (Z=0.004). The opacities used are described in [8]. The linear pulsation models allow a good description of the variability behaviour as long as non-linearity, as it is seen in classical Miras (compare [14]), plays a minor role. Fortunately, it seems that only a small fraction of the AGB variables in NGC 1846 is affected by non-linearity.

Fig. 1, taken from [8], shows the P-L-diagram for the AGB variables in NGC 1846. The best-fit model sequences (for various pulsation modes) are also given. The bending of the model sequences at the highest luminosities is due to the switch from O-rich to C-rich chemistry and the related change in opacities. As a result of this comparison the mass of the AGB stars are found to be close to $1.8\,M_{\odot}$. This leads to a cluster age of 1.4×10^{9} years (using [15]).

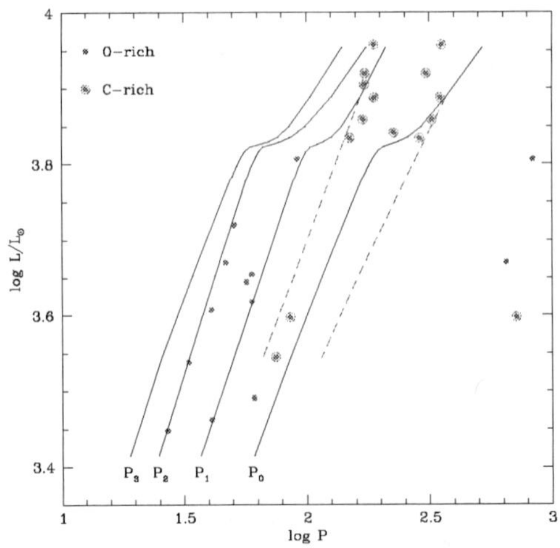

FIGURE 1. The $\log(L/L_{\odot})$ - log P diagram for NGC 1846 stars and models taken from [8]. The fundamental and the first three overtone modes are shown for models with L near the H-burning maximum of the He shall flash cycle (solid lines). For AGB stars in the He-burning phase of the flash cycle, only the fundamental and first overtone modes are shown (dashed lines).

An interesting feature of NGC 1846 is the possible existence of two populations in this cluster separated by about 300 Myr [16]. This aspect could not be tested from the variability data of [8].

3. C/O ALONG THE AGB

Having the mass of the stars constrained we can start to investigate the change of surface abundances due to third dredge-up. The fact that the most luminous AGB stars in this cluster are all carbon stars indicates that third dredge-up is taking place in these stars. We thus obtained high resolution near-infrared spectra of a sample of AGB stars in NGC 1846 to derive for each star the C/O and $^{12}C/^{13}C$ ratio, respectively. Spectra were obtained using the Phoenix spectrograph at Gemini South. The various abundances were determined by comparing the observed spectra with synthetic spectra from hydrostatic model atmospheres. For details on the data analysis we refer to Lebzelter et al. (submitted). Here we focus on the O-rich stars in the cluster.

We find a range of C/O values along the AGB of NGC 1846 as shown in Fig. 2. The C/O values and the $^{12}C/^{13}C$ are correlated with each other as expected (see Fig. 3). However, there is no simple relation between C/O and luminosity of the star. One might expect such a relation as the mean luminosity increases with the evolutionary status of a star along the AGB. We have to keep in mind that third dredge-up is linked to thermal pulses. Over the pulse cycle significant variations around the mean luminosity can be observed (see also Fig. 1).

In Fig. 3 we compare the measured C/O and $^{12}C/^{13}C$ values with model predictions.

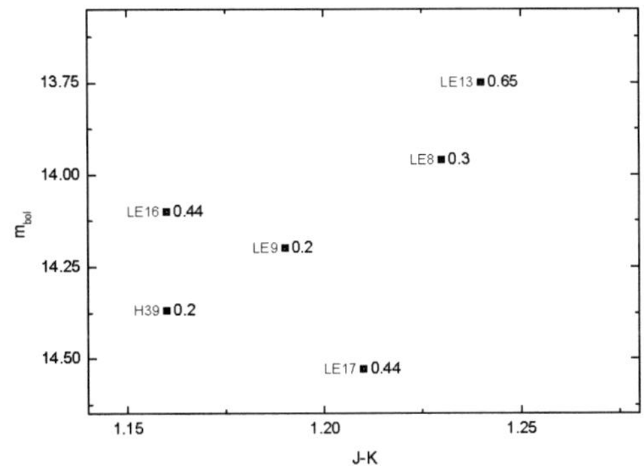

FIGURE 2. Colour-magnitude diagram of the O-rich stars in our sample. The C/O ratio of each data point is given. Taken from Lebzelter et al. (submitted).

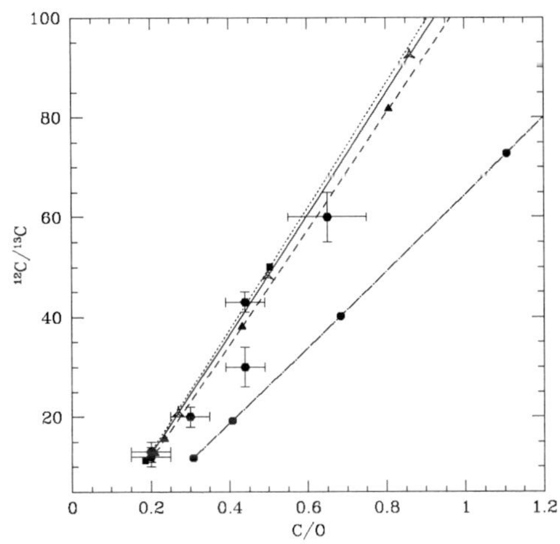

FIGURE 3. C/O versus $^{12}C/^{13}C$. For comparison predicted values from our evolutionary models are shown. The lines connect individual interpulse values for four sets of model parameters: Z=0.006, M=1.8 $M_{\odot}$, [O/Fe]=+0.2 (solid line); Z=0.003, M=1.8 $M_{\odot}$, [O/Fe]=+0.2 (dotted line); Z=0.006, M=1.5 $M_{\odot}$, [O/Fe]=+0.2 (dashed line); Z=0.006, M=1.8 $M_{\odot}$, [O/Fe]=+0.0 (dash-dotted line). Taken from Lebzelter et al. (submitted).

For this comparison we used the FRANEC stellar evolutionary code [17, 18, 19]. We find a good agreement with model predictions for both ratios assuming an enhancement in α-elements of +0.2 dex in the cluster. This is in agreement with earlier findings of an oxygen overabundance in the LMC [20].

Another factor influencing the C/O ratio on the AGB may be present. Comparing the C/O ratios with the findings from the stellar variability described above (Fig. 1) we find that the C/O ratio is higher for first overtone pulsators compared to higher overtone pulsators of the same luminosity. Thus there seems to be some kind of relation between the pulsation mode and the C/O ratio. At this point it is not clear how to explain such a relation. However, from the luminosities of the sample stars we can exclude a RGB contamination, all stars are on the AGB. The existence of two populations in this cluster suggested by [16] could be a possible explanation for this behaviour. AGB stars from both populations would differ in temperature, and there are indeed indications for such a temperature split among the AGB stars of NGC 1846. A mass difference may have also some influence on the pulsation mode. However, with our current data we cannot give a final explanation.

Beside the C/O measurements we could also derive values for the fluorine abundance in the sample stars. As we could use only one line blend for this investigation, the results have to be taken with some caution. We find a good correlation between F abundance and C/O ratio, confirming earlier findings (e.g. [21]) of an increasing F abundance with increasing C/O. This indicates a production of F on the AGB.

Finally, we want to briefly comment on LE13, a sample star classified as spectral type S in the literature. The star does not show a C/O ratio close to 1 as would be expected for a S-type star, but rather has a value close 0.65. This has likely two reasons: first, the α-element abundance is increased in this cluster. The classification as a S-type star indicates that a sufficient amount of s-process elements has been dredged up to the surface. For a given amount of dredged up material (including both ^{12}C and s-process elements) the higher starting abundance of oxygen leads to a lower C/O ratio. Second, the low metallicity stars observed here are hotter than typical disk S-stars, which is also allowing for a spectral type S with a C/O ratio lower than 1 [22].

4. THE ASPECT OF VARIABILITY ON THE ABUNDANCE DETERMINATION

Deriving stellar abundances from fitting the observed spectrum with a synthetic one naturally strongly depends on finding the appropriate model atmosphere. Beside the difficulties of deriving the correct global parameters from observations and of taking into account all sources of opacity correctly, AGB stars show a further difficulty, namely their atmospheric dynamics. Especially Miras show pronounced velocity fields in their atmospheres with velocity discontinuities where outward and inward moving layers collide (e.g. [23]). Furthermore, AGB stars show a significant mass loss, another dynamical phenomenon which has an effect on the spectrum. Thus hydrostatic models will in general not give a proper description of the atmospheric structure for highly evolved stars.

Due to this difficulty abundance measurements from Mira-like AGB stars are rare and most of the existing data have be taken with considerable caution (in the case of NGC 1846 the O-rich stars all show small amplitude variability only). In recent years dynamical model atmospheres of AGB stars are becoming available (e.g. [24]). While these models still exist only for a limited number of stellar parameters they have already proven to reproduce several observed features in this kind of stars (e.g. [25]). [26, 27] compared synthetic high resolution spectra with observations of C-rich Miras and found qualitative agreement.

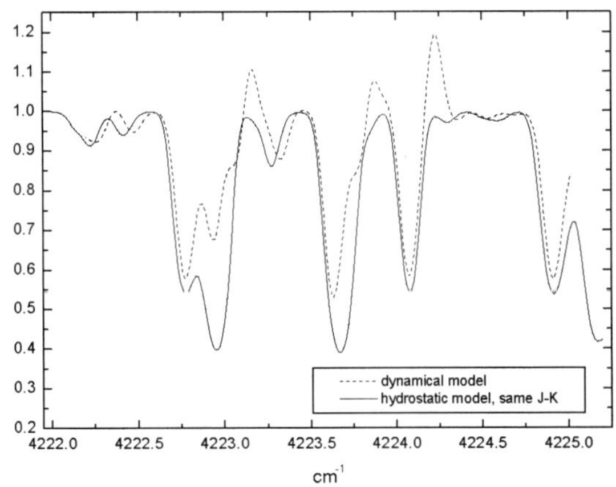

FIGURE 4. Comparison of synthetic spectra from a hydrostatic (solid line) and a dynamical model atmosphere (dotted line). The two models (respectively pulsation phase) were chosen to give the same $J-K$ value.

A detailed study of the effect of the atmospheric dynamics on the abundance determination is still missing. As a first step [28] presented a brief discussion of the change of the equivalent widths of some lines from various molecules over the pulsation cycle. Ratios between various lines are neither constant over the cycle nor do they agree with the hydrostatic case.

Here we present first results of our investigation of the influence of atmospheric dynamics on the abundance determination. For this, dynamical models for an O-rich, low metallicity AGB star, similar to a globular cluster Mira, were kindly provided by Susanne Höfner. In Fig. 4 we compare a small piece of the spectrum based on this dynamical model with a similar one based on a hydrostatic model with the same $J-K$ value and chemical composition. The difference is very obvious and, of course, variable over the pulsation cycle. For this dynamical model, the $J-K$ value varies between approximately 0.9 and 1.3. In the near future we plan to investigate this difference further to see which lines are the least affected ones and whether some kind of correction factor for the difference between a dynamical and a hydrostatic model approach can be found.

ACKNOWLEDGMENTS

TL and MTL received support from the Austrian Science Fund (FWF) under project P18171-N02. MTL acknowledges funding from the Austrian Academy of Science. WN was supported by FWF project P18939-N16.

REFERENCES

1. M. Busso, R. Gallino, and G.J. Wasserburg, *ARA&A* **37**, 239 (1999)
2. F. Herwig, *A&A* **360**, 952 (2000)

3. M. Busso, R. Gallino, D.L. Lambert, C. Travaglio, and V.V. Smith, *ApJ* **557**, 802 (2001)
4. T. Lebzelter and J. Hron, *A&A* **411**, 533 (2003)
5. S. van Eck, S. Goriely, A. Jorissen, and B. Plez, *Nature* **412**, 793 (2001)
6. P. Marigo, L. Girardi, and A. Bressan, *A&A* **344**, 123 (1999)
7. L. Girardi, C. Chiosi, G. Bertelli, and A. Bressan, *A&A* **298**, 87 (1995)
8. T. Lebzelter and P.R. Wood, *A&A* **475**, 643 (2007)
9. A.J. Grocholski, A.A. Cole, A. Sarajedini, et al., *AJ* **132**, 1630 (2006)
10. T. Lloyd Evans, *MNRAS* **193**, 87 (1980)
11. J.A. Frogel, J. Mould, and V.M. Blanco, *ApJ* **352**, 96 (1990)
12. T. Tanabé, S. Nishida, Y. Nakada, et al., *Ap&SS* **255**, 407 (1997)
13. M.W. Fox and P.R. Wood, *ApJ* **259**, 198 (1982)
14. T. Lebzelter and P.R. Wood, *A&A* **441**, 1117 (2005)
15. L. Girardi, A. Bressan, G. Bertelli, and C. Chiosi, *A&AS* **141**, 371 (2000)
16. A.D. Mackey and P. Broby Nielsen, *MNRAS* **379**, 151 (2007)
17. A. Chieffi, M. Limongi, and O. Straniero, *ApJ* **502**, 737 (1998)
18. O. Straniero, R. Gallino, and S. Cristallo, *Nucl. Phys. A* **777**, 311 (2006)
19. S. Cristallo, O. Straniero, M.T. Lederer, and B. Aringer, *ApJ* **667**, 489 (2007)
20. V. Hill, P. Francois, M. Spite, F. Primas, and F. Spite, *A&A* **364**, L19 (2000)
21. A. Jorissen, V.V. Smith, and D.L. Lambert, *A&A* **261**, 164 (1992)
22. J. Piccirillo, *MNRAS* **190**, 441 (1980)
23. K.H. Hinkle, D.N.B. Hall, and S.T. Ridgway, *ApJ* **252**, 697 (1982)
24. S. Höfner, *ASPC* **378**, 145 (2007)
25. R. Gautschy-Loidl, S. Höfner, U. G. Jørgensen, and J. Hron, *A&A* **422**, 289 (2004)
26. W. Nowotny, B. Aringer, S. Höfner, R. Gautschy-Loidl, and W. Windsteig, *A&A* **437**, 273 (2005)
27. W. Nowotny, T. Lebzelter, J. Hron, and S. Höfner, *A&A* **437**, 285 (2005)
28. T. Lebzelter, B. Aringer, and W. Nowotny, *ASPC* **304**, 111 (2003)

The Chemical Evolution in the Solar Vicinity and the Nucleosynthesis of Neutron Capture Elements

G. Cescutti*, F. Matteucci* and P. François †

*Dipartimento di Astronomia, Universitá di Trieste, via G.B. Tiepolo 11, I-34131
†Observatoire de Paris/Meudon, GEPI, 61 Avenue de l'Observatoire, 75014 Paris, France

Abstract. We propose an explanation for the considerable scatter of the abundances of neutron capture elements observed in low metallicity stars in the solar vicinity, compared to the small star to star scatter observed for the α-elements. We have developed a stochastic chemical evolution model, in which the main assumption is a random formation of new stars, subject to the condition that the cumulative mass distribution follows a given initial mass function. With our model we are able to reproduce the different spreads of neutron capture elements and α-elements in low metallicity stars. The reason for different observed spread in neutron capture elements and α-elements resides in the random birth of stars coupled with different stellar mass ranges from which α-elements and neutron capture elements originate. In particular, the site of production of α-elements is the whole range of the massive stars, from 10 to $80 M_{\odot}$ whereas the mass range of production for neutron capture elements lies between 12 and $30 M_{\odot}$.

Keywords: Stars: Abundances - Chemical Composition, Galactic Halo, Galaxy: Chemical Evolution
PACS: 98.35.Bd 98.35.Gi 97.10.Tk

1. INTRODUCTION

Early work by [1] first proposed that the stellar abundances of very heavy elements with respect to iron, particularly [Eu/Fe], show a large scatter at low metallicities. Their work suggested that this scatter appears to diminish with increasing metallicity. This was confirmed by the large spread observed later on in the [Ba/Fe] and [Eu/Fe] ratios in halo stars (e.g. [2, 3]). A more extensive study by [4] confirmed the very large star-to-star scatter in the early Galaxy, while studies of stars at higher metallicities, involving mostly disk stars [5, 6], show little scatter. In the last few years a great deal of observational work on galactic stars appeared: [7, 8, 9, 10, 11, 12]. All these works confirmed the presence of the spread for these elements and this spread can reach 2 dex at [Fe/H][1] ~-3. It is worth noting that such a high spread is not found for the [α/Fe] ratios in very metal poor stars (down to [Fe/H]=-4.0, [13]); in fact, the spread for these elements is around 0.5 dex at low metallicities. This fact suggests that the spread, if real, is a characteristic of these heavy elements and not only due to an inhomogeneous mixing in the early halo phases, as suggested by several authors [14, 15].

[1] We adopt the usual spectroscopic notations that [A/B]= $log_{10}(N_A/N_B)_{\star} - log_{10}(N_A/N_B)_{\odot}$ and that $log\epsilon(A) = log_{10}(N_A/N_H) + 12.0$, for elements A and B

CP1001, 9th Torino Workshop on Evolution and Nucleosynthesis in AGB Stars and the 2nd Perugia Workshop on Nuclear Astrophysics, edited by R. Guandalini, S. Palmerini, and M. Busso

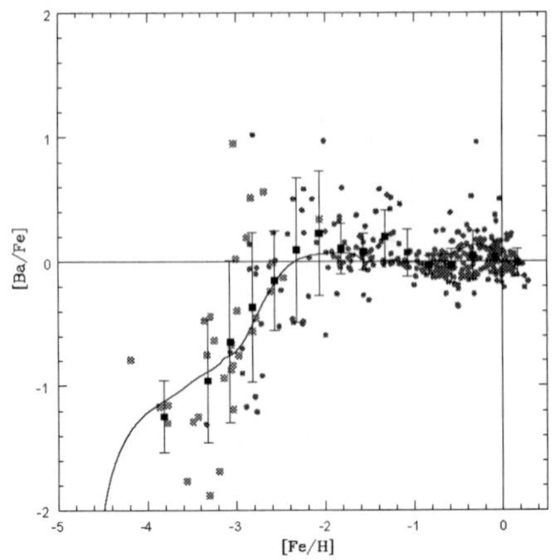

FIGURE 1. In the Figure is plotted the [Ba/Fe] versus [Fe/H] for the data by [16] (filled squares) and for the other observational data (filled circles). The squares with error bars are the mean values of the data bins and the error bars are the standard deviation. We show in solid line the results of by [19], model 1.

2. OBSERVATIONAL DATA

For the extremely metal poor stars (-4<[Fe/H]< -3), we adopted the recent results from UVES Large Program "First Star" [13, 16]. For the abundances in the remaining range of [Fe/H], we took published high quality data in the literature from various sources. We homogenized all of these data by normalizing them to the solar abundances by [17].

3. NUCLEOSYNTHESIS PRESCRIPTIONS

For the nucleosynthesis prescriptions of the Fe and Si we adopted those suggested by [18]. We underline that the site of production of the α-element Si is the whole range of massive stars. For the nucleosynthesis prescriptions of the r-process contribution, we used those of model 1 for Ba and Eu (see [19]). For Y we used the results shown in [20]. We have assumed that these neutron capture elements are produced by r-process in massive stars but only in the mass range 12-30 $M_{\odot}$. These empirical yields have been chosen in order to reproduce the surface abundances for all these neutron capture elements in low metallicity stars as well as the Y, Ba and Eu solar abundances, taking in account the s-process contribution at high metallicities by means of the yields of [21] for Ba and those of [22] for Y in the mass range 1.5-3 $M_{\odot}$. Eu is considered to be a purely r-process element produced in the same range of masses. This choice is extensively discussed in [19] and the nucleosynthesis prescriptions of the r-process contribution are summarized in Table1. We show in Fig. 1 the results by [19], model 1, in the case of Ba.

TABLE 1. The used prescriptions of the mass fractions of newly produced elements for neutron capture elements in massive stars (r-process).

M_{star}	X_{Ba}^{new}	X_{Eu}^{new}	X_{Y}^{new}
12.	$9.00 \cdot 10^{-7}$	$4.50 \cdot 10^{-8}$	$3.60 \cdot 10^{-7}$
15.	$3.00 \cdot 10^{-8}$	$3.00 \cdot 10^{-9}$	$2.10 \cdot 10^{-8}$
30.	$1.00 \cdot 10^{-9}$	$5.00 \cdot 10^{-10}$	$1.00 \cdot 10^{-9}$

4. INHOMOGENEOUS CHEMICAL EVOLUTION MODEL FOR THE MILKY WAY HALO

In order to have a comparison with a homogeneous chemical evolution model, we choose to use the same parameters of the chemical evolution (star formation rate, initial mass function, stellar lifetime, nucleosynthesis, gas density threshold) as those of the homogeneous model used in [19]. For this reason, we model the chemical evolution of the halo of the Milky Way for the duration of 1 Gyr. We consider that the halo consists of many independent cubic regions each with the same typical volume and each region does not interact with the others. We decided to have a typical volume of $2.8 \cdot 10^6 pc^3$. The number of assumed volumes is 100, in order to have good statistical results; we test that a larger number of volumes does not change significantly the statistic and it slows down the model. The dimension of the volume is large enough to allow us to neglect the interaction among different volumes, at least as a first approximation. In fact, for typical ISM densities, a supernova remnant becomes indistinguishable from the ISM (i.e. merges with the ISM) before reaching $\sim 50pc$ [23], less than the size of our cubic region. We do not take larger volumes because we would loose the stochasticity we are looking for; in fact, as we tested, for increasing volumes the model tends to an homogeneous model. We use timesteps of 1Myr, which is an interval of time shorter than any stellar lifetime considered here; in fact, the maximum stellar mass considered here is 80 $M_{\odot}$ with a lifetime of ~3 Myr. At the same time, this timestep is longer than the cooling time of the SN bubbles, which is normally ~ 0.1-0.2 Myr and at maximum 0.8 Myr. With this timestep we can say, in first approximation, that an homogenous mixing takes place and, at least for the first period, only few stars pollute the ISM; this is again important in the point of view of the stochasticity of the model. In each region we assume the same law for the infall of the gas with primordial composition and we consider a threshold in the gas surface density, which means that below this threshold the star formation is blocked, due to the too low gas density in the system. The star formation rate is set to match the one of the homogenous model during the halo phase. So, when for the first time the gas density threshold is reached, the mass which is transformed into stars in a timestep is defined as M_{stars}^{new}. Knowing the M_{stars}^{new}, we assign the mass to one star with a random function, weighted according to the initial mass function (IMF) of [24] in the range between 0.1 and 80 $M_{\odot}$. Then we extract the mass of another star and we repeat this cycle until the total mass of newly formed stars exceeds M_{stars}^{new}. In this way, in each region, at each timestep, the M_{stars}^{new} is the same but the total number and mass distribution of the stars are different and we are able to know the mass of each star contained in each region, when it is born and when it will die, assuming the stellar lifetimes of [25]. We compute the chemical evolution in the following way: at the end of its lifetime, each star enriches the interstellar medium with newly produced elements (see the next section), as a function of its mass and metallicity. The total mass of each element is determined at the end of the lifetime of each star, taking in account the enrichment due to that star and due to the already present mass of each element locked in that star when it is born. The model does not take into account the pollution produced by stars with mass $< 3M_{\odot}$ because their lifetimes exceed the duration of the simulation. We assume that the ejecta produced by the dying stars mix with the surrounding ISM only after that these ejecta cool down significantly. The cooling timescales are normally $\sim$ 0.1-0.2 Myr and reach at maximum 0.8 Myr in a set of 100 volumes; being the cooling timescales always smaller than the timestep of our simulation (1 Myr), we decided to neglect this delay time, so that the mixing in each simulated cubic volume is instantaneous. We show here the results concerning the chemical evolution of only 5 elements (Si, Fe, Y, Ba and Eu) in each region, even if the model is able to follow the chemical evolution of 28 elements. If the model is correct, our predictions will approximate the results of the homogeneous model, having the same parameters, as the number of stars increases. On the other hand, our model shows the spread that can be produced by different stars of various masses on the chemical enrichment at low metallicity, where the number of stars is low and the random effects in the birth of stellar masses can be important.

5. RESULTS

5.1. The Ratios of α-Elements and Neutron Capture Elements to Fe

We discuss here the results of our simulations compared to the observational data and to the prediction of the homogeneous model of [19]. We show the [Eu/Fe], [Ba/Fe] and [Y/Fe] ratios as functions of [Fe/H] in the Fig. 2, 3 and 4 respectively; for α-elements we show [Si/Fe] as functions of [Fe/H] in Fig. 5. In these Figures, we show the simulated stars still alive at the present time (blue points) together with the observations (red triangles).

It is worth noting that our model reproduces the large spread observed in the abundances of metal poor stars for the neutron capture elements and, at the same time, the small spread for the α-element Si. The chemical enrichment observed in very metal poor stars is due only to nucleosynthesis in massive stars. The site of production of the α-elements is different than the one of neutron capture elements (as described in the previous section): the α-elements and Fe are produced in the whole range of massive stars (10-80 $M_{\odot}$); on the other hand, the neutron capture elements are assumed to be produced in the smaller range between 12 and 30 $M_{\odot}$. Therefore, in regions biased randomly towards stars less massive than 30 $M_{\odot}$, the ratio of neutron capture elements over Fe is high. The opposite happens in regions where a large fraction of the stars are more massive than 30 $M_{\odot}$. This fact produces, in our inhomogeneous model, a large spread for the rates of neutron capture elements to Fe, but not for the ratio of the α-element Si to Fe, since the Si and Fe are produced in the same range. The model is not able to

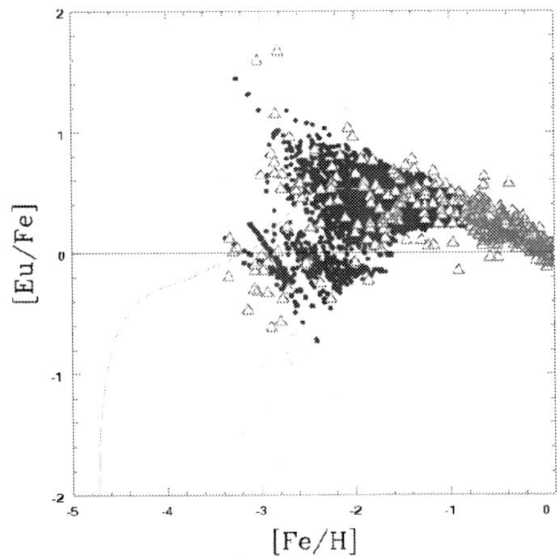

FIGURE 2. [Eu/Fe] vs [Fe/H]. The abundances of simulated stars still alive at the present time are indicated by the dots, the triangles are the observational data. The dashed line is the prediction of the homogeneous model ([19], model 1).

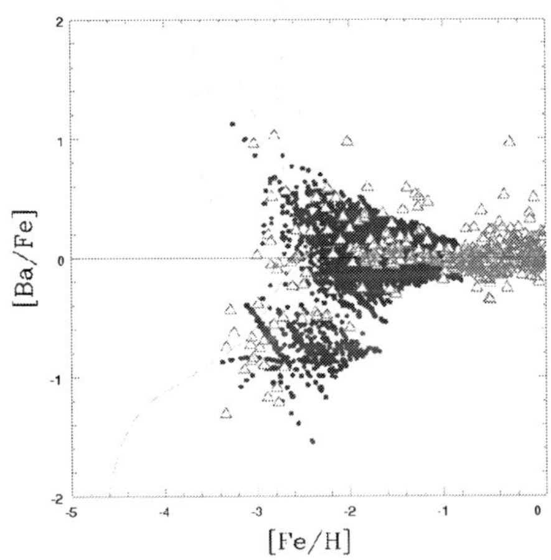

FIGURE 3. As in Fig. 2 but for [Ba/Fe]. The dashed line is the prediction of the homogeneous model ([19], model 1).

reproduce ultra metal poor stars, whereas stars with these metallicity are observed. In particular, the model predicts the formation of some metal free stars at the begin, but then the metallicity jumps to [Fe/H]$\sim$-3.6. This may be due to a too fast rise of [Fe/H], related to the used infall rate. Indeed, numerical simulations [26] show that the mixing timescale of the ejecta, although fast, is of the order of 10Myr, therefore larger than the assumed timestep. If included, this delayed mixing might increase the amount of very

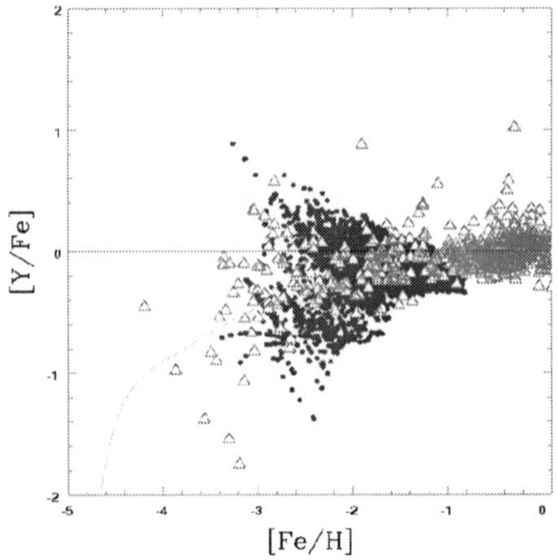

FIGURE 4. As in Fig. 2 but for [Y/Fe]. The dashed line is the prediction of the homogeneous model [20].

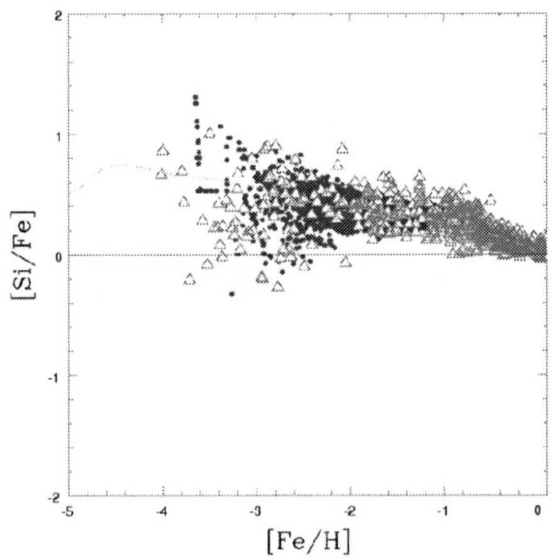

FIGURE 5. As in Fig. 2 but for [Si/Fe]. The dashed line is the prediction of the homogeneous model [18].

metal poor stars. Moreover, the model predicts that $\sim 5\%$ of the simulated stars are metal free. This is the fraction of stars formed in the first 5 Myr, in which no SN has yet exploded and enriched the ISM. We underline that the star formation can be slowed down for example by assuming a different infall rate or a strong outflow of gas from the halo, which would modulate a different star formation and this can solve these two problems. The problem of the metal free stars can be also solved by adopting a different IMF. The results of many works (see [27, 28, 29]) predict that the star formed in a metal free gas must be massive stars. The long living stars which are low mass stars, start to form when the ISM has been already enriched by these massive and metal free stars, the so-called Population III. The effect of the Pop III on the global chemical evolution should be negligible (see [30, 31]), being the total amount of recycled mass small. In fact, we tested a model with a top-heavy IMF, for the very first period, up to a metallicity $Z = 10^{-4}Z_{\odot}$; we decided to use this value because in literature $10^{-4}Z_{\odot}$ is considered the metallicity of the transition to have a normal IMF (see [32]). This stage lasts only about 5-10Myr, depending on the chosen cubic region. This top-heavy IMF has the same shape of the one we use normally [24] but we changed the lower limit from $0.1M_{\odot}$ to 1 $M_{\odot}$. We obtained in this way that the chemical evolution does not change but we eliminate the presence of metal-free stars, which are not observed. In fact all the stars, born during this stage will die before the present time and the stars born after will be not metal-free. We underline that we have decided to use the same parameters and conditions of the homogeneous model to test the validity of the inhomogeneous one. It is clear from the figures that this new model for a high number of star formation events well approximates the homogeneous model, which is shown in the Fig. 2–5 by the dashed line.

6. CONCLUSIONS

In this paper we tried to solve the problem of the large spread observed at low metallicity of the [el/Fe] ratios for neutron capture elements and, at the same time, the smaller spread observed in [α/Fe] ratios. We developed a new model for the chemical evolution of the halo of the Milky Way. In this model, we coupled the birth of stars with random masses, with different mass ranges for the production of α-elements and neutron capture elements. In particular, the site of production of α-elements is the whole range of the massive stars, from 10 to 80 $M_{\odot}$ whereas the mass range of production for neutron capture elements lies between 12 and 30 $M_{\odot}$. We showed that these two assumptions can explain the larger spread in the abundances of metal poor stars for neutron capture elements and the smaller spread for α-elements. We underlined that the parameters we used (described in the Section 4) are the same of the homogeneous model (see [19]). We adopted this point of view because these parameters have been already constrained to give good results compared to the observational data at higher metallicities and to the mean trends. In fact, toward high metallicities ([Fe/H]>-2.0 dex) the model naturally gives results compatible with the homogeneous model. However, this set of parameters is a starting point and the model still needs a better investigation of the parameter space. In particular, for what concerns the early galactic stages, the model is not able to reproduce ultra metal poor stars whereas stars with these metallicity are observed. This may be due to a too fast rise of [Fe/H], related to the used infall law. To avoid these problems a different infall law can be used or an outflow from the halo can be included, in order to provide a smoother increase of the metallicity. Moreover, this new model generates too large number of metal free stars, if we do not impose at least a slightly different IMF for the first episodes of stars formation to solve the problem of the metal free stars.

ACKNOWLEDGMENTS

GC would like to warmly thank the organizers of the workshop.GC also thanks Francesco Calura, Simone Recchi and Antonio Pipino for valuable conversations and useful comments. The work was supported by the Italian Ministry for the University and Research (MIUR) under COFIN03 prot.2003028039.

REFERENCES

1. K.K. Gilroy, C. Sneden, C.A Pilachowski, and J.J Cowan, *ApJ* **327**, 298 (1988)
2. A. McWilliam, G.W. Preston, C. Sneden, and L. Searle, *AJ* **109**, 2757 (1995)
3. S. Ryan, J.E. Norris, and T.C. Beers, *ApJ* **471**, 254 (1996)
4. D.L. Burris, C.A. Pilachowski, T.E. Armandroff, et al., *ApJ* **544**, 302 (2000)
5. B. Edvardsson, J. Andersen, B. Gustafsson, et al., *A&A*, **275**, 101 (1993)
6. V.M. Woolf, J. Tomkin, and D.L Lambert, *ApJ* **453**, 660 (1995)
7. J.P. Fulbright, *AJ* **120**, 1841 (2000)
8. L. Mashonkina and T. Gehren, *A&A* **364**, 249 (2000)
9. L. Mashonkina and T. Gehren, *A&A* **376**, 232(2001)
10. A. Koch and B. Edvardsson, *A&A* **381**, 500 (2002)
11. S. Honda, W. Aoki, T. Kajino, et al., *ApJ* **607**, 474 (2004)
12. Y. Ishimaru, S. Wanajo, W. Aoki, and S.G. Ryan, *ApJ*, **600L**, 47 (2004)
13. R. Cayrel, E. Depagne, M. Spite, et al., *A&A* **416**, 1117 (2004)
14. T. Tsujimoto, T. Shigeyama, and Y. Yoshii, *ApJ* **519L**, 63 (1999)
15. Y. Ishimaru and S. Wanajo, *ApJ* **511L**, 33 (1999)
16. P. François, E. Depagne, V. Hill, et al., *A&A* **476**, 935 (2007)
17. M. Asplund, N. Grevesse, and A.J. Sauval, *ASPC* **336**, 25A (2005)
18. P. François, F. Matteucci, R. Cayrel R., et al., *A&A* **421**, 613 (2004)
19. G.Cescutti, P. François, F. Matteucci, R. Cayrel, and M. Spite, *A&A* **448**, 557 (2006)
20. G. Cescutti, *PhD thesis, arXiv, 0708.4163*, (2007)
21. M. Busso, R. Gallino, D.L. Lambert, C. Travaglio, and V.V. Smith, *ApJ* **557**, 802 (2001)
22. C. Travaglio, R. Gallino, E. Arnone, J. Cowan, F. Jordan, and C. Sneden, *ApJ* **601**, 864 (2004)
23. K. Thornton, M. Gaudlitz, H.-Th. Janka, and M. Steinmetz, *ApJ* **500**, 95 (1998)
24. J.M. Scalo, *FCPh* **11**, 1 (1986)
25. A. Maeder and G. Meynet, A&A **210**, 155 (1989)
26. S. Recchi, F. Matteucci, and A. D'Ercole, *MNRAS* **322**, 800 (2001)
27. T. Abel, G.L. Bryan, and M.L. Norman, *ApJ* **540**, 39 (2000)
28. X. Hernandez and A. Ferrara, *MNRAS* **324**, 484 (2001)
29. J. Mackey, V. Bromm, and L. Hernquist, *ApJ* **586**, 1 (2003)
30. F. Matteucci and A. Pipino, *MNRAS* **357**, 489 (2005)
31. F. Matteucci and F. Calura, *MNRAS* **360**, 447 (2005)
32. R. Schneider, A. Ferrara, P. Natarajan, and K. Omukai, *ApJ* **571**, 30 (2002)

On the Fate of Extremely Low Metallicity Stars

P. Gil–Pons*, J.L. Gutiérrez*, E. García–Berro* and L. Althaus†

*Departament de Física Aplicada, Universitat Politècnica de Catalunya, Av. Canal Olímpic s/n, 08860 Castelldefels, Spain
†Facultad de Ciencias Astronómicas y Geofísicas, Universidad Nacional de La Plata, Paseo del Bosque s/n, B1900FWA, La Plata, Argentina

Abstract. We have computed and analyzed the evolution of intermediate-mass stars of metallicity $Z = 10^{-5}$ from the main sequence until the early stages of the TP-(S)AGB phase. In order to check the influence of mixing we have performed our calculations using two different evolutionary codes, `EVOLVE` — which does not include diffusion but allows for overshooting — and `LPCODE`, which includes diffusion and overshooting "à-la-Herwig". Important differences appear during the TP–AGB phase, as the calculations done with `EVOLVE` lead to an almost negligible third dredge-up, whereas the sequences computed with `LPCODE` show important third dredge-up and, therefore, significant enrichment in metals of the envelope. This fact has consequences on the strength of the winds during the TP-AGB phase and, ultimately, on the final fate of the considered stars. Also, the models computed with `LPCODE` show strong double flashes produced by the ingestion of protons on the helium burning shell.

Keywords: Stellar Structure, Abundances, Population III Stars, Supernovae
PACS: 97.10.Cv, 97.10.Tk, 97.20.Wt, 97.60.Bw

1. INTRODUCTION

Primordial and extremely low metallicity stars are drawing increasing attention, not only because we have now a harvest of new observations of more metal poor environments [1, 2], but also because of the recent progresses achieved in the theoretical calculations. Recent studies [3, 4] have found peculiar flashes that ensue during the evolution of extremely metal poor stars, which could dramatically affect — and be affected by — the mixing mechanisms that must be taken into account. We might be facing a problem not to be accounted for either by the standard mixing length recipes nor by diffusion. Instead, inputs from multidimensional hydrodynamic simulations might be required [4, 5]. Moreover, recent calculations of the evolution of intermediate-mass primordial stars [6, 7, 8], point out to the possibility that these stars might end their lives as SNeI1/2 [9], since the initial lack of metals of the envelope and the inefficiency of the dredge-up process favour very weak winds, that are unable to eject the envelope before the core reaches the Chandrasekhar mass.

In this work we explore the effects of the initial metallicity on the late stages and final fate of intermediate-mass stars of extremely low metallicity ($Z = 10^{-5}$), and we discuss whether the results obtained for their counterparts of primordial composition hold for these objects. We target as well the consequences of different mixing prescriptions. Finally, we also assess the impact of the numerical schemes adopted in the evolutionary codes, by computing evolutionary sequences with two different codes. In Section 2 we describe the codes and we outline the evolution during the main central burning stages.

CP1001, 9th *Torino Workshop on Evolution and Nucleosynthesis in AGB Stars and the 2nd Perugia Workshop on Nuclear Astrophysics,*
edited by R. Guandalini, S. Palmerini, and M. Busso

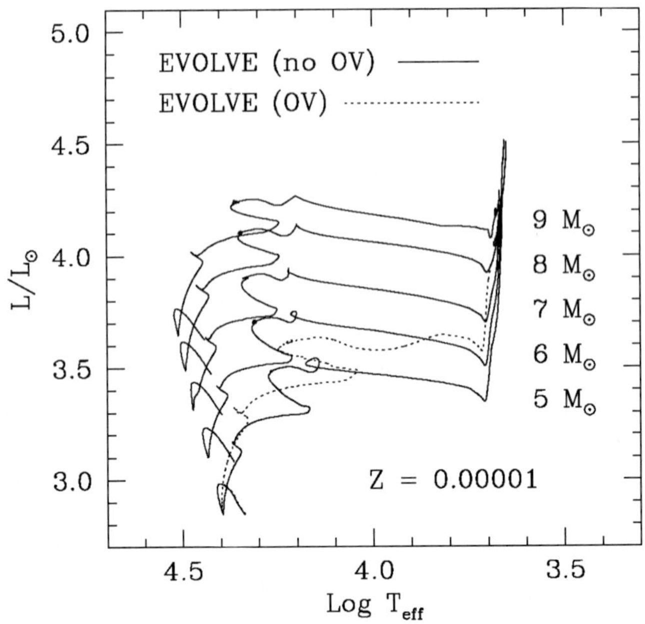

FIGURE 1. Evolution in the Hertzsprung-Russell diagram of our $Z = 10^{-5}$ models computed with EVOLVE.

In Section 3 we describe the early stages of the TP-AGB phase and focus on the problem of their final fate, whereas in the last section we summarize our findings and we draw the main conclusions of our work.

2. INPUTS, EVOLUTIONARY CODES AND SEQUENCES

In order to check how the mixing prescriptions affect the evolution of extremely low metallicity intermediate-mass stars, we have computed a series of sequences using two different codes: EVOLVE and LPCODE. The evolutionary code EVOLVE follows the H- and He-burning phases and has been updated to include carbon burning [10] and the new OPAL opacities [11], but does not include element diffusion. Recently, it has been used to analyze intermediate-mass primordial stars and how overshooting can affect their main properties [11, 7]. The treatment of overshooting has been implemented following [12, 13]. To be precise, convection sets in when $\nabla_{\rm rad} > \nabla_{\rm ad} - \delta$, where δ depends on the ratio $P_{\rm rad}/P_{\rm gas}$ and on a free parameter that has been calibrated in order to fit observations. Using EVOLVE without overshooting we have computed the evolution of $Z = 10^{-5}$ stars of initial masses between 5 and $9\,M_\odot$. For the sake of comparison, we have also computed the evolution of a $5\,M_\odot$ model star with overshooting. The tracks in the Hertzsprung-Russell diagram of these model sequences are shown in Fig. 1.

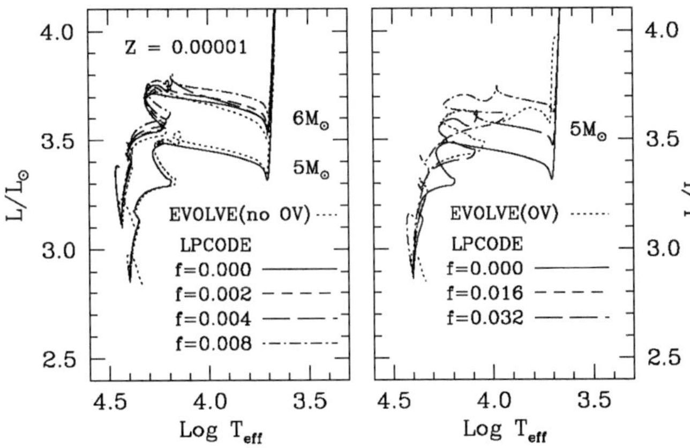

FIGURE 2. Evolution in the Hertzsprung-Russell diagram of our $Z = 10^{-5}$ models computed using LPCODE and several values of f.

LPCODE [14] includes diffusion and the treatment of overshooting follows the prescription of [15]. That is, convective mixing is extended beyond the frontier of stability according to the exponentially decaying law:

$$D_{\rm OV} = D_0 \exp\left(\frac{-2z}{H_v}\right) \tag{1}$$

where the parameters $D_0 = v_0 H_P$, $z = \mid r_{\rm edge} - r \mid$ and $H_v = f H_P$. The parameter f adopts values between 0 and 0.032, although the value recommended lately is 0.008 [16]. Larger values of f imply a more extended mixing. Unlike EVOLVE, LPCODE does not allow to compute the carbon burning phase.

We have analyzed the evolution of $5M_\odot$ and $6M_\odot$ stars, with composition $X = 0.755$, $Y = 0.245$ and $Z = 10^{-5}$, and with different values of f, ranging from 0.0 to 0.032. The left panel of Fig. 2 shows the tracks in the Hertzsprung-Russell diagram of the models computed with LPCODE and f between 0.0 and 0.008. The right panel of the same figure shows the evolution in this diagram of the $5M_\odot$ model computed with EVOLVE — with and without overshooting — and with LPCODE and values for f of 0.016 and 0.032. As can be seen in the figure, the amount of extra mixing due to overshooting obtained using EVOLVE corresponds to a fairly large value of f, intermediate between 0.016 and 0.032. On the other hand, the sequences computed with no overshooting show a fair agreement.

2.1. Evolution During the Main Central Burning Stages

Opposite to what happens in the case of primordial stars, stars with a metallicity of $Z = 10^{-5}$ develop core hydrogen burning (CHB) through the CNO cycle, as the initial amount of metals is more than enough to allow the onset of these reactions. However,

TABLE 1. Masses of the CO cores and central abundances of carbon and oxygen for the $6M_\odot$ model stars computed with EVOLVE and no overshooting and with LPCODE with several different values of f.

Code	f	M_{CO}	$X_c(^{12}C)$	$X_c(^{16}O)$
EVOLVE	—	0.96	0.31	0.69
LPCODE	0.000	0.85	0.41	0.59
LPCODE	0.002	0.87	0.32	0.68
LPCODE	0.004	0.88	0.26	0.74
LPCODE	0.008	0.92	0.36	0.64
LPCODE	0.016	0.98	0.34	0.66

the tracks of primordial and $Z = 10^{-5}$ stars show some aspects in common. For instance, they do not climb the red giant branch during hydrogen shell burning but instead, they do it for the first time during the helium shell burning phase. The reason for such behavior is that the cores are so hot and compact at the end of the CHB phase that the 3α reaction sets in before the hydrogen-burning shell is able to induce the expansion and cooling of the stellar envelope. Therefore, both CHB and core helium-core burning (CHeB) take place at the left hand side of the Hertzsprung-Russell diagram. This general behavior is independent of the code used and of the value adopted for f.

As for the evolutionary timescales, the $Z = 10^{-5}$ models computed with EVOLVE yield values longer than those corresponding to primordial stars obtained with the same code [7]. Also, the masses of the degenerate cores are larger and the central abundances of carbon (oxygen) are smaller (larger) than those of the corresponding primordial cases. When the sizes of the cores obtained with the two codes are compared, those resulting from the LPCODE sequences and $f = 0$ are about 12% less massive than the cores computed with EVOLVE — see Table 1. As f increases and reaches values ranging from 0.016 to 0.032, the CO core masses become similar for both codes. Finally, with regard to the ratio $X(\mathrm{C})/X(\mathrm{O})$, the $5M_\odot$ sequence computed with EVOLVE also yields a value similar to that of the corresponding LPCODE model and $f = 0.016$.

Carbon burning in our $Z = 10^{-5}$ models develops in conditions of partial degeneracy and, therefore, through a series of flashes in which the carbon luminosity, L_C, can reach values as large as $\sim 10^8 L_\odot$. This large release of energy causes a fast increase in the temperature gradient and the development of associated convective shells. The expansion and cooling of the layers just above causes the temporary switch-off of the helium-burning shell (HeBS), that recovers as soon as the carbon flash finishes.

Figure 3 shows the temporal evolution of the carbon (L_C) and helium (L_{He}) luminosities (upper panel) and the convective zones (lower panel) during the carbon-burning phase. As can be seen, the burning flame reaches the center of the star after the second flash. The most interesting feature we have found in this case is related to the evolution of convection during the last phases of carbon burning: convection advances inwards in the stellar envelope after the onset of the HeBS and stops its advance when carbon burning begins at a mass point of $0.19M_\odot$. But when the last convective zone associated to carbon burning is about to disappear and $L_C < 10^4 L_\odot$, the base of the convective en-

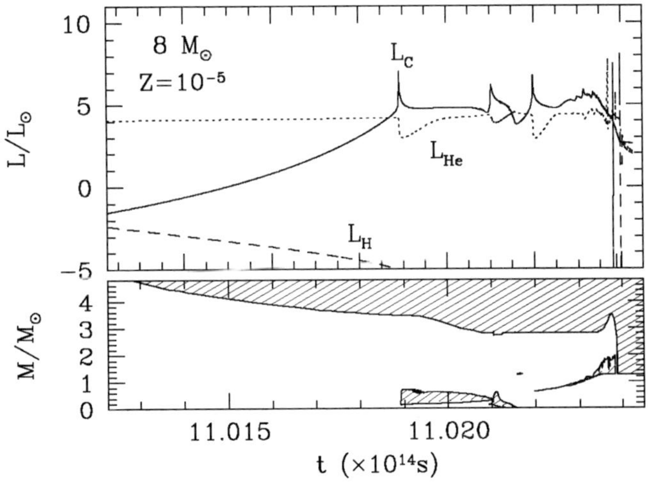

FIGURE 3. Evolution of our $8M_{\odot}$ model during the carbon burning phase. The upper panel shows the evolution of $L_{\rm H}$, $L_{\rm He}$ and $L_{\rm C}$ and the lower panel shows the evolution of the convective zones.

TABLE 2. Mass of the CO and ONe cores computed with EVOLVE and no overshooting.

$M_{\rm ZAMS}/M_{\odot}$	$M_{\rm ONe}/M_{\odot}$	$M_{\rm CO}/M_{\odot}$
5.0	—	0.91
6.0	—	0.97
7.0	1.02	1.05
8.0	1.21	1.22
9.0	1.39	1.40

velope resumes its advance inwards and is able to mix protons into the high temperature and density regions of the stellar interior. This causes a fast ignition of the protons, the occurrence of a hydrogen flash and the extension inwards of convection that allows the mixing of material processed during helium and carbon burning with the stellar envelope. This process was already observed and described in [17] and was later confirmed in [18], and has been referred to as dredge–out.

3. TP-AGB EVOLUTION AND THE EFFECTS OF VARYING f

Table 2 shows the sizes of the degenerate cores of the models computed with EVOLVE after the main central burning stages. The $7M_{\odot}$ model with $Z = 10^{-5}$ is already able to ignite carbon. For comparison, the minimum mass allowing carbon burning in the case

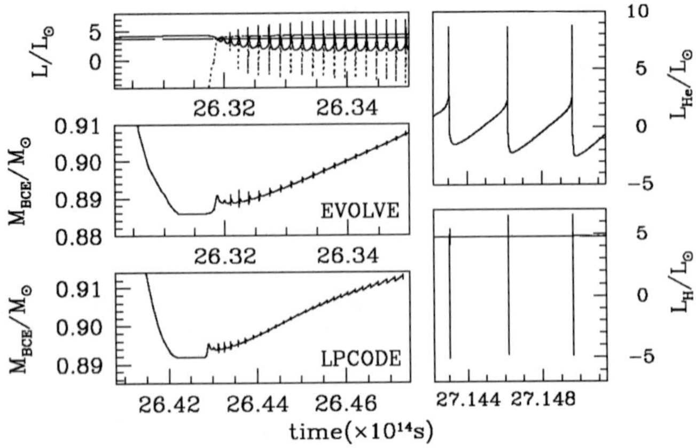

FIGURE 4. Left top panel: evolution of $L_{\rm H}$ and $L_{\rm He}$ during the early TP-AGB phase of the $5M_\odot$ star computed with EVOLVE. Left middle panel: position of the base of the convective envelope (BCE) computed with EVOLVE. Left lower panel: BCE for the same model star computed with LPCODE and f=0. Right upper panel: evolution of $L_{\rm He}$ during the double flash events obtained with LPCODE. Right lower panel: evolution of $L_{\rm H}$ during the double flash events obtained with LPCODE.

TABLE 3. Envelope CNO abundances and total metallicity of the $5M_\odot$ models computed with EVOLVE and LPCODE.

Code	f	$X_{\rm env}(^{12}{\rm C})$	$X_{\rm env}(^{14}{\rm N})$	$X_{\rm env}(^{16}{\rm O})$	$Z_{\rm env}$
EVOLVE	—	5.8×10^{-7}	3.2×10^{-6}	3.3×10^{-6}	1.10×10^{-5}
LPCODE	0.000	1.2×10^{-6}	1.4×10^{-6}	1.2×10^{-6}	1.02×10^{-5}
LPCODE	0.002	1.4×10^{-6}	1.7×10^{-6}	3.9×10^{-6}	1.06×10^{-5}
LPCODE	0.004	0.6×10^{-6}	1.5×10^{-6}	0.8×10^{-6}	1.83×10^{-5}
LPCODE	0.008	3.2×10^{-6}	3.0×10^{-6}	3.2×10^{-6}	1.32×10^{-5}
LPCODE	0.016	5.8×10^{-5}	1.1×10^{-5}	1.5×10^{-5}	1.10×10^{-5}

of primordial stars is $7.8M_\odot$.

Once the main central burning stages finish, the hydrogen burning shell (HBS) switches on again and the TP-AGB begins — see Fig. 4. On the left panels of this figure we compare the advance and retreat of the convective envelope in the evolutionary sequences calculated with EVOLVE and LPCODE. The agreement is quite close, but the importance of mixing induced by the third dredge-up (TDU) is much larger in the tracks computed with LPCODE. By this time the stellar envelope material has already been mixed with the products synthesized in the interior of the star.

We have followed the first thermal pulses of the $5M_\odot$ model star using both codes and different values of f. Table 3 shows the chemical composition of the envelope for this case during the TP-AGB phase. The models computed with LPCODE show larger C/N abundance ratios in the envelope. It must be stressed, however, that we have not followed the same number of pulses for each value of f, and hence the direct comparison of these

TABLE 4. Characteristic times (in years) needed for the cores to reach the Chandrasekhar mass and to remove the envelope using the mass-loss prescription of Reimers ($t_{\rm env}^{\rm R}$) and that of Blöcker ($t_{\rm env}^{\rm B}$).

$M_{\rm ZAMS}/M_{\odot}$	$t_{\rm Ch}$ (yr)	$t_{\rm env}^{\rm R}$ (yr)	$t_{\rm env}^{\rm B}$ (yr)
5.0	1.6×10^6	5.6×10^8	7.9×10^6
6.0	1.5×10^6	5.2×10^8	6.5×10^6
7.0	1.3×10^6	1.9×10^8	3.1×10^6
7.5	9.8×10^5	1.2×10^8	2.0×10^6
8.0	7.8×10^5	5.3×10^7	1.1×10^6
8.5	5.4×10^5	6.0×10^7	1.4×10^6

results can be somewhat misleading.

As to the characteristics of the TP-AGB phase, we have found that for the model computed with EVOLVE, after 15 pulses the effects of the TDU or hot bottom burning are negligible and that the envelope abundances remain constant. For the models computed with LPCODE, we have found that even for the $f = 0$ case TDU and hot bottom burning play an important role. After the first 20 pulses the metallicity of the envelope changes from $Z_{\rm env} = 1.00 \times 10^{-5}$ — the second dredge-up was almost negligible for this model — to $Z_{\rm env} = 2.33 \times 10^{-4}$. Another salient feature in the LPCODE sequences is the existence of double flashes, in which protons are ingested into the He-burning shell, thus producing a proton flash that quenches He shell burning. This can be seen on the right panels of Fig 4. Similar flashes have been found in [3, 4], but a firm observational test currently seems beyond of reach.

4. SUMMARY AND DISCUSSION

We have used two different evolutionary codes to compute the evolution of intermediate-mass extremely metal-poor stars. The two codes yield very different envelope enrichments during the TP-AGB and, therefore, point to very different possibilities for the final fates of the stars we are considering. According to the results obtained using EVOLVE, the envelope enrichment due to the TDU is almost negligible and, therefore, one can not assure that stellar winds will be powerful enough to remove the envelope before the degenerate core reaches the Chandrasekhar mass. As it was done in [11], an estimate of the final fate of these stars can be done. Using the prescription for stellar winds of Reimers [19] or that of Blöcker [20] and the sizes of the envelopes obtained from our sequences, we can estimate the time required by the stars to lose their envelopes ($t_{\rm env}$). Using the rates of core growth and the initial size of the cores, we can also estimate the time required to reach the Chandrasekhar mass ($t_{\rm Ch}$). These times are shown in Table 4. Even the values for $t_{\rm env}$ obtained with the prescription of Blöcker are larger than those for $t_{\rm Ch}$. Therefore, one can not discard that these stars might end their lives as SNeI1/2.

We have assumed that for the models computed using the evolutionary code EVOLVE the envelope abundances are almost constant along the TP-AGB phase. But, as already shown, this is not the case for the models computed with LPCODE. If, instead, this assumption is dropped, as it stems from the calculations performed with LPCODE, the fast increase in the metal content of the envelope seems enough to support stellar winds similar to those of solar metallicity stars. Therefore, they will allow a standard evolution for the $5M_{\odot}$ star and, therefore, it is foreseen that this model would lose its envelope and evolve to the white dwarf cooling track.

The problem of the evolution and fate of extremely low metallicity stars remains open and appears strongly dependent on the treatment of mixing in the model stars. The development of thermonuclear flashes that alter the mixing of isotopes in ways that do not occur in solar metallicity stars and the new observations make the subject of the evolution of extremely low metallicity stars increasingly interesting and challenging.

ACKNOWLEDGMENTS

Part of this work was supported by the MCYT grant AYA05-08013-C03-01, by the European Union FEDER funds and by the AGAUR.

REFERENCES

1. N. Christlieb, M.S. Bessell, T.C. Beers, B. Gustafsson, A. Korn, P.S. Barklem, T. Karlsson, M. Mizuno-Wiedner, and S. Rossi, *Nature* **419**, 904 (2002)
2. A. Frebel, W. Aoki, N. Christlieb, H. Ando, M. Asplund, P.S. Barklem, T.C. Beers, K. Eriksson, C. Fechner, M.Y. Fujimoto, S. Honda, T. Kajino, T. Minezaki, K. Nomoto, J.E. Norris, S.G. Ryan, M. Takada-Hidai, S. Tsangarides, and Y. Yoshii, *Nature* **434**, 871 (2005)
3. S. Campbell and J. Lattanzio, *ArXiV 0709.4567C* (2007)
4. P. Woodward, F. Herwig, D. Porter, T. Fuchs, A. Nowatzki, and M. Pignatari, in *"First stars III"*, AIP Conf Ser., Ed.: T. Abel, A. Heger and B. O'Shea, http://arXiv:0711.2091 (2008)
5. D. Arnett, this volume
6. A. Poelarends, Ph.D. Thesis (2006)
7. P. Gil-Pons, J.L. Gutiérrez, and E. García-Berro, *A&A* **464**, 667 (2007)
8. H. Lau, R. Stancliffe, and C. Tout, *MNRAS* **378**, 563 (2007)
9. D. Arnett, *Ap&SS* **5**, 180 (1968)
10. E. García-Berro and I. Iben Jr., *ApJ* **434**, 306 (1994)
11. P. Gil-Pons, et al., *A&A* **433**, 1037 (2005)
12. K.-P. Schröder, O. Pols, O., and P. Eggleton, *MNRAS* **285**, 696 (1997)
13. J. Eldridge and C. Tout, *MNRAS* **353**, 87 (2004)
14. L. Althaus, et al., *A&A* **404**, 593 (2003)
15. F. Herwig, *A&A* **360**, 952 (2000)
16. F. Herwig, *ApJ* **605**, 425 (2000)
17. C. Ritossa, E. García-Berro, and I. Iben Jr., *ApJ* **515**, 381 (1999)
18. L. Siess, *A&A* **476**, 893 (2007)
19. D. Reimers, *Mem. Soc. Roy. Sci. Liege* **8**, 369 (1975)
20. T. Blöcker, *A&A* **297**, 727 (1995)

SESSION III:

STELLAR NUCLEAR ASTROPHYSICS

Past and Present of the *s* Process: a Nuclear Physicist's View

F. Käppeler[†]

[†]*Forschungszentrum Karlsruhe, Institut für Kernphysik, Germany*

Abstract. The history of nuclear physics data for *s*-process calculations are briefly reviewed with emphasis on the actual status of the neutron capture cross sections. The remaining challenges will be illustrated and discussed in the light of new or optimized methods and state-of-the-art facilities, indicating the potential for accurate measurements and the possibility to study cross sections of radioactive isotopes. These opportunities will be considerably enriched by the enormous improvements provided by upcoming new facilities.

Keywords: Neutron Capture, *s*-Process, Radioactive Isotopes.
PACS: 25.40.Lw, 97.10.Cv.

1. INTRODUCTION

It is inspiring to see that the entire *s*-process panorama was sketched in its essential parts in the founding papers of Nuclear Astrophysics published 50 years ago by Burbidge, Burbidge, Fowler, and Hoyle [1] and by Cameron [2]. The visionary view of these papers is the more admirable as the basics were still pretty incomplete.

In spite of the fact that neither the neutron source reactions nor the neutron capture cross sections in the astrophysically relevant energy range were known, all essential features of the *s* process had been inferred from the available rudimentary information. The product of the stellar (n, γ) cross sections and of the resulting *s* abundances, $\langle\sigma\rangle N_s$, which represents the reaction flow, was found to be a smooth function of mass number A. From the composite slope of this function, two different *s* processes were postulated. The steep decline between $A \approx 63$ and 100 was interpreted as the result of an *s*-process site with not enough neutrons available per ^{56}Fe seed to build the nuclei to their saturation abundances. In the mass region beyond $A \approx 100$, the much smaller slope was suggesting that steady flow was achieved and that all of these nuclei reached their saturation abundances. It was concluded that "*the two different processes might have occurred in two different red-giant stars*" [1], completely consistent with what is now called the weak *s* process during core He [3, 4] and shell C burning [5-7] in massive stars on the one hand, and the main *s* component contributed by thermally pulsing low mass asymptotic giant branch (AGB) stars [8-10] on the other hand.

The main difficulty that prevented further conclusions was due to the lack of reliable neutron capture cross sections. In total, the cross sections of only 25 isotopes had been experimentally investigated at that time. Moreover, these data were uncertain by more than a factor of two in a number of cases. All the remaining data were extrapolated from thermal cross sections or from pile reactivity measurements using

CP1001, *9th Torino Workshop on Evolution and Nucleosynthesis in AGB Stars and the 2nd Perugia Workshop on Nuclear Astrophysics*, edited by R. Guandalini, S. Palmerini, and M. Busso

plausible assumptions on the energy dependence. The importance of a complete set of experimental data for the reliable description of the $\langle\sigma\rangle N_s$-curve is illustrated in Fig. 1, which corresponds to the situation obtained by the time of the cross section compilation of Bao *et al.* in 2000 [11], when most of the cross sections had been measured.

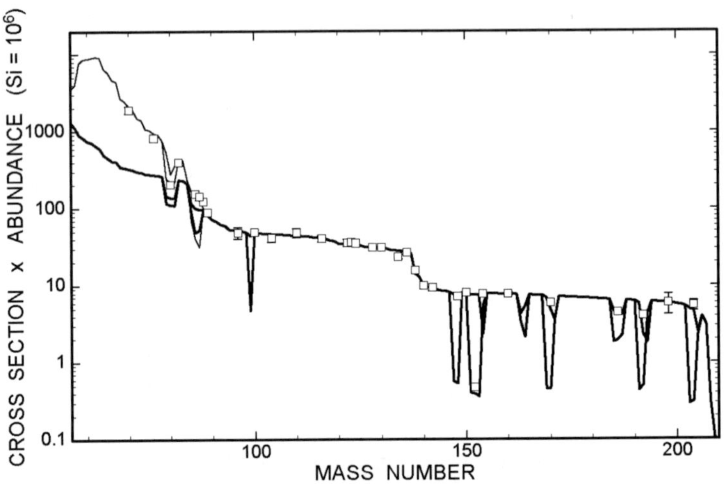

FIGURE 1 The characteristic product of cross section times s-process abundance, $\langle\sigma\rangle N_s$, plotted as a function of mass number. The solid line was obtained via the classical model, and the symbols denote the empirical products for the s-only nuclei. Some important branchings of the neutron capture chain are indicated as well.

Apart from the clear separation of the two *s*-process components, Fig. 1 shows also the pronounced effect of *s*-process branchings, which could not be addressed by B^2FH simply because the data were far too uncertain to reveal their signatures in the $\langle\sigma\rangle N_s$-curve. These branchings are the result of the competition between neutron capture and β-decay at unstable isotopes in the half-life range from a few months to a few years. For the same reason, also the effect of stellar temperatures on the β-decay half-lives had not been anticipated, i.e. for the branch point ^{79}Se. The branching at ^{79}Se is characterized by the strongly different $\langle\sigma\rangle N_s$ -values of the *s*-only isotopes ^{80}Kr and ^{82}Kr, although the terrestrial (ground state) half-life of ^{79}Se is about 10^5 years.

The lack of reliable stellar (n, γ) rates was clearly felt by B^2FH, who emphasized that *"unambiguous results would be obtained by measuring the total absorption cross sections, but it is very difficult experimentally. It is our view, however, that such measurements would serve as a crucial test of the validity of the s process"* [1].

2. EXPERIMENTAL NEUTRON CAPTURE CROSS SECTIONS

The role of neutron capture reactions for the nucleosynthesis of heavy elements was first reviewed in 1971 by Allen and co-workers [12], who published a list of recommended (experimental or semi-empirical) Maxwellian averaged cross sections at kT= 30 keV (MACS-30) in the mass range between C and Pu. The idea of a database for s-process studies was picked up 16 years later by Bao and Käppeler [13]. Their compilation included cross sections for (n, γ) reactions between ^{12}C and ^{209}Bi, some (n, p) and (n, α) reactions (from ^{33}Se to ^{59}Ni), and also (n, γ) and (n, f) reactions for long-lived actinides. A follow-up compilation by Beer *et al.* [14] was again updated in 2000 by Bao *et al.* [11], who included semi-empirical recommended values for nuclides without experimental cross section information. These estimates, which exhibit typical uncertainties of 25 to 30%, were derived by correcting the values obtained with the Hauser-Feshbach code NON-SMOKER [15] for known systematic deficiencies in the nuclear input of these calculations. Additionally, the database of [11] provided stellar enhancement factors and energy-dependent MACS for energies between kT= 5 and 100 keV.

Meanwhile, an extended electronic version of this database has been made available through the KADoNiS project [16], which provides a regularly updated compilation online at http://nuclear-astrophysics.fzk.de/kadonis. The current version as of January 2007 includes new recommended values for 38 improved and 14 new cross sections compared to [11]. In total, data sets are available for 353 isotopes, including 76 radioactive nuclei on or close to the s-process path. Experimental data are presently available for the (n, γ) cross sections of 265 out of the 277 stable isotopes, whereas only 12 of the radioactive nuclei have been measured so far. Most of the remaining cross sections are difficult to determine because they are not accessible by activation measurements or not available in sufficient amounts and/or enrichments for time-of-flight measurements.

The actual status of the cross sections required for s-process nucleosynthesis calculations is summarized in Fig. 2, showing the respective uncertainties as a function of mass number. Though the necessary uncertainties of 1 - 5% have been locally achieved, further improvements are clearly required, especially in the mass region below A = 120 and above A = 180, where the requested accuracy is far from being reached. Future efforts in this field are the more important as Fig. 2 reflects the situation for a thermal energy of 30 keV. In most cases, however, extrapolation to lower and higher temperatures implies still larger uncertainties.

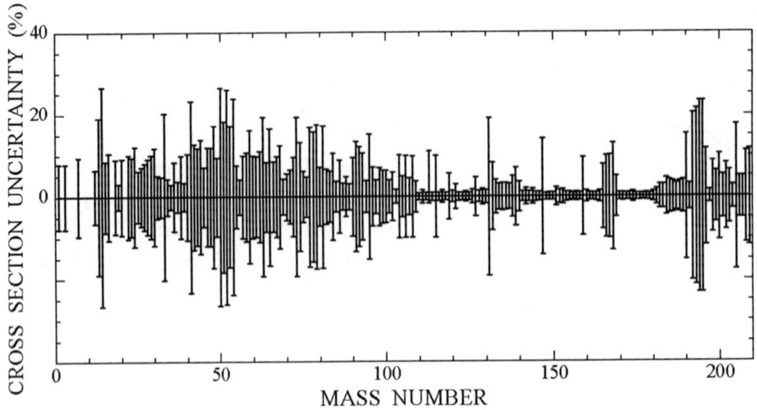

FIGURE 2. Present uncertainties of the stellar (n, γ) cross sections required for *s*-process nucleosynthesis. These values refer to a thermal energy of $kT = 30$ keV, but may be considerably larger at lower and higher temperatures.

3. STELLAR RATES FOR UNSTABLE ISOTOPES

3.1 Neutron Capture Measurements

As outlined above, most of the missing experimental information on stellar (n, γ) cross sections refers to unstable isotopes. The main difficulty in such measurements is clearly associated with the γ-activity of the sample, which restricts the amount of sample material that can be used in a measurement. Therefore, experiments with unstable samples are presently limited to cases where the specific γ-activity is rather low.

A second problem in this context is the production of sufficiently large and sufficiently clean samples. The specific γ-activity is difficult to suppress in measurements with liquid scintillators, e.g. with C_6D_6 detectors, because their poor energy resolution prevents the discrimination via pulse height analysis. Since the time-of-flight (TOF) technique is the only way of reducing the γ-background, these detectors can be applied only in cases of low specific activity. An example of such a measurement is the determination of the ^{151}Sm(n, γ) cross section performed by the n_TOF collaboration [17, 18]. This measurement took also advantage of the very low duty factor and unique luminosity of the n_TOF facility.

HPGe detectors represent an alternative in terms of energy resolution, but the difficulties in this case arise from the low peak efficiency. Consequently, the application of HPGe detectors appears to be useful only in cross section measurements on light isotopes, where the level scheme is comparably simple and well known, so that spectroscopy of the transitions between low lying states and the ground state in the reaction product can be used for the identification of capture events. In measurements on unstable samples HPGe detectors are further handicapped by their comparably large dead times.

The problems with C_6D_6 and HPGe detectors are to some extent avoided by the use of 4π BaF_2 arrays. In these highly segmented total absorption calorimeters the γ-background is distributed over many modules, thus reducing the dead time problem significantly. The intrinsic resolution in γ-ray energy of these detectors is important for discrimination of true capture events (characterized by capture cascades with sum energies of typically 5 to 10 MeV) from the γ-activity of the sample, which is normally limited to energies below 1 MeV. Moreover, the nearly 100% cascade efficiency of these arrays allows one to use much smaller sample masses than with other detectors (see also below).

Apart from a measurement on ^{151}Sm [19], measurements on unstable samples with such detector arrays have, so far, concentrated on the actinide region, but these examples showed that the technique can be successfully applied to unstable isotopes in general. As reported by the DANCE group, it was possible to perform a measurement with a sample of only 0.44 mg of ^{237}Np, the smallest sample ever used in an accelerator based TOF measurement of an (n, γ) cross section [20]. In a similar experiment, the n_TOF collaboration obtained high-resolution data with a sample of 44 mg of ^{237}Np [21].

Future improvements of the neutron flux will allow one to reduce sample masses even further. This is a crucial aspect, because the production of isotopically pure samples will remain a severe difficulty. Only the high intensities at future radioactive ion beam facilities will suffice to produce samples of about 10^{16} atoms with the required mass and purity.

3.2 Beta Decay Rates

A detailed evaluation of beta decay rates for *s*-process analyses has been presented by Takahashi and Yokoi [22] on the basis of a thorough classification of possible contributions from thermally excited states and by considering the relevant effects related to the high degree of ionization in the stellar plasma. The most spectacular consequence of ionization is the enormous enhancement of decays with small Q_β values, where the decay electrons can be emitted into unoccupied atomic orbits. This bound beta decay was eventually confirmed in storage ring experiments with fully stripped ^{163}Ho and ^{187}Re atoms at GSI Darmstadt, Germany [23, 24].

The quantitative assessment of the temperature dependent decay rates of the key branch point isotopes requires more experimental information on log *ft* values for the decay of excited states as well as more storage ring experiments to expand our knowledge of bound beta decay rates. Experimental possibilities have been discussed in [25], but must be extended by the recent application of (d, ^{2}He) reactions [26].

4. NEW AND IMPROVED EXPERIMENTAL TECHNIQUES

4.1 TOF Methods

With the TOF technique, neutron capture cross section can be measured as a function of neutron energy. From these results, MACSs can be determined for any stellar temperature of interest, provided that the data are determined in the proper energy range. Recent developments in both, pulsed neutron sources and detection techniques, have led to (n, γ) cross section measurements with improved accuracy, in many cases with uncertainties of a few percent. This progress is essential for obtaining the *s* abundances with sufficient accuracy to infer the physical conditions at the stellar site of the *s* process from the observed isotope patterns in solar material and in presolar grains.

The sum of the γ-ray cascade emitted in the decay of the compound nucleus represents the best signature of a capture event. Hence, 4π detectors with an efficiency close to 100% are the most direct way to unambiguously identify (n, γ) reactions and to determine capture cross sections. This calorimetric approach started with the use of large liquid scintillator tanks, which are meanwhile replaced by arrays consisting of BaF_2 crystals because of their superior resolution in γ-ray energy and correspondingly lower backgrounds. A detector of this type consisting of 42 modules was developed at Karlsruhe [27] and is also in use at the n_TOF facility at CERN [28]. In this design the BaF_2 crystals are shaped as truncated pyramids, forming a fullerene-type geometry, where each module covers the same solid angle. The segmentation of a 4π detector array is of particular advantage for separating true capture events from backgrounds. The state-of-the-art in this respect is the DANCE array with 162 BaF_2 modules that is operated at LANSCE [29]. A somewhat simpler approach was chosen at ORNL [30], where 12 hexagonal BaF_2 detectors are surrounding the sample in a cylindrical geometry.

The main problem in using 4π arrays is due to their response to neutrons, which are scattered in the sample. Although the scintillator is selected to consist of nuclei with small (n, γ) cross sections, about 10% of the scattered neutrons are captured in the scintillator. The resulting background can be attenuated by an absorber shell around the sample, which consists of ^{6}LiH [29] or a ^{6}Li containing compound [28], but it becomes prohibitive for measurements on neutron magic nuclei and on light isotopes with small capture cross sections, where the neutron scattering channel is orders of magnitude stronger.

In such cases, this problem can be avoided by using C_6D_6 liquid scintillators and the Pulse Height Weighting Technique (PHWT) [31]. Recently, these detectors were considerably improved by minimizing the construction materials and by replacing aluminum and steel by graphite or carbon fiber. In this way the sensitivity to scattered neutrons could be reduced to a practically negligible level [32, 33]. In parallel, the accuracy of the PHWT has been improved by detailed Monte Carlo calculations using precise and detailed computer models of the experimental setups [34, 35]. A dedicated set of measurements at the n_TOF facility confirmed that the refined simulations allow one to determine neutron capture cross sections with a systematic accuracy of better than 2% [35].

The use of new data acquisition systems has substantially contributed to the recent progress in experimental techniques. In particular, flash analogue-to-digital converters (FADCs), which provide at present the fastest and safest way to store and analyze analogue signals, made it possible to digitize the entire wave-form of the detector signals, instead of recording only their amplitude or integrated charge. Recording all detector signals during a measurement allows for a rigorous assessment of uncertainties related with the performance of the detectors, and distortions in the raw data such as baseline shifts, pileup, and noise are efficiently identified and can be corrected for during data analysis. The relevant parameters of the electronic signals such as time information, amplitude, area, are determined off-line via pulse shape analysis algorithms specifically designed for each type of detector. In this way, background sources can be better identified and corrected, particularly in the case of measurements performed with 4π BaF_2 arrays. An example for such a data acquisition system is the one used at the n_TOF facility [36].

4.2 Activation Method

The activation method has been used extensively in MACS measurements, once it was found that stellar neutron spectra can be produced in the laboratory. So far, three reactions have been invoked for this technique. The ^{7}Li(p, n)^{7}Be reaction yields a quasi-stellar spectrum for a thermal energy of kT = 25 keV [37, 38], very close to the 23 keV effective thermal energy in He shell flashes of low mass AGB stars, where neutrons are produced via the ^{22}Ne(α, n)^{25}Mg reaction. The main neutron source in these stars, the ^{13}C(α, n)^{16}O reaction, was simulated recently by means of the ^{18}O(p, n)^{18}F reaction [39].

While these two quasi-stellar spectra allow one to determine the MACSs necessary for studies of the main *s* component, the weak component associated with massive stars is characterized by higher temperatures, i.e. 26 keV thermal energy during core He burning and about 90 keV in the shell C burning phase. The situation during core He burning is again well described by the ^{7}Li(p, n)^{7}Be reaction, but the high temperatures during shell C burning are only roughly represented by means of the ^{3}H(p, n)^{3}He reaction, which provides a spectrum for kT = 52 keV [40]. In this case, the measured MACS have to be extrapolated by statistical model calculations.

The source intensities that can be presently achieved with these quasi-stellar neutron spectra are of the order of 10^9, 10^8, and 10^5 s^{-1} for the (p, n) reactions on ^{7}Li, ^{3}H, and ^{18}O, respectively. These numbers are obtained by assuming a proton beam current of 100 μA on target, a value that can be reached with electrostatic accelerators. Future developments, however, will provide much higher beam currents and correspondingly higher neutron fluxes.

The available neutron intensities for activation measurements exceed the fluxes for TOF measurements by orders of magnitude. Accordingly, activation represents the most sensitive method for (n, γ) measurements in the astrophysically relevant energy range. This feature is unique for the possibility to measure the MACSs of neutron poisons, abundant light isotopes with very small cross sections, as well as for the use of extremely small sample masses.

The latter aspect is most important for the determination of MACSs of unstable isotopes, where TOF measurements are challenged by the background due to the sample activity. An illustrative example in this respect is the successful measurement of the MACS of ^{147}Pm, which was performed with a sample of only 28 ng in order to compensate for the short half-life of 2.6 yr [41].

5. FUTURE STRATEGIES

The persisting challenges in neutron cross section data for *s*-process studies refer to the measurement of all MACSs with the required average accuracy of <5%, the reduction of the uncertainty to <2% for the cross sections of key isotopes, and the experimental determination of the cross sections of the unstable branch point nuclei.

The first goal of accurate measurements on stable isotopes can be achieved by present state-of-the-art C_6D_6 scintillators and total absorption BaF_2 calorimeters coupled with FADC data acquisition systems, in particular if experiments as well as data analysis are planned and performed by Monte Carlo simulations. The difficult measurements of small cross sections of neutron magic nuclei and of light neutron poison isotopes will benefit from the much higher intensities of the new neutron facilities, which will become available in the near future [42, 43]. The high neutron fluxes are also suited for measurements on very small amounts of unstable samples. Another aspect that will become important once higher neutron fluxes are available is that measurements can be repeated with modified experimental parameters, different detectors, and even with different methods. The redundancy obtained in this way is crucial for ensuring that corrections are properly treated and that systematic uncertainties are indeed reduced to better than 5% in general and to 1-2% for the isotopes of key importance.

REFERENCES

1. E. Burbidge, G. Burbidge, W. Fowler, and F. Hoyle, *Rev. Mod. Phys.* **29**, 547 (1957)
2. A. Cameron, Chalk River Report CRL-41, A.E.C.L. Chalk River, Canada (1957)
3. A. Heger, *Technical report,* http://www.ucolick.org/~alex/
4. M. Limongi, *Technical report,* http://www.mporzio.astro.it/~limongi/
5. C. Raiteri, M. Busso, R. Gallino, and G. Picchio, *ApJ* **371**, 665 (1991)
6. C. Raiteri, et al., *ApJ* **419**, 207 (1993)
7. M. Limongi, O. Straniero, and A. Chieffi, *ApJS* **129**, 625 (2000)
8. O. Straniero, et al., *ApJ* **440**, L85 (1995)
9. R. Gallino, et al., *ApJ* **497**, 388 (1998)
10. C. Arlandini, et al., *ApJ* **525**, 886 (1999)
11. Z. Bao, et al., *Atomic Data Nucl. Data Tables* **76**, 70 (2000)
12. B. Allen, J. Gibbons, and R. Macklin, *Adv. Nucl. Phys.* **4**, 205 (1971)
13. Z. Bao and F. Käppeler, *Atomic Data Nucl. Data Tables* **36**, 411 (1987)
14. H. Beer, F. Voss, and R. Winters, *ApJS* **80**, 403 (1992)

15. T. Rauscher and F.-K. Thielemann, *Atomic Data Nucl. Data Tables* **75**, 1 (2000)
16. I. Dillmann, et al., *Capture Gamma-Ray Spectroscopy and Related Topics*, AIP Conference Series 819, edited by A. Woehr and A. Aprahamian (AIP, New York, 2005), p.123, http://nuclear-astrophysics.fzk.de/kadonis
17. U. Abbondanno, et al. (n_TOF Collaboration), *Phys. Rev. Lett.* **93**, 161103 (2004)
18. S. Marrone, et al. (n_TOF Collaboration), *Phys. Rev. C* **73**, 034604 (2006)
19. K. Wisshak, et al.., *Phys. Rev. C* **73**, 015802 (2006)
20. R.C. Haight, *International Conference on Nuclear Data for Science and Technology*, Nice, France, April 22 - 27, 2007 (in preparation)
21. C. Guerrero and the n_TOF collaboration, *International Conference on Nuclear Data for Science and Technology*, Nice, France, April 22 - 27, 2007 (in preparation)
22. K. Takahashi and K. Yokoi, *Atomic Data Nucl. Data Tables* **36**, 375 (1987)
23. M. Jung, et al., *Phys. Rev. Lett.* **69**, 2164 (1992)
24. F. Bosch, et al., *Phys. Rev. Lett.* **77**, 5190 (1996)
25. F. Käppeler, *Prog. Part. Nucl. Phys.* **43**, 419 (1999)
26. D. Frekers, *Nucl. Phys.A* **752**, 580c (2005)
27. K. Wisshak, K. Guber, F. Käppeler, and F.Voss, *Nucl. Instr. Meth.A* **299**, 60 (1990)
28. M. Heil, et al., *Nucl. Instr. Meth. A* **459**, 229 (2001)
29. R. Reifarth, et al., *Nucl. Instr. Meth. A* **531**, 530 (2004)
30. K. Guber, R. Spencer, P. Koehler, and R. Winters, *Nucl. Phys.* A **621**, 254c (1997)
31. R. Macklin and J. Gibbons, *Phys. Rev.* **159**, 1007 (1967) [includes H. Maier-Leibnitz, priv. comm. (1963) and F. Rau, *Nukleonik* **5**, 191 (1963)]
32. P. Koehler, et al., *Phys. Rev. C* **62**, 055803 (2000)
33. R. Plag, et al., *Nucl. Instr. Meth. A* **496**, 425 (2003)
34. J. Tain, F. Gunsing, D. Cano-Ott, and C. Domingo, *J. Nucl. Sci. Tech. Suppl.* **2**, 689 (2002)
35. U. Abbondanno, et al. (n_TOF Collaboration), *Nucl. Instr. Meth. A* **521**, 454 (2004)
36. U. Abbondanno, et al. (n_TOF Collaboration), *Nucl. Instr. Meth. A* **538**, 692 (2005)
37. H. Beer and F. Käppeler, *Phys. Rev. C* **21**, 534 (1980)
38. W. Ratynski and F. Käppeler, *Phys. Rev. C* **37**, 595 (1988)
39. M. Heil, et al., *Phys. Rev. C* **71**, 025803 (2005)
40. F. Käppeler, A. Naqvi, and M. Al-Ohali, *Phys. Rev. C* **35**, 936 (1987)
41. R. Reifarth, et al., *ApJ* **582**, 1254 (2003)
42. http://j-parc.jp/, Technical report, JAERI and KEK (2004)
43. U. Ratzinger, et al., *18th Meeting of the International Collaboration on Advanced Neutron Sources*, April 25-29, 2007, Dongguan, Guangdong, PR China (2007)

Recent Results at n_TOF and Future Perspectives

S. Marrone[1 19], U. Abbondanno[20], G. Aerts[7], H. Álvarez[34], F. Alvarez-Velarde[30], S. Andriamonje[7], J. Andrzejewski[25], P. Assimakopoulos[16], L. Audouin[12], G. Badurek[1], P. Baumann[10], F. Bečvář[6], E. Berthoumieux[7], M. Calviani[18], F. Calviño[33], D. Cano-Ott[30], R. Capote[3,35], A. Carrillo de Albornoz[26], P. Cennini[36], V. Chepel[27], E. Chiaveri[36], N. Colonna[19], G. Cortes[32], A. Couture[40], J. Cox[40], M. Dahlfors[36], S. David[9], I. Dillmann[12], R. Dolfini[23], C. Domingo-Pardo[31], W. Dridi[7], I. Duran[34], C. Eleftheriadis[13], M. Embid-Segura[30], L. Ferrant[9], A. Ferrari[36], R. Ferreira-Marques[27], L. Fitzpatrick[36], H. Frais-Koelbl[3], K. Fujii[20], W. Furman[29], R. Gallino[22], I. Goncalves[27], E. Gonzalez-Romero[30], A. Goverdovski[28], F. Gramegna[18], E. Griesmayer[3], C. Guerrero[31], F. Gunsing[7], B. Haas[8], R. Haight[38], M. Heil[12], A. Herrera-Martinez[36], M. Igashira[24], S. Isaev[9], E. Jericha[1], Y. Kadi[36], F. Käppeler[12], D. Karamanis[16], D. Karadimos[16], M. Kerveno[10], V. Ketlerov[28,36], P. Koehler[39], V. Konovalov[29,36], E. Kossionides[15], M. Krtička[6], C. Lamboudis[13], H. Leeb[1], A. Lindote[27], I. Lopes[27], M. Lozano[35], S. Lukic[10], J. Marganiec[25], L. Marques[26], C. Massimi[21], P. Mastinu[18], A. Mengoni[3,36], P.M. Milazzo[20], C. Moreau[20], M. Mosconi[12], F. Neves[27], H. Oberhummer[1], S. O'Brien[40], J. Pancin[7], C. Papachristodoulou[16], C. Papadopoulos[14], C. Paradela[34], N. Patronis[16], A. Pavlik[2], P. Pavlopoulos[11], L. Perrot[7], M. Pignatari[22], R. Plag[12], A. Plompen[5], A. Plukis[7], A. Poch[32], C. Pretel[32], J. Quesada[35], T. Rauscher[37], R. Reifarth[38], M. Rosetti[17], C. Rubbia[23], G. Rudolf[10], P. Rullhusen[5], J. Salgado[26], L. Sarchiapone[36], I. Savvidis[13], C. Stephan[9], G. Tagliente[19], J.L. Tain[31], L. Tassan-Got[9], L. Tavora[26], R. Terlizzi[19], G. Vannini[21], P. Vaz[27], A. Ventura[17], D. Villamarin[30], M.C. Vincente[30], V. Vlachoudis[36], R. Vlastou[14], F. Voss[12], S. Walter[12], H. Wendler[36], M. Wiescher[40], K.Wisshak[12]

[1]Atominstitut der Österreichischen Universitäten,Technische Universität Wien, Austria,[2]Institut für Isotopenforschung und Kernphysik, Universität Wien, Austria,[3]International Atomic Energy Agency, NAPC-Nuclear Data Section, Vienna, Austria [4]Fachhochschule Wiener Neustadt, W iener Neustadt, Austria,[5]CEC-JRC-IRMM, Geel, Belgium,[6]Charles University, Prague, Czech Republic, [7]CEA/Saclay - DSM, Gif-sur-Yvette, France,[8]Centre National de la Recherche Scientifique/IN2P3 - CENBG, Bordeaux, France, [9]Centre National de la Recherche Scientifique/IN2P3 - IPN, Orsay, France, [10]Centre National de la Recherche Scientifique/IN2P3 IReS, Strasbourg, France,[11]Pôle Universitaire Léonard de Vinci, Paris La Défense, France,[12]Forschungszentrum Karlsruhe GmbH (FZK), Institut für Kernphysik Germany,[13]Aristotle University of Thessaloniki, Greece,[14]National Technical University of Athens Greece,[15]NCSR, Athens, Greece,[16]University of Ioannina, Greece,[17]ENEA, Bologna, Italy, [18]Istituto Nazionale di Fisica Nucleare(INFN), Laboratori Nazionali di Legnaro, Italy,[19]Dipartimento di Fisica and Istituto Nazionale di Fisica Nucleare, Bari, Italy, [20]Istituto Nazionale di Fisica Nucleare, Trieste, Italy,[21]Dipartimento di Fisica, Università di Bologna, and Sezione INFN di Bologna, Italy, [22]Dipartimento di Fisica, Università di Torino and Sezione INFN di Torino, Italy,[23]Università degli Studi Pavia, Pavia, Italy,[24]Tokyo Institute of Technology, Tokyo, Japan, [25]University of Lodz, Lodz, Poland[26]Instituto Tecnológico e Nuclear(ITN), Lisbon, Portugal,[27]LIP - Coimbra & Departamento de Fisica da Universidade de Coimbra, Portugal,[28]Institute of Physics and Power Engineering, Kaluga region, Obninsk, Russia,[29]Joint Institute for Nuclear Research, Frank Laboratory of Neutron Physics, Dubna, Russia,[30]Centro de Investigaciones Energeticas Medioambientales y Technologicas, Madrid, Spain,[31]Istituto de Fisica Corpuscolar, CSIC Universidad de Valencia, Spain,[32]Universitat Politecnica de Catalunya, Barcelona, Spain,[33]Universidad Politecnica de Madrid, Spain,[34]Universidade de Santiago de Compostela, Spain,[35]Universidad de Sevilla, Spain,[36]CERN, Geneva, Switzerland,[37]Department of Physics and Astronomy - University of Basel, Basel, Switzerland,[38]Los Alamos National Laboratory, New Mexico, USA,[39]Oak Ridge National Laboratory, Physics Division, Oak Ridge, USA,[40]University of Notre Dame, Notre Dame, USA.

[1] Corresponding author, address: via Orabona 4, 70126 Bari (Italy). Tel. +390805442511, e-mail stefano.marrone@ba.infn.it .

CP1001, 9th *Torino Workshop on Evolution and Nucleosynthesis in AGB Stars and the 2nd Perugia Workshop on Nuclear Astrophysics*,
edited by R. Guandalini, S. Palmerini, and M. Busso

Abstract. The neutron Time-of-Flight facility at CERN has been used for a series of neutron capture and fission cross sections measurements. The facility, detectors and data analysis techniques are described, followed by a brief review of the capture results and of the preliminary results for selected fission cross sections. The problem of photon strength functions is discussed and preliminary first results are presented. All these subjects are discussed with respect to their implications in Nuclear Astrophysics. The last section is dedicated to the future perspectives concerning the next experimental campaign and the upgrade of the facility.

Keywords: Neutron, Capture, Fission, Photon Strength Function, *s*-Process, *r*-Process.
PACS: 25.40.Lw, 26.20.+f, 26.50.+x, 27.60.+j, 97.10.Cv, 97.10.Tk.

1. INTRODUCTION

In nuclear astrophysics accurate data from neutron studies are necessary to clarify several issues such as the production of heavy elements through the rapid and slow neutron capture processes (*r*- and *s*-process for short) [1], the characteristics of stellar environments (Red Giants and Supernovae) and the age of the universe (Re/Os cosmo-chronometer). Related topics are the systematic of nuclear properties (masses and lifetimes) and the neutron reaction cross sections especially for isotopes far from the valley of stability [2]. In addition there are new concepts in nuclear technology for energy production, radioactive waste transmutation and nuclear medicine [3] that require particularly accurate neutron cross sections for a large number of isotopes. In order to fulfil these requests, an experimental program has been started for measuring capture and fission cross sections at the innovative neutron time-of-flight (n_TOF) facility at CERN. Recently, the particular features of n_TOF have been used to investigate the γ-rays following neutron capture for studies of photon strength functions.

The neutron beam at n_TOF is generated by bombarding a massive lead block with intense proton bunches (width 6 ns, intensity 7×10^{12} protons per bunch, proton energy 20 GeV) from the proton synchrotron (PS). The lead block ($80\times80\times60$ cm^3) [4] is surrounded by 5 cm of cooling water, which acts also as a moderator. The target is connected with the experimental area (EAR-1) by a 200 m long, evacuated neutron flight tube. The neutron beam is cleaned from charged particles by means of a sweeping magnet and beam-related backgrounds are suppressed by massive concrete and iron shieldings. Collimators, which are composed of iron and polyethylene layers, are located at 135 and 175 m from the target. The first collimator is fixed while the second can be modified according to the different beam diameters required for capture or fission measurements.

The neutron flux measured in the experimental area covers a wide energy range (Fig. 1). Thanks to the high proton densities and the short duration of the PS bunches, n_TOF represents the most luminous neutron source worldwide. In spite of the smaller beam diameter used in the capture measurements, the intensity in EAR-1 is still 6.2×10^5 neutrons per bunch with energies between 0.1 eV and several 100 MeV. In the fission configuration the flux is one order of magnitude higher. Flux measurements were performed with several detectors, particularly with a fission chamber, a ^{6}Li/silicon monitor, and via standard neutron capture resonances (^{197}Au, ^{56}Fe).

In order to monitor the neutron flux in the experimental area and to provide the normalization of the flux in measurements with different samples, the ^{6}Li/silicon monitor was extensively used throughout the various neutron capture campaigns [5]. The (n, γ) cross sections are measured with two types of detectors, a pair of C_6D_6 [6] scintillators and a Total Absorption Calorimeter (TAC). In C_6D_6 measurements, the samples are mounted on a remotely controlled sample changer made from carbon fibre, which can be directly integrated in the evacuated beam tube. The design of the 4π BaF_2 detector consists of a 42-element array [7] built from 30 hexagonal and 12 pentagonal crystals, which are forming a closed spherical BaF_2 shell with an inner diameter of 20 cm and a thickness of 15 cm. Given the high density of BaF_2 (4.88 g/cm^3), γ-ray cascades following neutron capture are detected with an efficiency of ~95%. Other important features of this detector are a good resolution in γ-ray (6% at 6.13 MeV) and an excellent time resolution (~ 500 ps).

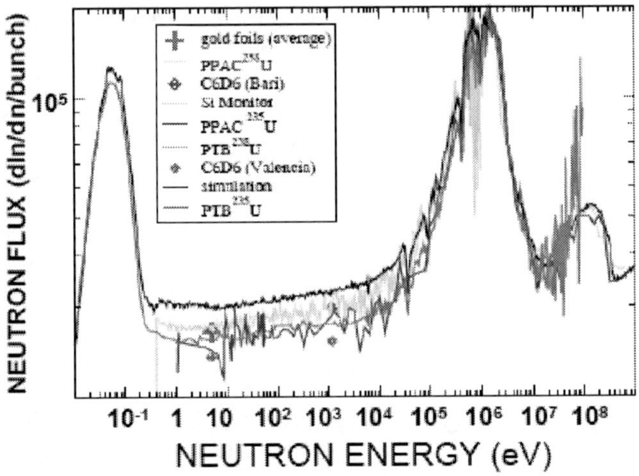

FIGURE 1. Neutron Flux of the n_TOF beam measured and simulated in the capture configuration. The legend indicates the different detectors used for the measurements.

Also in the fission cross section measurements, two types of detectors have been used, Parallel Plate Avalanche Counters (PPAC) and Fission Ionization Chambers (FIC) [8]. The PPACs can be built in large dimensions, are insensitive to radiation damage, and combine a high rate capability (more than 2 MHz frequency) with good time and position resolution. The Fission Chambers consist of a stack of parallel electrodes, accommodating up to 16 targets and 18 electrodes in a single detector.

Due to the high instantaneous neutron flux, several events are generally recorded in a single neutron bunch. In order to avoid pile-up and dead time problems, a data acquisition system based on high-frequency flash analogue to digital converters (FADC) has been developed [8]. The FADC modules can be operated with sampling rates up to 1 GSample/s and are equipped with 8 MByte of buffer memory for each channel. The raw data are recorded signal by signal for detailed off-line analysis, which allows one to derive the required information on timing, charge, amplitude, and particle identification to be stored on the Data Stream Tapes (DST). After extracting the DSTs, the data analysis consists of the calculation of the experimental yields for each sample and the subtraction of the respective backgrounds. The final steps are the correction for detector efficiency and for neutron flux normalization as indicated in the references for every specific measurement.

2. CAPTURE RESULTS AND IMPLICATIONS

The main published results of the capture measurements involve the cross sections of ^{139}La [9], ^{151}Sm [10,11], ^{186,187,188}Os [12], ^{204,207}Pb [13,14], ^{209}Bi [15] and ^{232}Th [16] measured with the C_6D_6 detectors. In the Resolved Resonance Region (RRR), the cross sections are given in terms of resonance parameters, which are determined by means of the R-matrix code SAMMY [17]. In the Unresolved Resonance Region, averaged cross sections are reported for some isotopes at higher energies up to about 1 MeV. The example illustrated in Fig. 2 shows the SAMMY fit of keV resonances in the ^{139}La cross section, essentially defining the resonance widths Γ_n and Γ_γ. In case of isotopes with higher level densities, the RRR ends at a few keV, so that only the averaged cross section is deduced at higher energies (Fig. 3). In both situations, it is possible to calculate the Maxwellian averaged cross section (MACS), which is the main input for stellar nucleosynthesis codes.

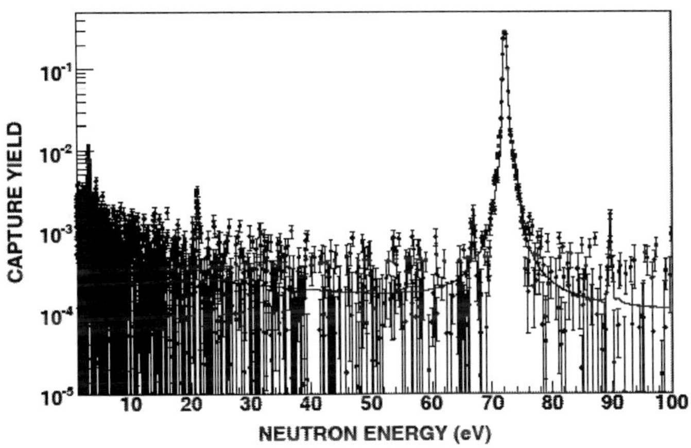

FIGURE 2. Resonance fits with the SAMMY code [17] in the resolved resonance region of the ^{139}La$(n,\gamma)^{140}$La reaction [9].

In recent years, La became increasingly important for stellar spectroscopy, since this element is easily observable in stellar atmospheres, it is relatively abundant, and consists essentially only of a single isotope, whose origin is well characterized. Because of the low capture cross section of ^{139}La and the characteristics of the backgrounds, the measurement could be analyzed only up to 9 keV in neutron energy. Nevertheless, MACS could be calculated with high accuracy, especially for low thermal energies (kT = 5-10 keV), where the stellar production of ^{139}La is most efficient. With these results, the s-process component, which accounts for 74% of the solar abundance, could be reliably determined. The r-process contribution to ^{139}La could then be improved by means of the residual method for comparison of the [La/Eu] spectroscopic ratio as a function of [Fe/H] in low metallicity stars [9].

The (n, γ) cross section of ^{151}Sm was measured over a wider energy range (1 eV-1 MeV) with an overall uncertainty of 5% [10,11]. The MACS at 30 keV of 3080 ± 150 mbarn was found to be systematically larger than theoretically predicted, resulting in a sizeable effect on the *s*-process branchings in the Sm/Eu/Gd region as illustrated by a sensitivity analysis with respect to the MACSs and β-decay rates of the involved isotopes.

The ^{186,187,188}Os capture cross sections were studied [12], because of their importance for cosmo-chronology. Due to the long half-life of ^{187}Re, the decay of this *r*-only isotope is particularly suited for determining the age of the Milky Way and hence of the Universe. Previous cross sections measurements [18,19] suffered from discrepancies and insufficient coverage of the astrophysically relevant energy range. The comparison with our results in Fig. 3 shows good agreement only for ^{187}Os, whereas differences of 9% and 30% were found for ^{186}Os and ^{188}Os, respectively. A preliminary estimate based on these results indicates a Galactic age of ~15 Gyr.

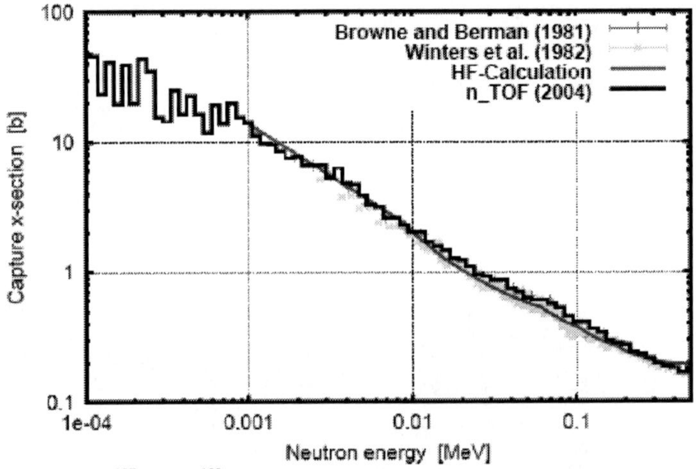

FIGURE 3. The averaged $^{187}Os(n,\gamma)^{188}Os$ cross section from 0.1 keV to 1 MeV compared with previous measurements [18,19] and with a theoretical calculation [12].

The Pb and Bi cross sections are particular relevant for describing the *s*-process termination point, where the *s* contributions of stars with about 10 times lower metallicity than the sun have a strong impact. The abundance pattern in the Tl/Pb/Bi region is affected by several branchings and by α-recycling from the unstable isotopes beyond ^{209}Bi. Accurate measurements of the (n, γ) cross sections of all stable Pb isotopes and of ^{209}Bi have been performed to improve the discussion of this important mass region. The results for ^{204}Pb [15] comprise 170 resonances between 1 and 90 keV and averaged cross section up to 400 keV, leading to significant changes in the MACS compared to previous measurements. The cross section of ^{207}Pb [16] was found to agree better with previous data, but – as in case of ^{204}Pb – the uncertainties could be reduced by factors of two to three. An important difference to previous results was found for ^{209}Bi [17], where the capture cross section and the corresponding MACS values are 15% lower than obtained in earlier experiments. As a consequence of the new MACS, the *s*-process fractions are now 95% and 77% of the solar ^{204}Pb and

^{207}Pb abundances compared to 99% and 82% predicted before, whereas the larger change of the ^{209}Bi cross section has almost no effect on the s abundance, which is now 19.0% instead of 18.7%.

Other capture measurements concern the isotopes ^{90,91,92,93,94,96}Zr, ^{24,25,26}Mg, ^{234,236}U, ^{243}Am, ^{237}Np and ^{240}Pu, which were partly measured with the C_6D_6 setup and partly with the TAC. The Mg and Zr results are important for their Nuclear Astrophysics aspects. The data analyses for these measurements are in an advanced stage and preliminary results have been presented at various conferences [20].

3. FISSION CROSS SECTIONS AND γ-RAY DATA

The preliminary results [21] for the ^{233}U(n, f) cross section are shown in Fig. 4. Since the higher α-activity required a higher pulse height threshold than for ^{235}U, the corresponding loss in efficiency was corrected by calculating an effective flux from the ^{235}U(n, f) cross section for the actual threshold. Small differences are observed in the amplitude of some ^{233}U resonances, and in the resonance minima the new data appear to be lower than in the evaluated data files, indicating a better background situation at n_TOF.

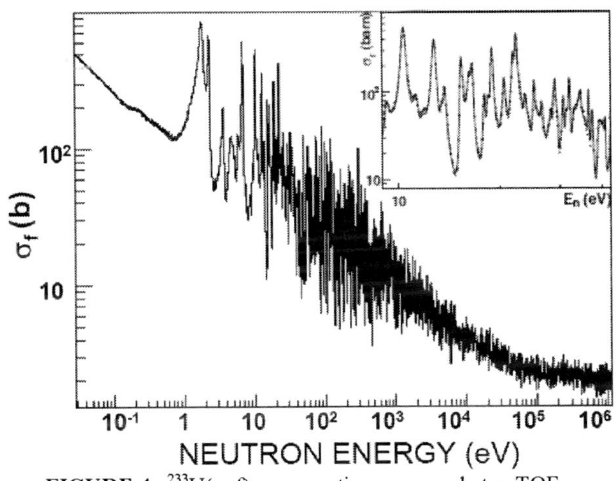

FIGURE 4. ^{233}U(n, f) cross section measured at n_TOF.

Another interesting isotope is ^{245}Cm since only few and discordant experimental measurement have been reported, and no data were given between thermal energies and 20 eV at all. In spite of the higher pulse height threshold, the experimental data had to be corrected for α-pile-up, because the relatively short half-life of 8500 yr implies a very large α-background. This correction was determined from runs performed without beam. Due to a remaining uncertainty in the detection efficiency, the cross section was normalized to a weighted mean of the available data at thermal energy. The yet preliminary n_TOF results are higher than the ENDF/B-VII evaluation, particularly in the energy range between 0.1 eV and 8 eV, where discrepancies of up to 10% were obtained. Compared with previous experiments, the fission resonances are well reproduced in general, except for some cases, where discrepancies of up to 20% are found in the minima and peak values.

In addition to the neutron cross sections, the n_TOF facility can also be used for γ-ray studies related to neutron capture reactions [22]. In this respect, the favourable features of the n_TOF facility provide γ-spectra with an excellent signal-to-noise ratio especially in the resolved resonance region as well as the possibility to resolve and identify levels with different quantum numbers. Certain disadvantages are mainly related to the characteristics of the γ-ray detectors. In fact the low efficiency and the poor energy resolution of the C_6D_6 scintillators are strongly limiting the quality of the measured γ-spectra. In future, these problems will be partially solved by using the TAC.

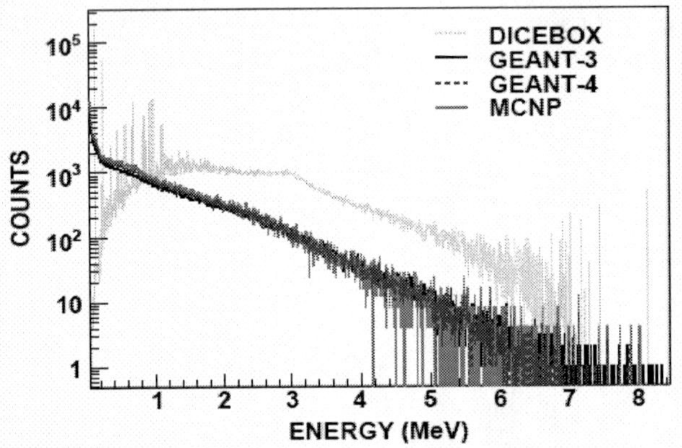

FIGURE 5. Photon strength function studies at n_TOF [22]. The various histograms indicate a DICEBOX simulation for the decay of ^{152}Sm (green) and the convolution of this primary spectrum with the response of the C_6D_6 detectors used by means of the tracking codes GEANT-3 (black), GEANT-4 (dashed-blue), and MCNP (red).

Decay spectra of ^{152}Sm were simulated using the DICEBOX algorithm within the extreme statistical model, embodying Bohr's idea of a compound nucleus, the paradigm of fragmentation of photon strength, and the validity of the Brink hypothesis. A response of the detection setup to each of these cascades was simulated using the tracking codes GEANT-3, GEANT-4, and MCNP. In these simulations, the geometry and the materials of the experimental set-up were carefully implemented and the points of origin of the photons from (n, γ) reactions were modelled according to the neutron beam profile at n_TOF. The γ-ray spectrum of ^{152}Sm obtained in the Monte Carlo simulations (Fig. 5) can be directly compared with its measured counterpart for deriving information on level density and photon strength functions.

4. FUTURE PERSPECTIVES AND CONCLUSIONS

Two goals are essentially planned at n_TOF for the next years. The first is the continuation of the experimental activities in neutron cross section measurements and photon strength function studies. There is a large set of isotopes to investigate, i.e. the isotopes of Mo, Ru, Pd, Fe, Ni, Zn, and Se as well as a number of actinide nuclei [23]. A part of these measurements (Fe, Zn and Se) is already approved by the Isolde/n_TOF Committee at CERN for 2008, when the neutron beam is expected to restart. From the experimental point of view, the detectors, samples, and the data acquisition system are ready for data taking. As in the first phase of the n_TOF project, Nuclear Astrophysics will play a fundamental role in the determination of the scientific programme at n_TOF.

The second task is the construction of a new experimental area (EAR-2). According to simulations, the best location for the additional beam line is at 90° with respect to the proton beam direction with a reduced flight-path of 20 m. The necessary modification of the spallation target will already be considered in the design of a new target that is planned for 2008. Compared to the present EAR-1, the neutron flux is expected to increase by almost a factor of 100 and the background due to ultra-relativistic particles and γ-rays will be strongly reduced. Since the EAR-2 will be particularly suited for measurements of very small samples, small cross sections, and radioactive isotopes, it will provide new horizons for the scientific programme at n_TOF.

ACKNOWLEDGMENTS

The n_TOF project is supported by the EC under the contract FIKW-CT-2000-00107 and by the participating National Institutions.

REFERENCES

1. F. Käppeler, *Prog. Nucl. Part. Phys.* **43**, 419 (1999)
2. NuPECC Long Range Plan, edited by M. Harakeh et al., in April 2004. (2004)
3. R. Terlizzi, et al., *Nucl. Instr. Methods B* **213**, 210 (2000)
4. n_TOF Collaboration, Performance Report, CERN-INTC-O-011 (2003)
5. S. Marrone, et al., *Nucl. Instr. Methods A* **517**, 389 (2004)
6. R. Plag ,et al., *Nucl. Instr. Methods A* **496**, 425 (2003)
7. D. Cano-Ott, et al., *AIP Conference Proceedings* **769**, 1442 (2004)
8. U. Abbondanno, et al., *Nucl. Instr. Methods A* **538**, 692 (2005)
9. R. Terlizzi, et al., *Phys. Rev. C* **75**, 035807 (2007)

10 U. Abbondanno, et al., *Phys. Rev. Lett.* **93**, 161103 (2001)
11 S. Marrone, et al., *Phys. Rev. C* **73**, 034604 (2006)
12 M. Mosconi, et al., *Prog. in Part. and Nucl. Phys.* **59**, 165 (2007)
13 C. Domingo-Pardo, et al., *Phys. Rev. C* **75**, 015806 (2007)
14 C. Domingo-Pardo, et al., *Phys. Rev. C* **74**, 055802 (2006)
15 C. Domingo-Pardo, et al., *Phys. Rev. C* **74**, 025807 (2006)
16 G. Aerts, et al., *Phys. Rev. C* **73**, 054610 (2005)
17 N. M. Larson, Report ORNL/TM 2000/252, Oak Ridge National Laboratory (2000)
18 J. Browne and B. Berman, *Phys. Rev. C* **23**, 563 (1980)
19 R. Winters, et al., *Phys. Rev. C* **21**, 034604 (2006)
20 Visit website www.cern.ch/n_TOF
21 M. Calviani, et al., Proc. of Nuclear Data for Science and Technology, Nice April 11-16 2007 (2007)
22 S. Marrone, et al., Proc. of Photon Strength Function and Related Topics, Prague June 17-20 2007 (2007)
23 n_TOF Collaboration, "The n_TOF phase 2 initiative", CERN-INTC-205-21 (2006)

Indirect Study of $^{13}C(\alpha,n)^{16}O$

F. Hammache*, M. G. Pellegriti*, P. Roussel*, L. Audouin*, D. Beaumel*, S. Fortier*, J. Kiener†, A. Lefebvre-Schuhl†, V. Tatischeff†, P. Descouvemont**, L. Gaudefroy‡ and M. Stanoui§

*IPN-Orsay, IN2P3-CNRS, Université Paris XI,Orsay, 91406, France
†CSNSM, IN2P3-CNRS, Université Paris XI,Orsay, 91405, France
**Physique Nucléaire Théorique et Physique Mathématique, ULB CP229, B-1050 Brussels, Belgium
‡GANIL, IN2P3-CNRS, Caen, France
§GSI, Postfach 110552, D-64220, Darmstadt, Germany

Abstract. The reaction $^{13}C(\alpha,n)^{16}O$ is considered as the main neutron source for s-process in low-mass asymptotic giant branch stars. At low energies of astrophysical interest, the contribution of the subthreshold state 6.356 MeV of ^{17}O to the $^{13}C(\alpha,n)^{16}O$ cross section should be taken into account. However, the results of previous studies of this contribution lead to different conclusions.

Hence, we investigated the effect of this resonance on the astrophysical S-factor through a new precise measurement of the alpha spectroscopic factor, S_α, of the 6.356 MeV state using the transfer reaction $^{13}C(^7Li,t)^{17}O$ at two different incident energies. The measured angular distributions and the obtained spectroscopic factors will be presented as well as their impact on $^{13}C(\alpha,n)^{16}O$ cross section and reaction rate.

Keywords: AGB, Neutron Source, s-process, α-transfer, $^{13}C(\alpha,n)^{16}O$
PACS: 26.20.-f,26.20.Fj,26.20.Kn

1. INTRODUCTION

Nearly half of the elements heavier than iron are produced by a slow sequence of neutron capture reactions, the so-called s-process. For the AGB stars of 1-3 solar masses at low temperatures, the $^{13}C(\alpha,n)^{16}O$ reaction is considered as the main neutron source for the s-process [1, 2, 3]. In these stars, ^{13}C is synthesized in the He intershell of the star by the CNO cycle $^{12}C(p,\gamma)^{13}N(\beta^+)^{13}C$. This leads to the formation of the so-called ^{13}C pocket on the top of the intershell region, where the neutrons necessary for the s-process, are produced through the burning reaction $^{13}C(\alpha,n)^{16}O$. Hence, all the models describing the s-process nucleosynthesis in these AGB stars depend critically on the neutron flux from the $^{13}C(\alpha,n)^{16}O$ reaction and so on the cross section of this reaction which occurs in these stars at temperatures around 10^8 K, i.e around the Gamow peak $E_{cm} \sim$190 keV.

Direct measurement of $^{13}C(\alpha,n)^{16}O$ reaction at this energy is extremely difficult because the cross section decreases drastically when the incident α energy decreases. Thus, direct measurements [4] have only been performed down to 270 keV (see Fig. 1) too far away from the energy range of interest around 190 keV. The estimation of the cross section of $^{13}C(\alpha,n)^{16}O$ reaction at the energy of astrophysical interest depends critically on the way extrapolations are made. First one has to extrapolate the measured cross sections from high energies and consider in the same time the likely effect of the subthreshold state at 6.356 MeV of the compound nucleus ^{17}O in this energy region.

CP1001, 9th Torino Workshop on Evolution and Nucleosynthesis in AGB Stars and the 2nd Perugia Workshop on Nuclear Astrophysics, edited by R. Guandalini, S. Palmerini, and M. Busso

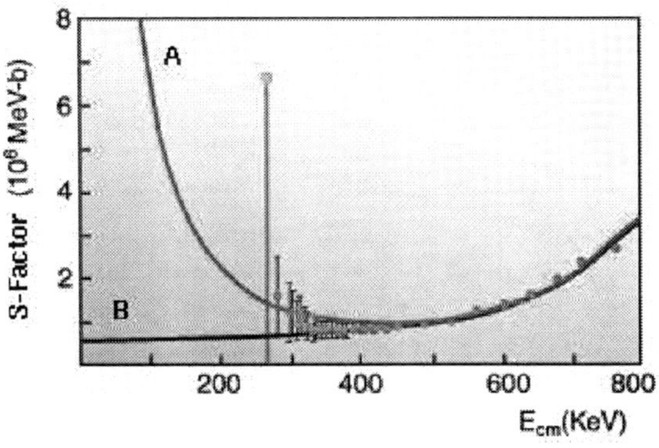

FIGURE 1. $^{13}C(\alpha,n)^{16}O$ astrophysical S-factors as a function of center of mass energy. The data points are the direct measurements of [4]. Curves A and B inferred from R-matrix calculations using an α spectroscopic factor for the subthreshold state at 6.356 MeV of about 0.7 and 0 respectively.

R-matrix extrapolations [5], which consider a large value of 0.7 for the α spectroscopic factor (S_α) of the subthreshold state at 6.356 MeV lead to a rapid increase of the S-factor when the energy decreases to zero (see curve A in Fig. 1), while they lead to constant extrapolated S-factor at low energies when they use a null S_α (see curve B in Fig.1).

If the experimental points in Fig.1 suggest a rise of the S-factor when the energy decreases as expected with the usually used spectroscopic factors [6, 7] in the s-models ($S_\alpha \sim$ 0.3-0.7(curve A)), the error bars are too large to well discriminate curves A and B.

Kubono et al. [8] measured then the S_α of the subthreshold state at 6.356 MeV through the α transfer reaction $^{13}C(^6Li,d)^{17}O$ at the laboratory energy of 60 MeV. Through a finite-range DWBA analysis of their transfer angular distributions, a very small value of $S_\alpha \sim$ 0.01 was deduced for the 6.356 MeV $1/2^+$ state of ^{17}O. This value leads to a very small contribution of this subthreshold state to the cross section of $^{13}C(\alpha,n)^{16}O$ at energies of astrophysical interest and thus to the S-factor. This result has dramatic effects on the neutron flux and subsequently on the onset of the s-process nucleosynthesis. However, a reanalysis of the same data by Keeley et al. [9] ends up to a drastically different conclusion since they found a value of S_α of about 0.4, forty times higher than Kubono's value and consistent with the values ($S_\alpha \sim$ 0.3-0.7) usually used in astrophysical calculations. In addition to these controversial results, the new determination of the contribution of the $1/2^+$ state to the total $^{13}C(\alpha,n)^{16}O$ astrophysical S-factor, by the recent $^{13}C(^6Li,d)^{17}O$ ANC measurement [10], was found 10 times smaller than Nacre [5] adopted value (based on S_α=0.1-0.7) but 5 times larger than the result of Kubono et al. at the energy of 190 keV.

Therefore, it appeared highly desirable to perform a new determination of this α spectroscopic factor through another transfer reaction $^{13}C(^7Li,t)^{17}O$ and an improved DWBA analysis.

2. EXPERIMENT DESCRIPTION

The experiment was performed using a $^{7}Li^{3+}$ beam provided by the Orsay TANDEM. Two self-supporting enriched ^{13}C targets, with a ^{13}C thickness of 0.072 ± 0.004 mg/cm^2 and 0.133 ± 0.007 mg/cm^2 and an initial purity of about 90% were used. A ^{12}C target of 0.080 ± 0.004 mg/cm^2 was also used for calibration purposes and background subtraction. The absolute amount of ^{13}C and the final amount of ^{12}C in the ^{13}C target were deduced, respectively, from an α-energy loss measurement and by comparing the measured ratio between 28 MeV elastically scattered particles at 21° from the ^{12}C nuclei in ^{12}C and ^{13}C targets. The reaction products were analyzed with an Enge Split-pole magnetic spectrometer and detected at the focal plane by a 50 cm long position-sensitive gas chamber and a ΔE proportional gas-counter. The particle identification was made unambiguously using ΔE versus position measurements. The tritons were detected at angles ranging from 0 to 31 degrees corresponding to angles up to 43 degrees in the center of mass frame. The beam and ^{12}C amount were continuously monitored with a telescope of silicon detectors mounted inside the scattering chamber at $\theta_{lab}=35^o$.

Due to the presence of ^{12}C impurities, spectra coming from the $^{13}C(^{7}Li,t)^{17}O$ reaction were contaminated by the excited levels of ^{16}O, produced via the reaction $^{12}C(^{7}Li,t)^{16}O$. So the (^{7}Li,t) reaction was measured on both ^{13}C and ^{12}C targets at each angle with the same setup.

3. RESULTS

The experimental $^{13}C(^{7}Li,t)^{17}O$ differential cross sections measured for the 6.356, 3.055, 4.55 and 7.38 MeV, at the two incident energies of 34 and 28 MeV, are displayed in Fig.2a and Fig.2b, respectively. The data points displayed for 3.055 MeV in the 34 MeV left-column are Clark's measurements [11] at 35.5 MeV. The accuracy assigned to our measured cross sections includes the uncertainties on the peak yield, the number of target atoms, the solid angle and the integrated charge except for the zero degree run (no charge measurement) where the number of counts in the silicon monitor detector placed at 35° was used.

Finite-range DWBA calculations, using the FRESCO code [12], were performed. For the triton channel, the optical potential parameters used were taken from [13]. Concerning the ^{7}Li channel, we used for the transfer data at 34 MeV, the optical potential parameters of [14] who performed ^{7}Li elastic scattering measurements on ^{13}C at the same energy. Concerning the data at 28 MeV, we used both those from Schumacher et al. [14] as well as those provided from fitting our elastic scattering cross sections. The optical potential parameters finally selected are those giving the best fit for all the studied transitions in the (^{7}Li,t) reaction.

For the α+^{13}C Wood-Saxon optical potential parameters, we adopted the radius value R=4 fm and the diffuseness value a=0.76 fm. These values provide a good fit of the angular distributions measured for the four studied states at both incident energies. The depth was adjusted to reproduce the binding energy of each considered ^{17}O states and the number of radial nodes N were fixed by the oscillator energy conservation relation. The calculated angular distributions normalized to the data are shown in Figure 2. For

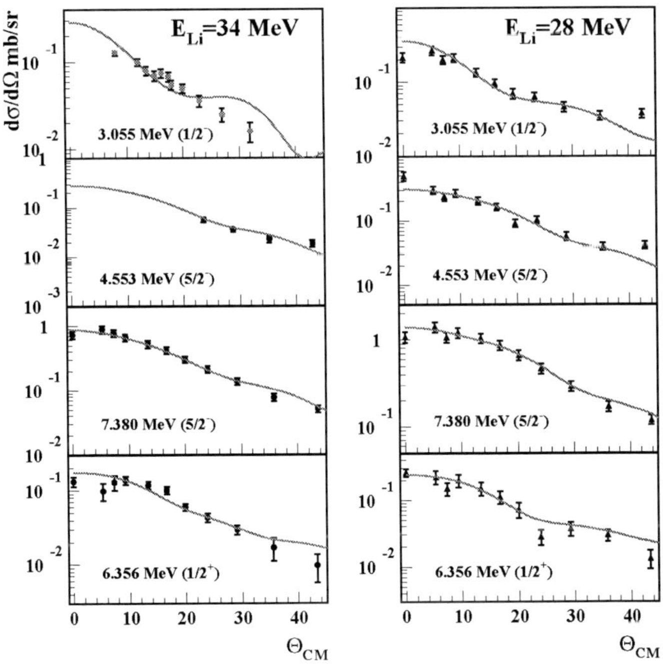

FIGURE 2. Experimental differential cross sections of the $^{13}C(^{7}Li,t)^{17}O$ reaction obtained at 28 and 34 MeV, compared with finite-range DWBA calculations normalized to the data.

both incident energies, the calculated curves agree fairly well with all the measured angular distributions of the different populated states. The α-spectroscopic factors were extracted from the normalization of the finite-range DWBA curves to the experimental data, $S_\alpha=\sigma_{exp}/\sigma_{DW}$. The spectroscopic factor for the overlap between α+t and ^{7}Li was taken to be 1.0. With this value, the α-spectroscopic factor of the reference state at 3.055 MeV is found to be $S_\alpha=0.26\pm0.06$, similar to that obtained in nuclear model calculations $S_\alpha=0.25$ [15].

The good agreement between the DWBA calculations and the measured differential cross sections of the different excited states of ^{17}O at the two bombarding energies of 28 MeV and 34 MeV respectively, gives strong evidence of the direct nature of the (^{7}Li,t) reaction populating these levels and confidence in our DWBA analysis. An S_α mean value of 0.29 ± 0.11 is deduced for the state of interest at 6.356 MeV of ^{17}O, which is in good agreement with that obtained by [9] and those used earlier ($S_\alpha \approx 0.3$-0.7) in the s-process models. The uncertainty on the extracted α spectroscopic factor for our state of interest (6.356 MeV) was evaluated from the dispersion of the deduced S_α values at the two incident energies and using different sets of optical potentials in the entrance [14] and exit channels [13] and different α-^{13}C well geometry parameters.

The α-reduced width γ_α^2 of about 13.5±4.3 keV for the 6.356 MeV state was obtained by using the expression $\gamma_\alpha^2 = \frac{\hbar^2 R}{2\mu} S_\alpha |\phi(R)|^2$ [16], where μ is the reduced mass and $\phi(R)$ is the radial part of the α-^{13}C cluster wave function calculated at R=7.5 fm. This radius was chosen in order to reach the coulomb asymptotic behavior of the radial part of the α-^{13}C wave function. The contribution of the $1/2^+$ state to the astrophysical S-factor when using our deduced γ_α^2 is shown in dashed curve in Figure 3.

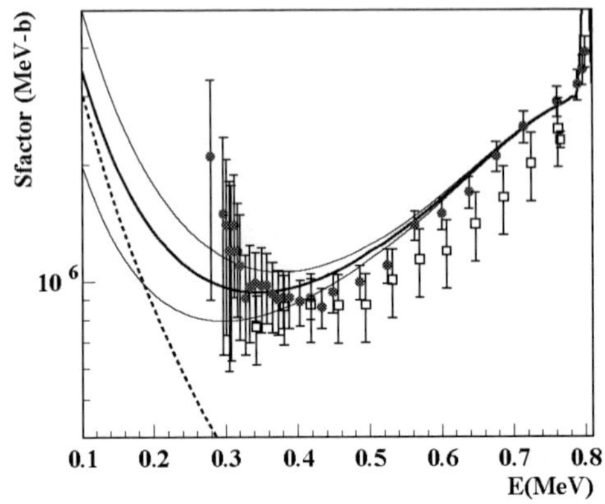

FIGURE 3. Astrophysical S-factor for the ^{13}C(α,n)^{16} reaction with R-matrix calculations. The data are taken from [4, 18]. The contribution of the 6.356 MeV state is shown as the dashed curve. The thick curve correspond to our recommended γ_α^2 values, and the thin curves to the lower and upper limits.

Its value at 190 keV is 9.6×10^5 MeV-b, which is five times larger than in reference [10]. The present value for γ_α^2 (6.356 MeV) has been used to evaluate the ^{13}C(α,n)^{16}O S-factor at astrophysical energies. All ^{17}O resonances up to 7.38 MeV have been included in an R-matrix calculation. As the resonant structure near the $3/2^+$ resonance at 0.84 MeV is not well reproduced when using Tilley's resonance parameters [17], we have readjusted the partial widths of this resonance. We find $\Gamma_\alpha = 0.09$ keV, and $\Gamma_n = 400$ keV. These results are slightly different from the recommended values [17] ($\Gamma_\alpha = 0.07$ keV, $\Gamma_n = 280$ keV), but are consistent with a ^{16}O+n measurement of [19]. Note, that in [10] this problem has been addressed by including a constant term of $0.4\pm0.2\times10^6$ MeV-b, expected to simulate a non-resonant contribution in the S-factor. Our resulting S-factor is shown in Fig.3 as a solid curve, with the uncertainty associated with the error bar on γ_α^2 (6.356 MeV). At the energy of astrophysical interest, E_{cm}=0.19 MeV, the contribution of the $1/2^+$ subthreshold state to the total S-factor is dominant ($\approx 70\%$).

Our calculated ^{13}C(α,n)^{16}O reaction rate, $5.41^{+1.8}_{-1.8}\times10^{-15}$cm^3 mol^{-1} s^{-1}, at temperature T=0.09 GK important for the s-process in low mass AGB stars, is found to be 1.3 times less than adopted in NACRE compilation and 2.3 times larger than Jonhson's deduced rate (see Fig. 4).

The discrepancy of a factor as large as 5 between our $1/2^+$ contribution to the reaction rate and the one of [10] is found reduced to a factor of about 2.3 in the total reaction rate because of the questionable role of the non-resonant term which was found to be dominant in [10]. According to our result, the non-resonant term is the main source of error in the evaluation of the reaction rate. This points out the importance to

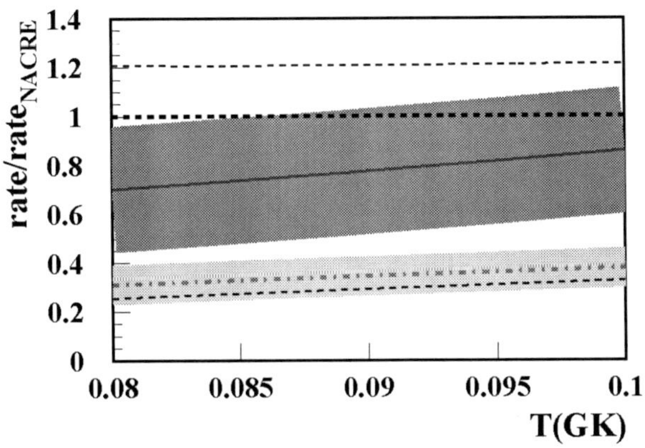

FIGURE 4. The ratio of the $^{13}C(\alpha,n)^{16}O$ reaction rate with respect to the Nacre adopted rate is shown as a function of the temperature. The blue solid curve and the dark band represent this experiment deduced values and uncertainties. The thick dashed black line and the thin ones correspond to the NACRE [5] adopted values and uncertainties respectively. The dotted-dashed curve and the light band represent the deduced values and uncertainties of [10].

determine experimentally the non-resonant contribution from the angular distributions of $^{13}C(\alpha,n)^{16}O$. However the difference with the ANC measurement of [10] shows that further work is necessary before drawing definite conclusions.

4. CONCLUSION

In summary, the reaction $^{13}C(\alpha,n)^{16}O$ was investigated through the direct α transfer reaction (^{7}Li,t) at 28 and 34 MeV incident energies. The spectroscopic factor and hence the α-width of the ^{17}O subthreshold state at 6.356 MeV was deduced from a DWBA analysis using different sets of selected optical and well parameters. These parameters were selected in order to fit the measured angular distributions of the various populated states at both incident energies. This procedure leads to a good evaluation of the uncertainties on S_α and α-reduced width. The α-reduced width was used in the calculation of the astrophysical S-factor and the $^{13}C(\alpha,n)^{16}O$ reaction rate at stellar temperatures. The latter was found 1.3 times lower than the NACRE adopted value but with a lower uncertainty. With our precise measurement the contribution of the $1/2^+$ state is found dominant at astrophysical energies like Keeley's conclusion and contrary to Kubono's and Johnson's results.

REFERENCES

1. R. Gallino, et al., *ApJ* **497**, 388 (1998)
2. M. Busso, et al., *ARAA* **37**, 239 (1999)
3. W. Aoki, et al., *ApJL* **536**, L97 (2000)

4. H.W. Drotleff, et al., *ApJ* **414**, 735 (1993)
5. C. Angulo, et al., *Nucl. Phys. A* **656**, 3 (1999)
6. G.M. Hale, *Nucl. Phys. A* **621**, 177c (1997)
7. P. Descouvemont, *Phys. Rev. C* **36**, 2206 (1987)
8. S. Kubono, et al., *Phys. Rev. Lett.* **90**, 062501 (2003)
9. N. Keeley, et al., *Nucl. Phys. A* **726**, 159 (2003)
10. E.D. Johnson, et al., *Phys. Rev. Lett.* **97**, 192701 (2006)
11. M.E. Clark, K.W. Kemper, and J.D. Fox, *Phys. Rev. C* **18**, 1262 (1978)
12. I.J. Thomson, et al., *Comp. Phys. Rep.* **7** 167 (1988)
13. J.D. Garrett, et al., *Nucl. Phys. A* **212**, 600 (1973)
14. P. Schumacher, et al., *Nucl. Phys. A* **212**, 573 (1973)
15. H. Furutani, et al., *Prog. Theor. Phys. Suppl.* **68**, 193 (1980)
16. F.D. Becchetti, et al., *Nucl. Phys. A* **305**, 293 (1978)
17. D.R. Tilley, et al., *Nucl. Phys. A* **564**, 1 (1993)
18. C.R. Brune, I. Licot, and R.W. Kavanagh, *Phys. Rev. C* **48**, 3119 (1993)
19. D. Lister, et al., *Phys. Rev.* **143**, 745 (1966)

Indirect Measurement of $^{15}N(p,\alpha)^{12}C$ and $^{18}O(p,\alpha)^{15}N$. Applications to the AGB Star Nucleosynthesis

M. La Cognata*, C. Spitaleri*, R. Tribble†, T. Al-Abdullah†, A. Banu†, S. Cherubini*, V. Crucillà*, C. Fu†, V. Goldberg†, M. Gulino*, L. Lamia*, A. Mukhamedzhanov†, R.G. Pizzone*, S.M.R. Puglia*, G.G. Rapisarda*, S. Romano*, M.L. Sergi*, G. Tabacaru†, L. Trache†, A. Tumino* and Y. Zhai†

*Istituto Nazionale di Fisica Nucleare, Laboratori Nazionali del Sud and Università degli Studi di Catania, Catania, Italy

†Cyclotron Institute, Texas A&M University, College Station, TX 77843, USA

Abstract. The Trojan Horse Method has been recently applied to the study of reactions involved in fluorine nucleosynthesis inside AGB stars. Fluorine abundance is important since it allows to constrain mixing models from the comparison of the observed fluorine abundances with the ones predicted by models. Anyway direct measurements of the cross section do not extend down to the Gamow peak, which is the astrophysically relevant energy region. In particular the study focuses on the $^{15}N(p,\alpha)^{12}C$ and the $^{18}O(p,\alpha)^{15}N$ reactions which can influence fluorine yield as they are part of ^{19}F production/destruction network.

Keywords: Nuclear Reactions, Indirect Techniques, AGB Star Nucleosynthesis
PACS: 24.10.-i – 26.20.-f

1. ASTROPHYSICAL BACKGROUND

^{19}F is one of the few naturally occurring isotopes whose nucleosynthesis is still uncertain. SNe, AGB and WR stars are its most likely sources [1]. In particular it was shown that it is not possible to explain the observed abundances of ^{19}F in Milky Way giants without taking into account its production during the AGB phase [2]. In recent years it has been pointed out that fluorine abundances observed in giants can constrain AGB star models since they are sensitive to the environmental conditions in the intershell region and to the dynamics of the mixing phenomena taking place in such stars [2]. Large efforts were then devoted to understand how ^{19}F nucleosynthesis proceeds in AGB's and its production chain is extensively discussed in [3]. Anyway large uncertainties characterize ^{19}F nucleosynthesis since experimental ^{19}F abundances are not reproduced by current AGB models [2].

In this context it has been shown that a key isotope for ^{19}F production in AGB stars is ^{15}N and a crucial role is played by the $^{15}N(p,\alpha)^{12}C$ reaction, removing both protons and ^{15}N nuclei from fluorine synthesis chain in the AGB intershell environment. The available reaction rates give at least a 8% uncertainty on the fluorine surface abundance [2], due to a factor 2 difference between NACRE [4] and CF88 [5] reaction rates usually employed in the calculations. On the other hand the $^{18}O(p,\alpha)^{15}N$ reaction can affect

CP1001, *9th Torino Workshop on Evolution and Nucleosynthesis in AGB Stars and the 2nd Perugia Workshop on Nuclear Astrophysics*, edited by R. Guandalini, S. Palmerini, and M. Busso

fluorine yield from AGB stars since it produces ^{15}N nuclei which are later burnt to ^{19}F through the $^{15}N(\alpha,\gamma)^{19}F$ capture reaction during a thermal pulse [3].

Besides fluorine observations, the study of meteorite grains can also provide useful clues to understand the processes taking place inside AGB stars as they allow to investigate the isotopic ratios for a number of key elements (C, N, O, Al), which are also strongly affected by the internal structure of AGB stars [6]. If the experimental ratios are compared with the ones from AGB star models, it is found that only the largest $^{14}N/^{15}N$ ratios can be reproduced even if the $^{12}C/^{13}C$ ratio for both MS and A+B grains (see [7] for a detailed discussion) can be justified according to present AGB models (including hot bottom burning for A+B grains). If a sub-solar initial abundance is assumed, the MS full range of the $^{14}N/^{15}N$ ratio can be retrieved, though the extremely low values displayed by A+B grains do not agree with the theoretical predictions [6]. The mentioned reactions can play a very important role also in this case since they both can change the $^{14}N/^{15}N$ ratio in opposite ways. Indeed a possible solution of these observational problems can come from nuclear physics studies.

The measurement of nuclear cross sections at ultra-low energies is a very difficult task because of the presence of the Coulomb barrier exponentially suppressing the cross section. Therefore extrapolation is necessary when data at astrophysical energies are unavailable. Even when measurements are available inside the Gamow window, electron screening makes the bare-nucleus cross section inaccessible. Thus indirect techniques such as the Trojan Horse Method (THM) [8] have been developed to extract low-energy cross sections with no need of extrapolation. The THM is extensively discussed in [9, 10].

2. THE INDIRECT STUDY OF THE $^{15}N(p,\alpha)^{12}C$ REACTION

The $^{15}N(p,\alpha)^{12}C$ reaction is studied through the $^{2}H(^{15}N,\alpha\ ^{12}C)n$ TH reaction and its astrophysical S(E)-factor is extracted. The $^{15}N(p,\alpha)^{12}C$ reaction proceeds in the astrophysically relevant energy region through a $J^{\pi}=1^{-}$ state in the intermediate ^{16}O nucleus at $E_x = 12.44$ MeV (Γ=91 keV) [11], associated with the resonance at $E_{c.m.}$=312 keV in the $^{15}N-p$ excitation function. To extract the astrophysical factor the plane wave approximation for the initial and final state scatterings of the nuclei in the TH reaction is used. This approximation leads to a factorized form of the TH double differential cross section. In the THM the transferred particle is virtual (off-energy-shell). However, we neglect the off-shell character of the transferred particle and use the on-shell approximation. A detailed discussion of the experimental setup and of the data analysis is given in [10]. The relevant details and the results are discussed hereafter.

2.1. Data Analysis

The first step of the analysis is the identification of the events corresponding to the $^{2}H(^{15}N,\alpha^{12}C)n$ TH reaction. This is accomplished first through a selection of the carbon locus in the ΔE-E 2D-spectra. Since isotopic discrimination of ^{12}C from ^{13}C is not possible, the experimental kinematical locus is compared to a Monte Carlo simulation

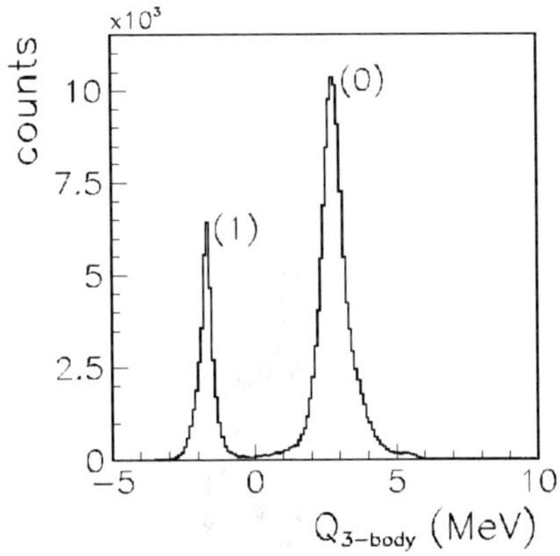

FIGURE 1. Typical $Q-$value spectrum for the ^{2}H(^{15}N,α^{12}C)n three-body process. The peak labelled with (0) corresponds to the events in which ^{12}C is left in its ground state, the one labelled with (1) refers instead to the events in which the ^{12}C first excited state at 4.44 MeV is excited.

for the ^{2}H(^{15}N,α_0^{12}C)n and ^{2}H(^{15}N,α_1^{12}C)n reaction channels. From the comparison, we can conclude that no additional channels contribute. The Q-value spectrum for these events is given in Fig. 1. Two peaks show up corresponding to the ground state and first excited state of ^{12}C (Q-values 2.74 MeV and -1.70 MeV respectively). The good agreement between experimental and theoretical Q-values confirms the identification of the reaction channel and the accuracy of the calibration. The ^{15}N(p,α_1)^{12}C reaction is of negligible astrophysical interest since its cross section is much lower than the ^{15}N(p,α_0)^{12}C reaction inside the astrophysical relevant energy range [12]. Thus, only events corresponding to the Q-value region $0 \leq Q_{3-body} \leq 6$ MeV are considered in the following analysis to extract the ^{15}N(p,α_0)^{12}C cross section.

Afterwards, the quasi-free (QF) reaction mechanism is singled out by comparing the shape of the experimental momentum distribution of the undetected neutrons with the theoretical one. In order to reconstruct the experimental p_{sx} distribution, the energy sharing method [9, 13] is applied choosing a narrow $E_{^{15}Np}$ relative energy window. Dividing the resulting quasi-free coincidence yield by the kinematic factor, a quantity which is proportional to the product of the p_{sx} momentum distribution and the differential ^{15}N$-p$ binary cross section is deduced

$$|\varphi_u(p_{sx})|^2 \left(\frac{d\sigma}{d\Omega_{c.m.}}\right)^{HOES}_{E_0} \propto \left[\frac{d^3\sigma}{d\Omega_\alpha d\Omega_{^{12}C} dE_{c.m.}}\right] [KF]^{-1}, \qquad (1)$$

where E_0 is the mean energy of the relative energy window. In a restricted $E_{^{15}Np}$ relative energy range selected by choosing the energy region between the 12.44 MeV and the 13.09 MeV ^{16}O resonance, the differential binary cross section $(d\sigma/d\Omega_{c.m.})^{HOES}$ of the ^{15}N$+p$ reaction can be considered constant. Thus Eq. 1 provides the experimental p_{sx} momentum distribution in arbitrary units. This result is compared with the square of the

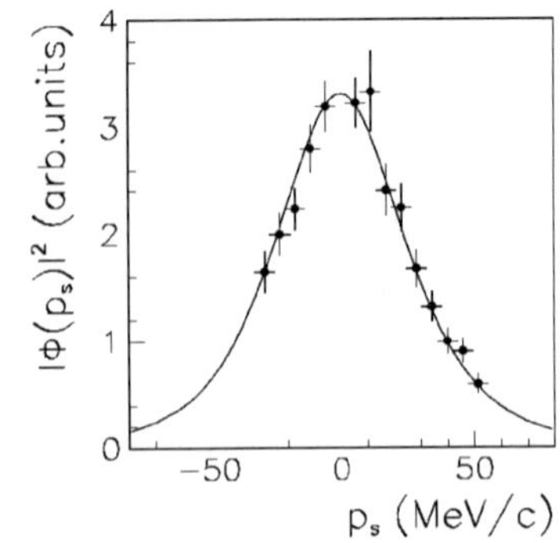

FIGURE 2. The momentum distribution deduced from the $^2H(^{15}N,\alpha^{12}C)n$ three-body process compared with the behaviour of the Hulthén function.

Hulthén wave function in momentum space [9, 14]

$$\varphi_a(p_{sx}) = \frac{1}{\pi}\sqrt{\frac{ab(a+b)}{(a-b)^2}}\left[\frac{1}{a^2+p_{sx}^2} - \frac{1}{b^2+p_{sx}^2}\right] \tag{2}$$

with standard parameters $a = 0.2317$ fm^{-1} and $b = 1.202$ fm^{-1} [14] for the deuteron, with the normalization constant fixed at the experimental maximum (Fig. 2). The good agreement between the two makes us confident that the QF mechanism gives the main contribution to the $^{15}N+d$ reaction at an energy of 60 MeV in the selected kinematical experimental regions.

2.2. Results

The astrophysical S-factor is finally deduced and compared to the ones in literature [15, 16]. Normalization to direct data is deduced by equating the areas subtended by the 314 keV peak in both data sets. The result for the astrophysical S(E)-factor is shown in Figure 3. THM and direct data are displayed as full red and black symbols respectively. Vertical error bars include both statistical errors and a systematic error due to the normalization procedure. The black and red lines represent instead Breit-Wigner best-fits to direct data from [15] and to the THM S-factor. In our fit the following

TABLE 1. Fitting parameters for S(E), Eq. 3. The centroid of the Breit-Wigner is fixed at the 1^- resonance energy of 312 keV. In the last column the χ^2 per degree of freedom is also reported.

a_0 (MeVb)	a_1 (MeVb/MeV)	a_2 (MeVb/MeV^2)	Γ (MeV)	W (MeV^{3}b)	χ^2/N
-55 ± 10	-459 ± 140	1928 ± 260	0.118 ± 0.005	11.7 ± 0.3	0.1

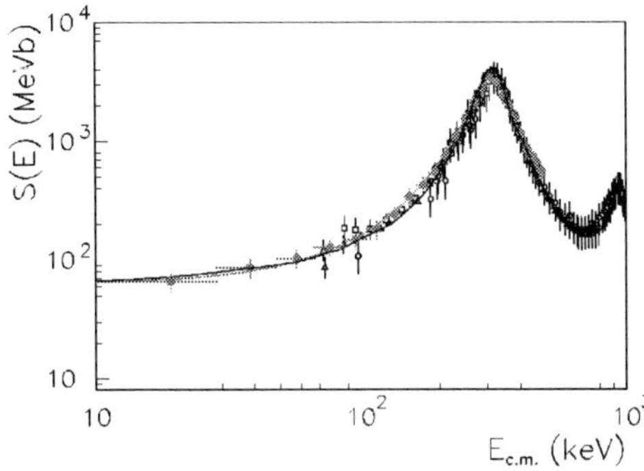

FIGURE 3. Astrophysical S(E)-factor for the $^{15}N(p,\alpha)^{12}C$ reaction extracted via THM (red dots) compared with direct data (open symbols) from [15].

expression has been adopted:

$$S(E) = a_0 + a_1 E + a_2 E^2 + \frac{W}{(E - E_R)^2 + \frac{\Gamma^2}{4}} \tag{3}$$

with parameters a_i ($i = 0,1,2$), W, resonance energy E_R and width Γ; $E \equiv E_{^{15}N\,p}$. The best fit parameters are reported in Table 1. A good agreement is apparent in the energy region where direct data are available. At lower energies, THM data support the extrapolation reported by [15], which is the one adopted to evaluate the reaction rate [4].

3. THE INDIRECT STUDY OF THE $^{18}O(p,\alpha)^{15}N$ REACTION

The $^{18}O(p,\alpha)^{15}N$ reaction has been subject of extensive studies both in direct [18, 19] and indirect [20] ways. Nevertheless, the resonance parameters of some important peaks, namely the 20 and the 656 keV resonances, are not well established yet [4]. In particular, the width and strength of the 676 keV resonance are poorly known, though this resonance provides a non negligible contribution to the reaction rate both at high and low temperatures (around about 10^8 K), being hundreds of keV wide [4]. In the following, we focus on the extraction of the resonance parameters of the 656 keV resonance.

3.1. Experimental Procedure

The $^{18}O(p,\alpha)^{15}N$ S(E)-factor has been deduced from the analysis of the $^2H(^{18}O,\alpha^{15}N)n$ three-body process (see [17] for a more detailed discussion). The experimental procedure is similar to the one reviewed in the previous sections. After detector calibration, the first step in the analysis is the selection of the $^2H(^{18}O,\alpha^{15}N)n$ channel. To this purpose the three-body kinematics is studied in order to disentangle the

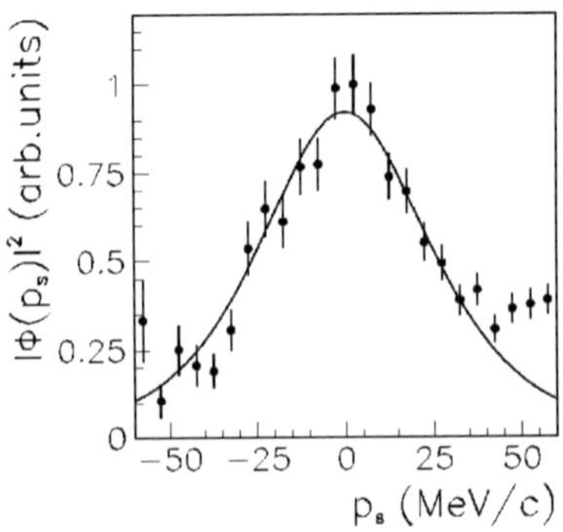

FIGURE 4. The momentum distribution deduced from the ^{2}H(^{18}O, α^{15}N)n reaction compared with the behaviour of the Hulthén function.

contribution of the channel of interest from others. The Q-value spectrum is deduced, showing a single peak, which demonstrated that the ^{2}H(^{18}O, α^{15}N)n channel only contributes to the three-body coincidence yield.

A three-body process can proceed through different reaction mechanisms besides the quasi-free one. For instance, in the ^{18}O+d interaction the same exit channel can be fed through a two-step sequential decay (SD) process involving an intermediate compound nucleus. Possible SD channels in the ^{2}H(^{18}O, α^{15}N)n reaction are:

$$^{18}\mathrm{O}+d \to {}^{16}\mathrm{N}^* + \alpha \to {}^{15}\mathrm{N}+\alpha+n \quad (4)$$

$$^{18}\mathrm{O}+d \to {}^{5}\mathrm{He}^* + {}^{15}\mathrm{N} \to {}^{15}\mathrm{N}+\alpha+n \quad (5)$$

$$^{18}\mathrm{O}+d \to {}^{19}\mathrm{F}^* + n \to {}^{15}\mathrm{N}+\alpha+n\,. \quad (6)$$

In the last case ^{19}F excited states can be populated both via a QF and a SD mechanism. Thus, while the contribution from ^{16}N* and ^{5}He* decays can be distinguished by investigating the 2D relative energy plots, a more thorough analysis is required for the ^{19}F$^* + n$ channel, focusing on reaction dynamics. A quantity which turns out to be strongly dependent on the reaction mechanism is the shape of the momentum distribution of participant - spectator relative motion in the THM nucleus (here, $p-n$ in the deuteron). As in the previous case, a good agreement is found between the theoretical and experimental momentum distributions (see Figure 4), making us confident about the presence of the QF mechanism, which is the dominant process at the energies of interest.

Finally, the astrophysical S(E)-factor for the ^{18}O(p,α)^{15}N reaction is extracted. The result is shown in Figure 5 for $\theta_{c.m.} = 46°$. Similar spectra were derived for other $\theta_{c.m.}$ angles in the $30° - 150°$ range. The resonance parameters were deduced for each $\theta_{c.m.}$ angle in this preliminary analysis by fitting the THM S(E)-factor (in arbitrary units) with the sum of two Breit-Wigner functions, taking into account the 656 keV and 799 keV resonances, plus an interference term, according to the discussion in [19]. The strength of the 656 keV resonance was inferred by comparison to the well-established strength of the 799 keV resonance for each $\theta_{c.m.}$ angle. Similar results were deduced for each $\theta_{c.m.}$ after taking into account the effect of the angular distributions.

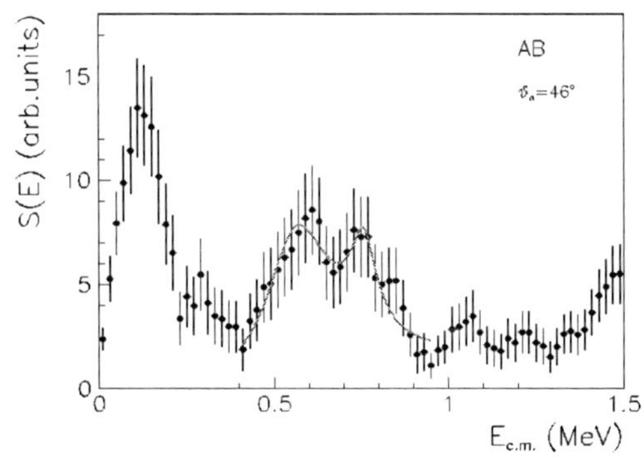

FIGURE 5. Indirect astrophysical S-factor for $\theta_{c.m.} = 46°$ (black dots). Experimental data has been averaged out to reduce statistical fluctuations. The red line represents a fit of the two levels at 656 keV and 799 keV including interference effects.

Finally the resonance parameters from the best-fit were averaged out to deduce our best estimate. This procedure provides a resonance energy of 590 ± 20 keV and a FWHM of 208 ± 26 keV. Regarding the strength of the 656 keV resonance, the mentioned approach leads to: $(\omega\gamma)_{656}/(\omega\gamma)_{799} = 14.8 \pm 2.5$ (where the cited error is only statistical), to be compared with ~ 0.5 [4], ~ 3 [19] and ~ 0.4 [21]. Since we evaluate the ratio of the two strengths, no systematic uncertainties from the normalization procedure affects the strength parameter $(\omega\gamma)_{656}$ derived via the THM. On the other hand a careful analysis of the half-off-energy-shell effects is important since these might affect in different ways the $^{18}O + p$ partial widths for the two resonances (see [10, 22] for a detailed study). For this reason a more accurate R-matrix analysis of the experimental data is in progress.

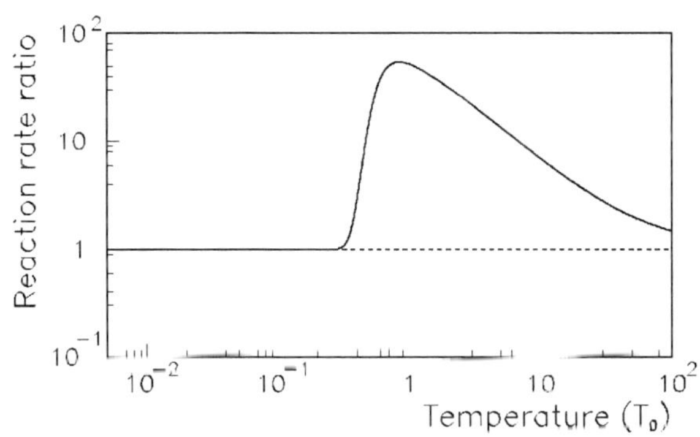

FIGURE 6. The ratio of the THM $^{18}O(p,\alpha)^{15}N$ reaction rate to the one from [4] is displayed as a function of temperature (T_9=T/10^9 K) (full line). For comparison, the line for equal rates is also shown as a dashed line.

3.2. Some Astrophysical Remarks

In order to evaluate the astrophysical implications of the present result, the reaction rate was calculated using the new resonant parameters, as determined in this study. As a first approximation the contribution of the 656 keV peak was evaluated by assuming the narrow resonance limit [4]. The ratio of the THM rate to the one from [4] is given in Fig. 6. The plot clearly shows that the THM rate is about 50 times larger than the NACRE one at about 1 GK, while good agreement is found in the temperature region of interest for ^{19}F nucleosynthesis inside AGB stars (about 10^8 K). Therefore no change is expected in ^{19}F calculated yields though this is a preliminary result providing only an estimate of the expected correction.

REFERENCES

1. A. Renda, et al., *MNRAS* **354**, 575 (2004)
2. M. Lugaro, et al., *ApJ* **615**, 934 (2004)
3. A. Jorissen, et al., *A&A* **261**, 164 (1992)
4. C. Angulo, et al., *Nucl. Phys. A* **656**, 3 (1999)
5. G.R. Coughlan and W.A Fowler, *At. Data Nucl. Data Tables* **40**, 283 (1988)
6. K.M. Nollett, et al., *ApJ* **582**, 1036 (2003)
7. S. Amari, et al., *ApJ* **546**, 248 (2001)
8. C. Spitaleri, et al., *Phys. Rev. C* **63**, 055801 (2001)
9. C. Spitaleri, et al., *Phys. Rev. C* **69**, 055806 (2004)
10. M. La Cognata, et al, *Phys. Rev. C* **76**, 065804 (2007)
11. D.R. Tilley, et al., *Nucl. Phys. A* **564**, 1 (1993)
12. C. Rolfs and W.S. Rodney, *Nucl. Phys. A* **235**, 450 (1974)
13. M. La Cognata, et al., *Phys. Rev. C* **72**, 065802 (2005)
14. M. Zadro, et al., *Phys. Rev. C* **40**, 181 (1989)
15. A. Redder, et al., *Z. Phys. A* **305**, 325 (1982)
16. A. Schardt, et al., *Phys. Rev.* **86**, 527 (1952)
17. M. La Cognata, et al., *J. Phys. G: Nucl. Part. Phys.*, in press (2007)
18. H.B. Mak, et al., *Nucl. Phys. A* **304**, 210 (1978)
19. H. Lorentz-Wirzba, et al., *Nucl. Phys. A* **313**, 346 (1979)
20. A.M. Champagne and M.L. Pitt, *Nucl. Phys. A* **457**, 367 (1986)
21. K. Yagi, et al., *J. Phys. Soc. Japan*, **17** 595 (1962)
22. A.M. Mukhamedzhanov, et al., *J. Phys. G: Nucl. Part. Phys.*, in press (2007); arXiv-nucl-th 0708.0658 (2007)

Indirect Measurements for (p,α) Reactions Involving Boron Isotopes

L. Lamia*,†, C. Spitaleri*,†, S. Romano*,†, N. Carlin**, S. Cherubini*,†, V. Crucillà*,†, M.G. Del Szanto**, M. Gulino*,†, M. La Cognata*,†, R. Liguori Neto**, M.M. De Moura**, M.G. Munhoz**, R.G. Pizzone*,†, S.M.R. Puglia*,†, M.L. Sergi*,†, F.A. Souza**, A.A.P. Suaide**, E. Szanto**, A. Szanto de Toledo**, S. Tudisco*,† and A. Tumino*,†

Laboratori Nazionali del Sud, Catania, Italy
†*Dipartimento di Metodologie Fisiche e Chimiche per l'Ingegneria, Università di Catania, Catania, Italy*
***Departamento de Fisica Nuclear, Universitade de São Paulo, São Paulo, Brasil*

Abstract. Light elements lithium, beryllium and boron (LiBeB) were used in the last years as "possible probe" for a deeper understanding of some extra-mixing phenomena occurring in young Main-Sequence stars. They are mainly destroyed by (p,α) reactions and cross section measurements for such channels are then needed. The Trojan Horse Method (THM) allows one to extract the astrophysical S(E)-factor without the experience of tunneling through the Coulomb barrier. In this work a résumé of the recent results about the ^{11}B(p,α_0)^{8}Be and ^{10}B(p,α)^{7}Be reactions is shown.

Keywords: Nuclear Astrophycsics; Indirect Methods
PACS: 24.10.-i, 24.50.+g

1. INTRODUCTION

Stellar structure and mixing mechanisms for low-mass Main-Sequence stars have been better understood in the last years thanks to the amount of work regarding the "depletion problem" of light elements lithium, beryllium and boron (LiBeB) [1]. Different authors ([1] and references therein) suggest that the study of LiBeB abundances in young F and G stars can give useful informations about different mixing processes presently not well known. In particular for such stars, the Standard Stellar Model does not take into account the possibility of "communication" between the convective zone and the nuclear destruction zone where the burning of such elements occurs mainly via (p,α) reactions induced at temperatures of about $T_6 \sim 10$. This means that the residual LiBeB abundances in such stars would have to reflect the abundances of the original interstellar gas [2]. In contrast, the observational evidence suggests that for F main sequence young stars there is a depletion of lithium and beryllium as the observations on Hyades, Praesepe ($\sim$600 My) and on other young clusters reveal, while there is not evidence of this depletion in F pre-main sequence stars according to the observations on Pleiades cluster ($\sim$70 My) [3]. The evidence of a Li-dip connected with a less pronounced Be-dip and the constancy of boron abundance as the stellar temperature varies together with the Li-Be and Be-B correlation are interpreted as a signature of non-standard mixing mechanisms acting inside these stars: in particular, *slow-mixing mechanisms* induced

CP1001, *9th Torino Workshop on Evolution and Nucleosynthesis in AGB Stars and the 2nd Perugia Workshop on Nuclear Astrophysics,* edited by R. Guandalini, S. Palmerini, and M. Busso

by stellar rotation seem to give the largest contribution for the description of the current observational status [4, 5].
Moreover these processes (and mixing mechanisms in general) transport the surface material into the regions where LiBeB are mainly destroyed via (p,α) reactions induced at a Gamow energy E_G of few keV's and then precise cross-section measurements focused in the astrophysically-relevant energy window are needed. Since both Coulomb barrier and electron screening effects make it difficult to directly determine the cross-section for charged-particle induced reactions in the energy domain relevant for astrophysics, the Trojan Horse Method (THM) provide an alternative way. It allows to extract the cross section for a charged-particle induced two body reaction by selecting the quasi-free (QF) contribution to a suitable three body reaction. Due to its theoretical apparatus the method allows to extract the information down to the astrophysical energies, overcoming the problem connected with both Coulomb barrier penetration and electron screening effects.
The present work reports on the recent investigations of the ^{11}B(p,α)^{8}Be and ^{10}B(p,α)^{7}Be reactions through the THM applied to the ^{2}H(^{11}B,α^{8}Be)n and ^{2}H(^{10}B,α^{7}Be)n three-body processes respectively. Both (p,α) reaction were then studied by using the deuteron (^{2}H) as "Trojan-Horse" nucleus: figuratively, this "nuclear Trojan Horse" brings directly the proton into the nuclear field of the boron isotopes inducing the (p,α) reactions on ^{11,10}B without those feel the presence of the Coulomb barrier "Troy's walls" effects.

2. THEORY OF THE THM

The main idea of the THM [6, 7] is to extract the cross section for the astrophysical reaction a+x→C+c through the selection of the QF contribution of a suitable three-body reaction a+A→C+c+s, being A the cluster structure A=x⊕s. A very intuitive theoretical description of such mechanisms can be given by using the *impulse approximation (IA)* [8]. In particular, in the framework of the Plane Wave Impulse Approximation (PWIA), the three-body cross section can be factorized into three different terms [9] by the relation:

$$\frac{d^3\sigma}{dE_c d\Omega_c d\Omega_C} \propto (KF) \mid \Phi(\vec{p_n}) \mid^2 \left(\frac{d\sigma}{d\Omega}\right)_{cm} \tag{1}$$

The term $\left(\frac{d\sigma}{d\Omega}\right)_{cm}$ is the differential two-body cross section induced at energy E_{cm} given in post-collision prescription by $E_{cm}=E_{ax}=E_{Cc}-Q_{2body}$. The variable E_{Cc} is the relative energy between the outgoing particles and Q_{2body} is the Q-value of the virtual two body reaction. If the energy of the incoming nucleus is chosen high enough to overcome the Coulomb barrier in the entrance channel of the three-body reaction, the decay of the "TH nucleus" *A* in its constituents *a* and *x* occurs in the nuclear field and both Coulomb barrier penetration and electron screening effects are negligible. Moreover, the effect of the binding energy of the cluster structure compensates for the energy of the incoming projectile, inducing the 2-body reaction even at energy relevant for astrophysics. Since the two-body interaction occurs without the effects of penetration through the Coulomb barrier, it is necessary to introduce an appropriate penetration function P_l in order to take into account these effects affecting the direct data below the Coulomb barrier. After the normalization to the direct data, the astrophysical S(E)-factor is extracted following the usual definition; however it is necessary to keep in mind that the TH S-factor is "bared" from the electron screening effects present in the direct data. The comparison between the indirect "bared" and the direct "shielded" data represents an independent experimental approach to extract the experimental value for the electron screening potential for the studied reaction (more details on the method can be found in [7, 11]).

3. THE $^{11}B(p,\alpha_0)^8Be$ REACTION STUDIED THROUGH THE THM

3.1. The Experiment and the Selection of the Three-Body Channel

The ^{2}H(^{11}B,α^8Be)n experiment was performed at the Laboratori Nazionali del Sud in Catania and as a first stage of our investigation only the ^{2}H(^{11}B,$\alpha_0{}^8$Be)n channel was investigated [12]. The SMP Tandem Van de Graaf accelerator provided a 27 MeV ^{11}B beam with a spot size on target of about 1.5 mm and intensities up to 2-3 nA. Deuterated polyethylene targets (CD_2) of about 170 μg/cm^2 were placed at 90° with respect to the beam direction. The detection set up consisted of a Dual Position Sensitive Detector (DPSD), made of two 50x10 mm^2 silicon detectors mounted one above the other and separated by 1 mm, and three 50x10 mm^2 standard Position Sensitive Detectors (PSD1, PSD2 and PSD3).
Due to its instability against α-decay, the ground state of ^{8}Be was selected by reconstructing the relative energies between two-alpha particles hitting in coincidence the upper and lower part of the DPSD. The trigger for the event acquisition was given by the triple coincidences between the upper and lower part of the DPSD and one of the three PSD's. This allowed for the kinematic identification of ^{8}Be in the DPSD and its coincident detection with an α particle.
After the calibration of the involved detectors, the first step of a typical "TH analysis" is to discriminate the channel of the three-body reaction of interest from the ones induced, in this case, by the interaction of ^{11}B on CD_2. Under the hypothesis that the third *undetected particle* had mass number 1, all the variables of interest were further calculated. In particular, by means of the energy conservation law, the Q-value spectrum for the selected events was also reconstructed and reported in Fig. 1. The presence of a well separated peak around 6.4 MeV must be compared with the theoretical Q-value of 6.36 MeV for the ^{2}H(^{11}B,$\alpha_0{}^8$Be)n reaction. The agreement, within the experimental uncertainties, is a signature of our good calibration and a precise selection of the three-body channel.

3.2. Selection of the Quasi Free Mechanism

As we mentioned above, the basic idea of the method is to extract the two-body cross section of interest for astrophysics by selecting properly the Quasi-Free (QF)

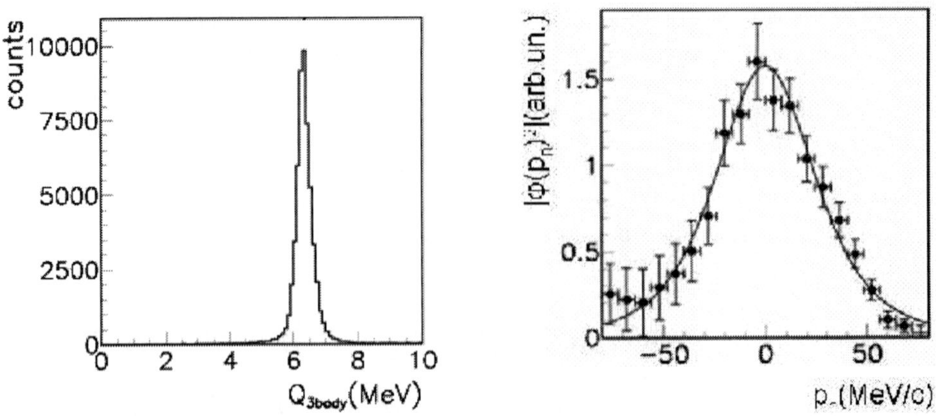

FIGURE 1. Experimental Q-value (left panel) and momentum distribution (right panel) for the ^{2}H(^{11}B,α^8Be)n reaction.

contribution on the three-body data: only after this strict selection we can move from the ^{2}H(^{11}B,α^8Be)n reaction to the astrophysically relevant ^{11}B(p,α_0)^{8}Be one. However, since in the THM approach one works with a three-body channel, it is expected that the same particles of interest (α, ^{8}Be, n) could come from the so called Sequential Mechanism (SM), possible formations/de-excitation of intermediate compound nucleus state. In order to evaluate the contribution of such "competing" mechanisms, the relative energies between the outgoing particles were then reconstructed. In particular the study of the $E_{\alpha Be}$, $E_{\alpha n}$ and E_{Ben} relative energies allows to obtain information on the presence of excited states of ^{12}C, ^{5}He and ^{9}Be respectively. From such analysis different states of ^{12}C were recognized; in particular the 16.106 MeV (J^π=2$^+$) one corresponds to a l=1 resonance in the ^{11}B-p system at E_{cm}=150 keV and then its contribution to the total S(E)-factor must be carefully evaluated. Moreover, different contribution come from the 1.68 MeV, 2.43 MeV and 3.30 MeV levels of ^{9}Be right in the energy region of interest for astrophysics. These events represent a "noise" for the THM and a careful determination of their contribution on the total yield is necessary. After this study, we can proceed with the further steps of the analysis and, in particular, with the selection of the events coming from the QF-mechanism.

The QF-mechanism is connected with the behavior of the undetected third particle in the exit channel. A necessary check is performed by studying the experimental momentum distribution of the neutron in the exit channel. In the "quasi-free" hypothesis, the neutron should maintain in the exit channel the same impulse distribution for the p-n relative motion inside the deuteron that it had before interaction with the impinging particle. Selecting then a small energy region where the two-body cross section can be assumed almost constant, the three-body coincidence yield (Eq.1) correct for the phase-space factor will be proportional to the momentum distribution. The experimental result is shown in Fig.1, where the coincidence yield for the events in the energy window E_{cm}=0.15±0.05 MeV correct for the phase-space factor is reported. The good agreement between the experimental data and the theoretical Hulthèn function for the p-n motion inside the deuteron represents a further experimental evidence that the neutron acted as a "spectator" during the break-up occurred in the ^{2}H(^{11}B,α^8Be)n reaction.

4. THE $^{10}B(p,\alpha)^8Be$ REACTION STUDIED THROUGH THE THM

4.1. The Experiment and the Selection of the Three-Body Channel

The ^{2}H(^{10}B,α^7Be)n experiment was performed at the Pelletron-Linac laboratory (Departamento de Fisica Nuclear (DFN)) in São Paolo (Brazil) and more details about can be found in [13]. The Tandem Van de Graaf accelerator provided a 27 MeV ^{10}B beam with a spot size on target of about 2 mm and intensities up 1 nA. A deuterated polyethylene target (CD_2) of about 192 μg/cm^2 was placed at 90° with respect to the beam axis direction. The detection setup consisted of a 1000 μm PSD detector (PSD1) and a "telescope" system, having a proportional counter (PC) as ΔE and an E-detector made of a standard 500 μm PSD detector (PSD2). In order to evaluate the contribution of the ^{2}H(^{10}B,α^7Be)n reaction, a selection of the ^{7}Be particles was made by using the standard ΔE-E technique while no identification was used for α particles on PSD2. After particle identification and under the assumption of mass number 1 for the undetected third particle, the Q-value was calculated. The selected events correspond to the experimental Q-value spectrum (Fig.2) centered at about $Q_{exp}\sim$-1 MeV, according to the expected theoretical value (Q_{theor}=-1.08 MeV). Events within this peak were then selected as belonging to the ^{2}H(^{10}B,α^7Be)n and used for the further analysis.

4.2. Selection of the Quasi Free Mechanism

As before, the presence of the QF-mechanism has to be evaluated also. A necessary check is performed by studying the experimental momentum distribution of the neutron in the exit channel. A small energy region where the two-body cross section can be assumed almost constant was then selected and the three-body coincidence yield (Eq.1) correct for the phase-space factor. This quantity will be proportional to the momentum distribution. The experimental result is shown in Fig.2, where the coincidence yield for the events in the energy window $E_{\alpha Be}$=1.15±0.05 MeV corrected for the phase-space factor is reported. The good agreement between the experimental data an the theoretical Hulthèn function for the p-n motion inside the deuteron represents a further experimental evidence that the neutron acted as a "spectator" during the break-up occurred in the ^{2}H(^{10}B,α^7Be)n reaction. After these experimental evidence and the further check shown in [12], we can conclude that the QF-mechanism is present in the energy region relevant for the astrophysics.

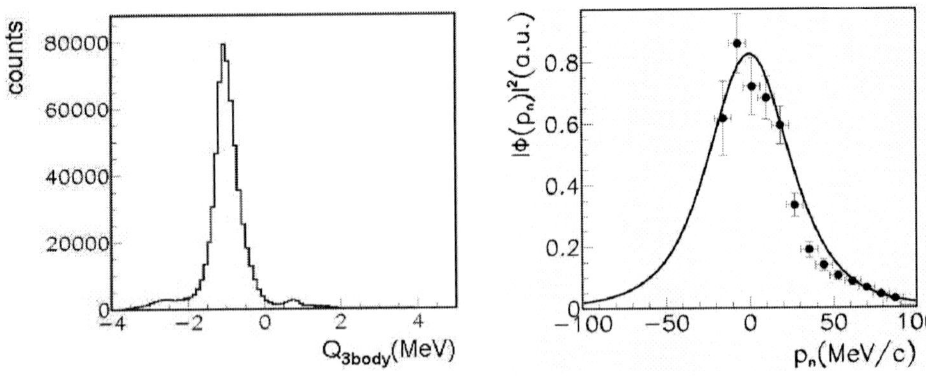

FIGURE 2. Experimental Q-value (left panel) and momentum distribution (right panel) for the ^{2}H(^{11}B,α^{8}Be)n reaction.

5. RESULTS AND DISCUSSION

5.1. S(E)-Factor for the $^{11}B(p,\alpha_0)^{8}Be$ and the $^{10}B(p,\alpha)^{7}Be$ Reaction

After the selection of the QF-mechanism on the three-body ^{2}H(^{11}B,$\alpha_0{}^{8}$Be)n channel and using the PWIA approach (Eq.1), it was possible to extract the nuclear part of the two body ^{11}B(p,α)^{8}Be cross section in the energy range $0<E_{cm}<1000$ keV. In particular events corresponding to $|p_n|<40$ MeV/c condition were selected, being the kinematical region where the QF-mechanism is dominant. Moreover, due to the different energy resolution of the involved detectors, in this recent analysis only data coming from the pair DPSD-PSD2 were further selected, showing an energy resolution of about 40 keV. The bare-nucleus S(E)-factor was then extracted following the usual definition [10]. For the normalization to the available direct data [14] the relative weights between the resonant (l=1) and the non-resonant (l=0) contributions, known from the direct measurements, were taken into account. The preliminary result for this new-approach is shown on the left side of Fig.3 where both data set, direct and indirect ones, are reported with the same energy resolution ($\sim$20 keV). The reaction mainly proceeds through the formation of the 16.106 MeV ($J^{\pi}=2^{+}$) level of ^{12}C, representing a l=1 resonance in the ^{11}B-p channel; a good agreement is clearly evident in the whole energy range and, in particular, around the resonant state at about E_{cm}=150 keV. A preliminary evaluation on these data gives a value of S(0)=1.98$\pm$0.31 (MeV b).

Following the same steps of the previous case, it was possible to extract the S(E)-factor for the ^{10}B(p,α)^{7}Be reaction. The preliminary experimental data (points) are shown in the right panel of Fig.3. The histogram reproduces above $E_{cm}\sim$30 keV the behavior of direct data smeared out at the same energy resolution (about 45 keV) while, below this value, the same histogram reports the extrapolation of the bare-nucleus S(E)-factor [15]. The increase of the S(E)-factor down to 300 keV is mainly due to the contribution of

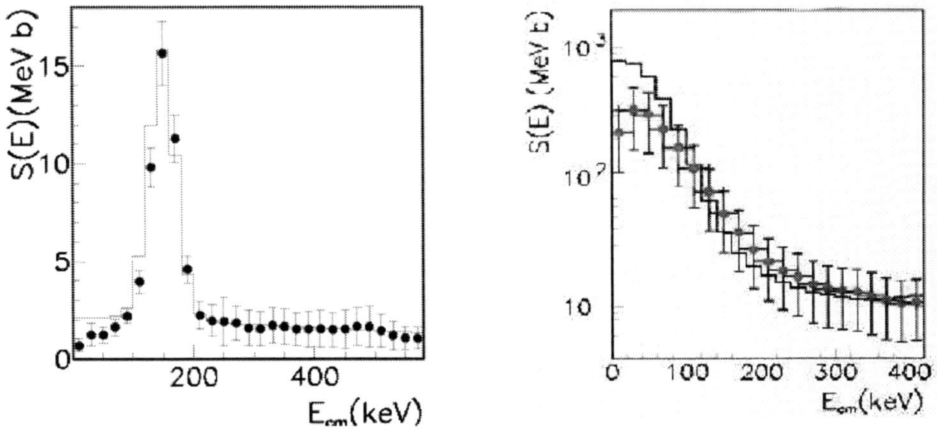

FIGURE 3. Astrophysical S(E)-factor for the $^{11}B(p,\alpha_0)^8Be$ (left panel) and the $^{10}B(p,\alpha)^7Be$ (right panel) reaction. The indirect TH-data are reported as points while the available direct data as histogram.

the 8.701 MeV level of ^{11}C (J^π=5/2$^+$), that represents a l=0 resonant state in the ^{10}B-p channel. The FWHM obtained for the resonance was about 110 keV, while independent measurements of such level report a value of about 16 keV ([15] and ref. ther.). Due to the poor resolution, it is not possible to extract any information about the value of the astrophysical S(0)-factor or about the electron screening potential. However, this first approach to the study of the present reaction confirms the power of the THM to reach the ultra-low energy values. This is an important feature of the method, since its application allows to study the resonant contribution even in the energy region, usually reached through the extrapolations, where both Coulomb penetration and electron screening effects are dominant.

In conclusion, the main channels for the boron destruction in the energy range relevant for astrophysics have been studied by means of the Trojan Horse Method. The (p,α) reactions involving boron isotopes show a clear contribution of resonant levels. These are well reproduced by the present indirect investigation, confirming that the Trojan Horse is a powerful "tool" for the investigation of reactions relevant for astrophysics. However, both set of data are still under investigation and, in particular, the analysis of new experimental data available from two different experiments recently performed at LNS of Catania is presently being undertaken.

REFERENCES

1. A.M. Boesgaard, *"The light elements lithium, beryllium and boron"*, Carnegie Obs. Astroph. Ser., Vol. 4, Origin and evolution of the elements, ed. A. Mc William, (2004)
2. A. Stephens, et al., *ApJ* **491**, 339 (1997)
3. A.M. Boesgaard, arXiv:astro-ph/0412198 v1, (2004)
4. A.M. Boesgaard, et al., *ApJ* **605**, 864 (2004)
5. A.M. Boesgaard, et al., *ApJ* **621**, 991 (2005)

6. G. Baur, et al., *Nucl. Phys.* **A458**, 188, (1986)
7. C. Spitaleri, et al., *Phys. Rev. C* **60**, 055802 (1999)
8. G.F. Chew and G.C. Wick, *Phys. Rev.* **85**, 636 (1952)
9. M. Jain, et al., *Nucl. Phys.* **A153**, 49:81 (1970)
10. C. Rolfs and W. Rodney, *"Cauldrons in the Cosmos"*, The University of Chicago press, Chicago (1988)
11. A.M. Mukhamedzhanov, et al., arXiv:nucl-th/0602001 v1 1 Feb, (2006)
12. C. Spitaleri, et al., *Phys. Rev. C* **69**, 055806 (2004)
13. L. Lamia, et al., *Nucl. Phys. A* **787**, 309c (2007)
14. H.W. Becker, et al., *Z. Phys.* **A327**, 341 (1987)
15. C. Angulo, et al., *Z. Phys.* **A345**, 231 (1993)

SESSION IV:

NUCLEOSYNTHESIS COMPUTATIONS

Carbon- and s-Process-Rich Post-AGB Stars: Theoretical Interpretations

R. Gallino[*,†], S. Bisterzo[*] and L. Husti[*]

[*]Dipartimento di Fisica Generale, Universitá di Torino, Via Pietro Giuria 1, 10125 Torino, Italy
[†]Center for Stellar and Planetary Astrophysics, School of Mathematical Sciences, Monash University, PO Box 28M, Victoria 3800 Australia

Abstract. The study of post-AGB stars gives important informations on the s-process inside AGB stars, as the s-enrichment of post-AGB stars is due to an intrinsic process. A sample of thirteen C- and s-rich post-AGB stars were analyzed with AGB nucleosynthetic models at different masses, metallicities and s-process efficiencies. We present and discuss the theoretical fits obtained for 3 selected stars of our sample. A summary of the results obtained by comparing spectroscopic abundances and AGB theoretical models for all the considered stars is reported.

Keywords: Stars: Abundances - Stars: Population II - Stars: Carbon and s Elements
PACS: 97.10.Cv, 97.10.Tk, 97.30.Kn

1. INTRODUCTION

Post-AGB stars are suitable objects for the study of the changes in chemical composition during the AGB phase. The spectra of post-AGB stars are relatively simpler with respect to the ones of AGB stars and of planetary nebulae (PNe). In the case of AGB stars, the atmosphere pulsations, the large mass loss rates and the presence of numerous molecular lines in the spectra make it difficult to derive accurate abundances. Meanwhile the central star of PNe, to which post-AGB stars evolve, are so hot that many elements do not provide lines in the optical spectra. In post-AGB spectra atomic transitions prevail, allowing a better determination of a wide range of elements [1].

During the TP-AGB phase, the star gets its energy from two alternately burning shells of H and He. The He produced by H burning sinks onto the He intershell, increasing the temperature and density in the He-rich zone. When the conditions to ignite He are reached, it burns producing a strong instability in the He shell, which becomes convective for a short period of time (thermal pulse, TP). During a thermal pulse the outer layers expand and the H shell becomes inactive. About a thousand years after the quenching of a TP, while the stellar envelope is recontracting, the bottom of the convective envelope penetrates into the top layers of the He intershell, mixing with the envelope freshly synthesized C and s-elements (third dredge up episode, TDU). Furthermore, during a TDU the H-rich convective envelope and the He- and C-rich radiative He intershell come into contact, allowing a small amount of protons to be mixed into the top layers of the He intershell. At H reignition, these protons captured by the abundant ^{12}C present in the He intershell, forming a thin layer of ^{13}C, the so-called ^{13}C pocket. This is the major neutron source in low mass AGB stars, via the $^{13}C(\alpha, n)^{16}O$ reaction, which takes place later on still in radiative conditions when compression

CP1001, 9th Torino Workshop on Evolution and Nucleosynthesis in AGB Stars and the 2nd Perugia Workshop on Nuclear Astrophysics, edited by R. Guandalini, S. Palmerini, and M. Busso

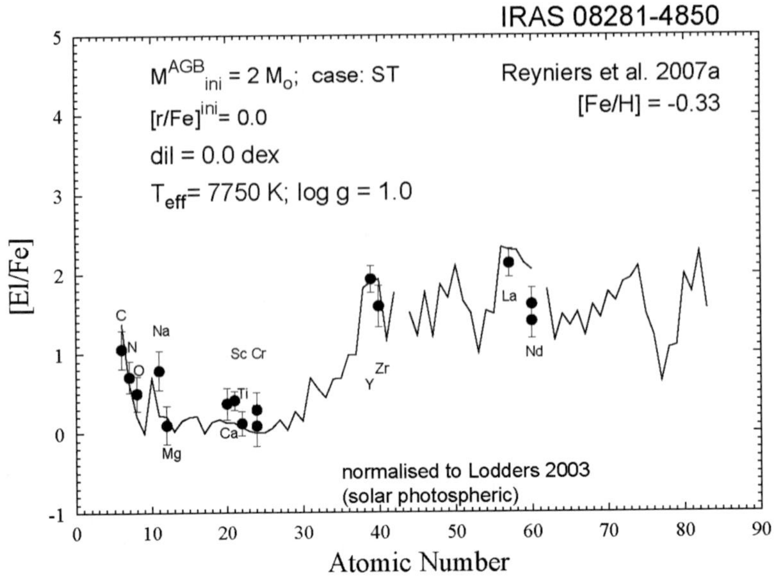

FIGURE 1. Theoretical interpretation of the post-AGB star IRAS 08281-4850 by Reyniers et al. (2007a) [2], with $M_{\rm ini}^{\rm AGB} = 2\ M_\odot$, case ST.

and heating increase the temperature in the pocket to $T \sim 0.9 \times 10^8$ K [3]. Notice that the TDU mechanism ceases to operate when the mass of the envelope is reduced by mass loss below $M^{env} \sim 0.5\ M_\odot$ [4].

A second minor neutron source is driven by the ^{22}Ne(α, n)^{25}Mg reaction, which is marginally activated during the TP. There, a large abundance of ^{22}Ne is present, due to the conversion of all CNO nuclei into ^{14}N in the H-burning shell followed by the α chain ^{14}N(α,γ)^{18}F($\beta^+\nu$)^{18}O(α,γ)^{22}Ne in the early phases of the thermal instability. Additional production of ^{22}Ne derives from the primary ^{12}C mixed with the envelope by previous TDU episodes. The impact of this primary source of ^{22}Ne becomes predominant in very metal-poor AGB stars, contributing both as neutron poison and as source of neutrons [5].

AGB stellar models predict a minimum initial mass for the TDU to occur, which at metallicities close to solar is estimated at around 1.3 $M_\odot$. Thus there is an important fraction of low-mass stars with mass below this minimum and having had the possibility along the time of the Galaxy to evolve off the main sequence, reaching the AGB phase and later become post-AGB stars without showing in their photosphere any enhancement of C and s-process elements [4]. This may explain the dichotomy in the s-process enhancement shown by post-AGB stars [1]. In our sample, only s-enhanced single stars were considered. The enhancement is intrinsic, and not due to mass transfer of s-rich material from a companion (extrinsic AGB), as in the case of the Barium stars or the CEMP-s stars, so no dilution factor must be applied to explain the observations. This is why post-AGB stars are a very useful tool to test the validity of the AGB theoretical models used, including the choice of the initial mass and the global treatment of mass

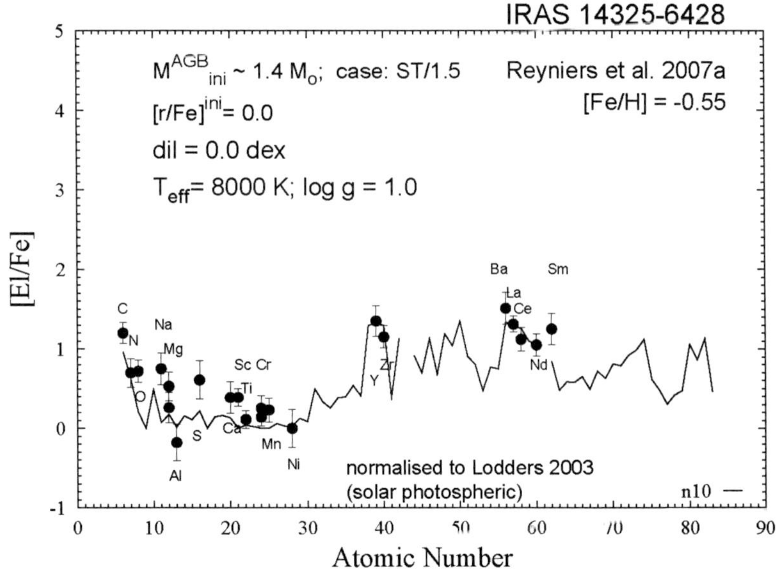

FIGURE 2. Theoretical interpretation of the post-AGB star IRAS 14325-6428 by Reyniers et al. (2007a) [2], with $M_{\rm ini}^{\rm AGB} = 1.4\ M_\odot$, case ST/1.5.

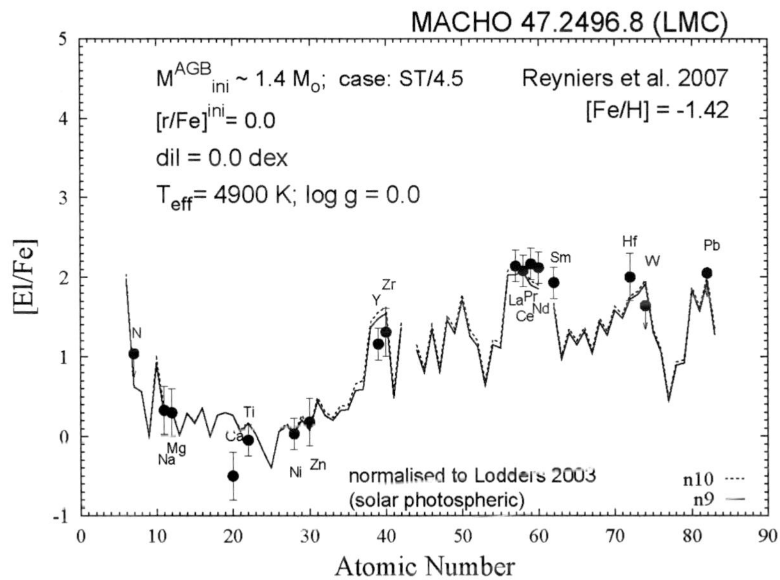

FIGURE 3. Theoretical interpretation of the post-AGB star MACHO 47-2496-8 by Reyniers et al. (2007b) [6], with $M_{\rm ini}^{\rm AGB} = 1.4\ M_\odot$, case ST/4.5. (n9 and n10 are the numbers of thermal pulses followed by TDU in two AGB models of initial mass differing by only $0.25\ M_\odot$, see the text).

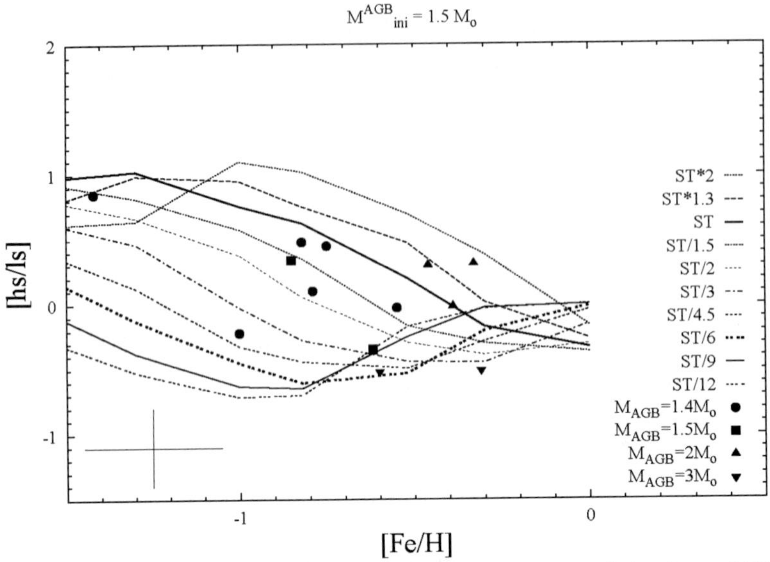

FIGURE 4. The average neutron exposure indicator [hs/ls] versus metallicity for an AGB model of initial mass 1.5 $M_\odot$.

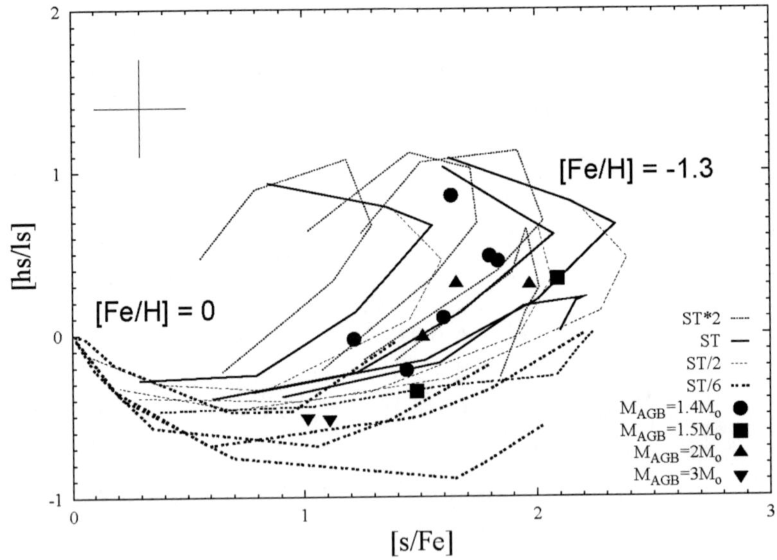

FIGURE 5. The average neutron exposure indicator [hs/ls] versus [s/Fe].

loss rate, TDU efficiency versus pulse number, choice of the ^{13}C pocket. Our models are based on FRANEC (Frascati Newton-Raphson Stellar Evolutionary Code), using the prescriptions of Straniero et al. [4].

2. COMPARISON BETWEEN THEORY AND OBSERVATIONS

We computed AGB nucleosynthetis models of different initial masses ($M_{\rm ini}^{\rm AGB}$ = 1.3, 1.4, 1.5, 2, 3 $M_\odot$), metallicities ($-1.5 \leq$ [Fe/H] ≤ 0) and different choices of the ^{13}C-pocket efficiency in the range ST/12 to ST*2, where ST is the standard case of Gallino et al. [7].

Among the heavy neutron capture elements beyond the iron peak, one can distinguish two separate s-peaks, the first s-peak at Sr, Y, Zr (light-*s*, or 'ls") and the second s-peak at Ba, La, Ce, Pr, Nd, Sm (heavy-*s*, or 'hs"). A third s-process peak is at Pb, at the termination of the s-process. These abundance peaks correspond to isotopes with neutron magic numbers N = 50, 82 and 126, which have a very low neutron capture cross section and consequently may accumulate at high abundances during the s-process flow.

We adopt for ls the usual average of the Y and Zr abundances, and for hs the average of La, Nd and Sm. The relative ratio of ls and hs elements produced in an AGB star depends on the average neutron exposure released during the s-process, that is on the strength of the ^{13}C-pocket efficiency and on the amount of Fe-seed nuclei (see the review of Busso et al. [8]).

We then compared these theoretical predictions with a sample of post-AGB stars analysed with spectra obtained at very high resolution by [9, 10, 11, 2, 6]. We discuss here in detail the two A-type post-AGB stars most recently discovered: IRAS 08281-4850 and IRAS 14325-6428 [2], and the LMC post-AGB star MACHO 47.2496.8 of low metallicity [6]. The other eight Galactic post-AGB stars previously observed by [9, 11] were already compared with AGB model prescriptions of our Group in Busso et al. [12] and in Reyniers et al. [11]. We will resume in Table 1 the results of an updated AGB model analysis for all the C- and s-enhanced post-AGB stars of our sample.

IRAS 08281-4850 has [Fe/H]=-0.33 with an estimated errobar of 0.2 to 0.3 dex. Among the ls elements, the abundance of Y is based on 9 lines of YII, while for Zr only one line of ZrII was detectable. Among the hs elements, the abundance of La was based on 12 lines, while the Nd estimate is of poor quality, based on only one line of NdII and one line of NdIII. No Sm lines have been detected. Analysing AGB models calculated at [Fe/H]=-0.3 for various initial massed and ^{13}C-pocket efficiencies we could not find a satisfactory solution. However, at [Fe/H]=-0.52 we found a satisfactory fit for an AGB models of initial mass $M_{\rm ini}^{\rm AGB}$ = 2 $M_\odot$, and the ST case (see Fig. 1).

IRAS 14325-6428 has [Fe/H]=-0.55. Among the ls elements, 15 lines of YII and 14 lines of ZrII were detected. Among the hs elements, the abundance of La is based on 5 LaII lines, of Nd on 3 NdIII lines and of Sm on only 1 SmII line. Besides, 1 line of BaII and 2 lines of CeII were detected. As illustrated in Fig. 2 we found a satisfactory fit with an AGB model on initial mass 1.4 $M_\odot$ and ST/1.5. The errorbars indicated in the plot for the ls and hs elements are the line-to line scatter reported in [2].

MACHO 47.2496.8 has [Fe/H] = -1.4. It belongs to the Large Magellanic Cloud and is the C(N type) post-AGB star with the lowest metallicity discovered so far [6].

This is a variable star of very high luminosity, of the class of RV Tauri stars. Contrary to what is found in Galactic RV Tauri stars, this object shows an important enhancement in s-process elements. The observed metallicity is unusually low for a field LMC star. Due to its lower temperature as compared with all other post-AGB stars ($T_{\rm eff}$ = 4900 $\pm$ 250 K), strong C_2 bands and molecular lines dominate the spectrum, making it difficult to get a precise determination of the atmospheric parameters. Despite this difficulty, 4 lines of YII were detected for an abundance determination using the spectrum synthesis analysis, one line of ZrII, and a several lines of elements belonging to the hs peak (9 lines of LaII, 23 lines of CeII, 12 lines of PrII, 20 lines of NdII and 12 lines of SmII). Also one line of HfII, WI, and PbI were detected, providing an estimate of the Hf abundance and upper limits for W and Pb. In Fig. 3, we compare the spectroscopic abundances of MACHO 47.2496.8 with the predictions of an AGB stellar model of initial mass $M_{\rm ini}^{\rm AGB}$ = 1.4 $M_\odot$, ST/4.5 case, which best reproduces the data. As discussed in [6], the errorbars of the observational abundances were difficult to assess: if only one line was detected an error of 0.3 dex was adopted, while errors of 0.2 dex were assumed for values derived from more than one line. The s-process distribution of two slightly different AGB models are reported in Fig. 3, allowing for a slight variation in the choice of the initial mass of the AGB model, labeled as $n9$ and $n10$ by the number of thermal pulses followed by TDU suffered during the AGB phase.

In order to monitor the s-process efficiency in stars, the average [ls/Fe] and [hs/Fe] indexes are often used. In case of missing data for a single element, we took the theoretical [X/Fe] value of the best fit we obtained for the given star. In Table 1 we report the effective temperature, the surface gravity, the metallicity, the [ls/Fe] and [hs/Fe] index that we obtained, as well as the ^{13}C-pocket efficiency and the initial mass of the AGB used to obtain the fits for all the stars of the sample. The method of comparison of AGB model predictions with the distribution of all the observed elements for a given star provides a more reliable analysis than the one simply based on the average ls and hs abundances. The stars already presented in [11] were revised (see Table 1) and in some cases the parameters of the fits obtained with our new models are not exactly the same as the ones reported previously. In Table 1 the two carbon-rich post-AGB stars analysed by Reddy et al. [10]: IRAS 05113+1347 and IRAS 22272+5435 of about the same metallicity [Fe/H] ~ -0.8 are also reported, showing also quite similar ls and hs abundances, except for a discrepant Sm abundance in IRAS 22272+5435, whose [Sm/Fe], based of 5 detected line of SmII, is 0.4 dex higher than [La/Fe], while for IRAS 05113+1347 [Sm/La] = -0.1 dex. The expected [Sm/La] in AGB stars of various initial masses and ^{13}C-pocket efficiencies stays in the range 0 to -0.4. Discrepant [Sm/La] were also found in IRAS 23304+6147 and IRAS 06530-0213. For them [Sm/Fe] is 0.9 dex lower than [La/Fe]. For these discrepant cases, in order to calculate the [hs/Fe] average, we used the theoretical value of [Sm/Fe] deduced from the best fit for each star.

For the sample of post-AGB stars considered, the choice of the ^{13}C pocket efficiency is peaked at ST/1.5, with a spread between ST/4.5 and ST*1.3, similarly to the result previously found by Busso et al. [12], which were based on different classes of s-enhanced stars, including the MS, S, SC, C(N), Ba and CH stars.

In Fig. 4 we plot the average neutron exposure indicator [hs/ls] with respect to [Fe/H]. Note that [hs/ls] is independent of the fact that the s-enhanced star considered is an

TABLE 1. References: Van Winckel and Reyniers (2000) [9], Reddy et al. (2002) [10], Reyniers et al. (2004) [11], Reyniers et al. (2007a,b) [2, 6].

Star	$T_{\rm eff}$(K)	log *g*	[Fe/H]	[ls/fe]	[hs/fe]	^{13}C-pocket	$M_{\rm ini}^{\rm AGB}/M_\odot$	Ref.
IRAS 04296+3429	7000	1.0	-0.62	1.66	1.31	ST/1.5	1.5	[9]
IRAS 05341+0852	6500	1.0	-0.85	1.92	2.26	ST/1.5	1.5	[9]
IRAS 07134+1005	7250	0.5	-1.00	1.55	1.33	ST/3	1.4	[9]
IRAS 19500-1709	8000	1.0	-0.60	1.37	0.85	ST/3	3	[9]
IRAS 22223+4327	6500	1.0	-0.31	1.27	0.76	ST/2	3	[9]
IRAS 23304+6147	6750	0.5	-0.79	1.55	1.65	ST/2	1.4	[9]
IRAS 05113+1347	5250	0.25	-0.75	1.61	2.06	ST/1.5	1.4	[10]
IRAS 22272+5435	5750	0.5	-0.82	1.56	2.04	ST/1.5	1.4	[10]
IRAS 06530-0213	7250	1.0	-0.46	1.82	2.12	ST*1.3	2	[11]
IRAS 08143-4406	7050	1.2	-0.39	1.52	1.50	ST*1.3	2	[11]
IRAS 08281-4850	7750	1.0	-0.33	1.50	1.81	ST	2	[2]
IRAS 14325-6428	8000	1.0	-0.55	1.23	1.20	ST/1.5	1.4	[2]
MACHO 47.2496.8	4900	0.0	-1.42	1.21	2.06	ST/4.5	1.4	[6]

intrinsic AGB or post-AGB star, or an extrinsic AGB in a binary system where the observed star got polluted in C and s-process element by the winds of a more massive companion while on the AGB phase (now a white dwarf). Since for Galactic disc metallicities with [Fe/H] ≥ -0.8 the predicted values of [hs/ls] at a given ^{13}C pocket efficiency are only slightly sensitive to the choice of the initial AGB mass, we plot the theoretical curves of the 1.5 $M_\odot$ model. In general, good agreement is found between the theoretical predictions and the observational data and the and the need of a spread in the ^{13}C-pocket efficiency as already outlined in Busso et al. [12]. Two post-AGB stars, IRAS 04296+3429 and IRAS 08281-4850 (see Fig. 1), have few observational data in the hs region and for them the theoretical fit should be taken with caution.

A very interesting plot is illustrated in Fig. 5, where the average neutron exposure indicator [hs/ls] is plotted versus the average s-process observed abundance [s/Fe] in the stellar photosphere, with [s/Fe] = ([ls/Fe] + [hs/Fe])/2. In intrinsic AGB or post-AGB stars, the [s/Fe] value depends on the effective overabundance reached in the envelope, which in turns depends on the ^{13}C-pocket efficiency but also on the number of thermal pulses followed by TDU that the stars experienced, thus on the initial mass of the AGB model adopted. Choosing for example the ST case, predicted [s/Fe] will increase with increasing [Fe/H] and also with increasing the initial mass of the AGB model. The first observational evidence by Van Winkler and Reyniers [9] was for an empirical quite strict linear relationship of [hs/ls] versus [s/Fe]. With the more ample sample of post-AGB stars presently available, a general rise of [hs/ls] versus [s/Fe] is still observed but with a more ample spread, in accordance with the range of theoretical expectations.

3. CONCLUSIONS

We analyzed a sample of post-AGB stars collected from the literature. We investigated a range of masses and ^{13}C-pocket efficiencies in order to obtain the theoretical fits. All post-AGB stars have intrinsic s-process enhancements, so no dilution factor is needed. A unique interpretation is obtained for each star in the range of all AGB models of various initial mass models and of ^{13}C-pocket efficiencies considered. In case of IRAS 08281-4850 we could not find a satisfactory fit using the models with closest metallicity, and relaying on the uncertainty of [Fe/H], we used a 0.2 dex lower metallicity model to obtain the fit. MACHO 47.2496.8 is the only star in the sample that has a determined upper limit of lead, which is well matched by the corresponding value of the theoretical fit. In a few stars, a large discrepancy between the observational value of [Sm/Fe] and [La/Fe] with respect to theoretical predictions can be noticed. In these cases, we concentrated our efforts to fit La and Nd, for which usually a large number of detected lines concur to the abundance estimate, at the detriment of Sm. In Table 1 we report a summary of the observational data and the fit parameters for all the stars of the considered sample. Considering the indicator [hs/ls] we also found a good agreement between theory and observations, and the need of a spread in the ^{13}C-pocket efficiency is confirmed.

ACKNOWLEDGMENTS

We acknowledge for su pport the Italian MIUR-PRIN06 Project 'Late phases of Stellar Evolution: Nucleosynthesis in Supernovae, AGB stars, Planetary Nebulae".

REFERENCES

1. H. Van Winckel, *ARA&A* **41**, 391 (2003)
2. M. Reyniers, et al., *A&A* **471**, 247 (2007a)
3. O. Straniero, et al., *ApJ* **478**, 332 (1997)
4. O. Straniero, et al., *PASA* **20**, 389 (2003)
5. R. Gallino, et al., *POS(NIC-IX)* 100 (2006)
6. M. Reyniers, et al., *A&A* **461**, 641 (2007b)
7. R. Gallino, et al., *ApJ* **497**, 388 (1998)
8. M. Busso, R. Gallino, and G.J. Wasserburg, *ARA&A* **37**, 239 (1999)
9. H. Van Winckel and M. Reyniers, *A&A* **354**, 135 (2000)
10. B.E. Reddy, et al., *ApJ* **564**, 482 (2002)
11. M. Reyniers, et al., *A&A* **417**, 269 (2004)
12. M. Busso, et al., *ApJ* **557**, 802 (2001)
13. K. Lodders, *ApJ* **591**, 1220 (2003)
14. R. Gallino, et al., *Mem. S.A.It.* **77**, 786 (2006)

s–Process in Low Metallicity Stars: CEMP–s and CEMP–s+r

S. Bisterzo and R. Gallino

Dipartimento di Fisica Generale, Università di Torino, Via Pietro Giuria 1, 10125 Torino, Italy

Abstract. A sample of 74 very metal–poor, C– and s–rich rich stars were collected from the literature. We give here a summary of the results obtained by comparing spectroscopic abundances of CEMP–s stars and AGB theoretical models, using different ^{13}C–pocket efficiencies, initial masses and initial r–enrichments. They acquired the C and s enrichments by mass transfer in a close binary system from the more massive companion while on the AGB (now a white dwarf). In general, reasonable solutions are obtained for all the stars. About half of these objects have a relevant number of elements detected at high resolution spectroscopy, while only 10 of them were obtained at moderate resolution. When few elements were detected, several solutions are possible. For most of the main–sequence turnoff stars, due to the low Na and, in some cases, the low ls peak (Sr–Y–Zr), a good interpretation can only be obtained with lower initial AGB masses ($M^{\rm AGB}_{\rm ini}$ ≤ 1.4 $M_\odot$), and the absence of dilution factors suggests that no efficient thermohaline mixing had occurred. Among the stars with r–elements measured, an important number is r–enhanced (∼ 46%), including six objects which need a strong initial r–enrichment ([r/Fe]^{ini} = 2.0 dex). We show here some examples.

Keywords: Stars: Abundances - Stars: Population II - Stars: Carbon and s Elements
PACS: 97.10.Cv, 97.10.Tk, 97.20.Tr, 97.30.Hk, 98.62.Bj, 98.80.Ft

1. INTRODUCTION

During the last years, an increasing number of CEMP–s stars were studied at high resolution spectroscopy. These halo stars, which are observed today, cannot intrinsically synthesize the huge amount of neutron capture elements detected in the envelope: a very low initial mass (M ≤ 0.9 $M_\odot$) must be considered, and no third dredge–up (TDU) is predicted by AGB models at these low metallicities [1]. Moreover, all these stars are dwarfs or giants, far from the AGB phase. The only possible explanation of the s–enrichment is pollution from a companion with higher initial mass, which has already gone through the TP–AGB phase. The low mass AGB stars, during their thermal pulse phase, are considered to be the main site of carbon and s–process nucleosynthesis [2]. During the thermal pulse phase, the shells of H and He burn alternately, making strong instabilities in the He–shell, which begins convective. Subsequently, the bottom of the convective envelope penetrate into the top layers of the He–intershell, the region comprised between the H-shell and the He-shell, mixing with the envelope freshly synthesized ^{12}C and s-process elements (TDU episode). Since the envelope is hydrogen rich, a small amount of protons can penetrate in the top layers of the He–intershell. When the temperature increases, these protons interact with the ^{12}C present in the He–intershell, forming a thin layer of ^{13}C, the so–called ^{13}C–pocket, which is the major neutron source via ^{13}C(α, n)^{16}O. This reaction burns in radiative conditions [3].

CP1001, 9th Torino Workshop on Evolution and Nucleosynthesis in AGB Stars and the 2nd Perugia Workshop on Nuclear Astrophysics, edited by R. Guandalini, S. Palmerini, and M. Busso

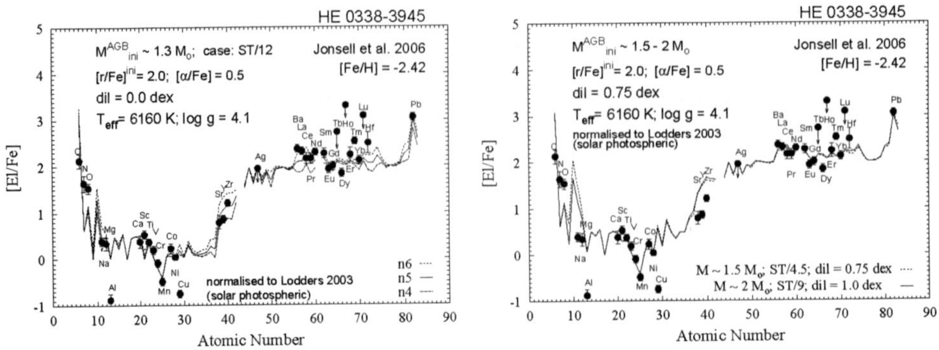

FIGURE 1. Theoretical interpretation for the dwarf HE 0338−3945 [7], with an AGB model of $M \sim 1.3\ M_{\odot}$, a case ST/12 and an initial r−process enrichment of $[r/Fe]^{ini} = 2.0$ (*left panel*). For this star, we used three thermal pulses n4, n5, and n6. In *right panel*, two AGB models are shown: $M \sim 1.5\ M_{\odot}$, ST/4.5, dil = 0.75 dex (dotted line), and $M \sim 2\ M_{\odot}$, ST/9, dil = 1.0 dex (full line). These last solutions cannot reproduce the observed Na and the ls elements.

The second neutron source is the ^{22}Ne(α, n)^{25}Mg, which is partially activated during the thermal pulses. At low metallicities, ^{22}Ne is efficiently produced in a primary way, contributing both as neutron poison and as source of neutrons [4].

We analyzed in detail AGB model results at different masses (M = 1.3, 1.5, 2, 3 $M_{\odot}$), metallicities ($-3 \leq$ [Fe/H] ≤ -1) and s−process efficiencies starting from the standard case ST of Gallino et al. [5] (ST/150 $\leq$ ^{13}C−pocket $\leq$ ST*2). We compared these theoretical results with a sample of 74 very metal−poor, C− and s−rich rich stars collected from the literature (see [6]). In the next §, we present three of these stars, the dwarf HE 0338−3945, the subgiant CS 22183−015, and the giant HD 196944. Moreover, the peculiar abundance pattern of the RV Tauri star V453 Oph is discussed.

2. COMPARISON BETWEEN THEORY AND OBSERVATIONS

HE 0338−3945 is a star close to the main−sequence turnoff, analyzed by Barklem et al. [10], and, afterward, the same authors performed a detailed analysis at higher spectral quality [7]. A high enhancement in both s− and r−elements is observed. Vanhala and Cameron [11] showed through numerical simulations how the supernova ejecta may interact with a surrounding molecular cloud, polluting it with freshly synthesized material and likely triggering the formation of a binary system. Starting from this hypothesis, to reproduce stars with both s and r enhancements, we assumed that the parental cloud, from which the binary system formed, was already enriched in r−elements. After the formation of the system, the s−process enrichment is explained with the C− and s−rich material accreted onto the observed star from the more massive AGB companion (now a white dwarf). This scenario might explain very high observed overabundance of r−process elements, as detected for this star ([r/Fe] $\sim$ 2). Such a large r−process enrichment is very rare in very metal−poor stars: the enhancement of Eu found among pure

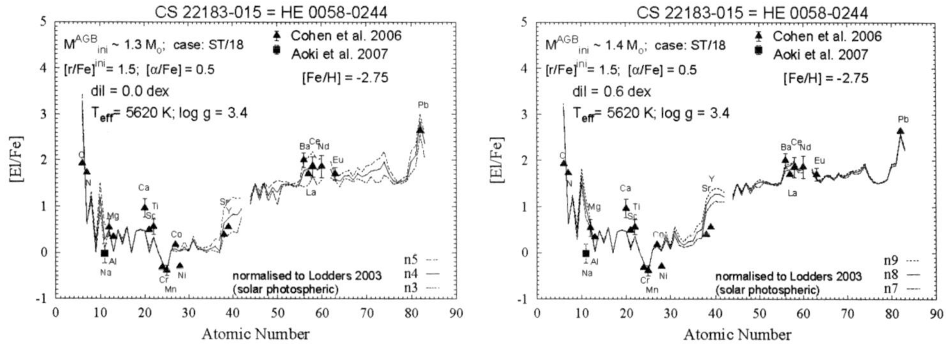

FIGURE 2. *Left panel*: theoretical interpretations for the subgiant CS 22183−015 using the spectroscopic abundances of [8, 9], with $M \sim 1.3\ M_{\odot}$, ST/18 and no dilution. Three thermal pulses are shown, n3, n4 and n5. *Right panel*: the same as left panel, but for higher AGB initial mass (M = 1.4 $M_{\odot}$), ST/18 and dil = 0.6 dex. With the increase of the thermal pulses (n7, n8, and n9), the observed Na and the ls elements are overestimated by the models. In both cases an enhancement in r−process elements of $[r/Fe]^{ini}$ = 1.5 dex is applied.

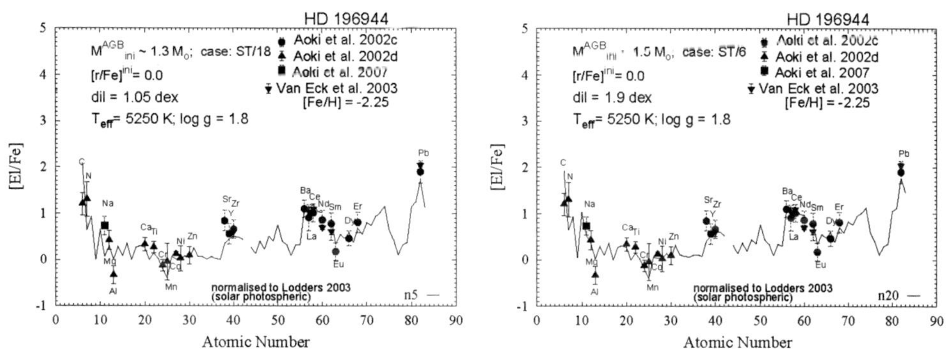

FIGURE 3. *Left panel*: the giant HD 196944 is interpreted with an AGB model of M = 1.3 $M_{\odot}$, ST/18 and dil = 1.05 dex. *Right panel*: the same as left panel, but for M = 1.5 $M_{\odot}$, ST/6 and dil = 1.9 dex. No initial r−enrichment is necessary for this object. The high Na measurement favors the solution with higher initial mass.

r−process−enhanced stars is slightly lower[1]. A good match of the observed abundances is obtained for an AGB stellar model of 1.3 $M_{\odot}$ and a ST/12, including Na, Sr, Y and Zr (see Fig. 1, left panel). Instead, for M_{ini}^{AGB} = 1.5 and 2 $M_{\odot}$, Na and the ls elements are overestimated by the models, as shown in Fig. 1, right panel. This suggest that the

[1] The higher values found in the literature are [Eu/Fe] $\sim$ 1.8 recently measured in HE 1523−0901 by [12], and [Eu/Fe] $\sim$ 1.6, e.g. CS 22892−052 by [13], and CS 29497-004 by [14].

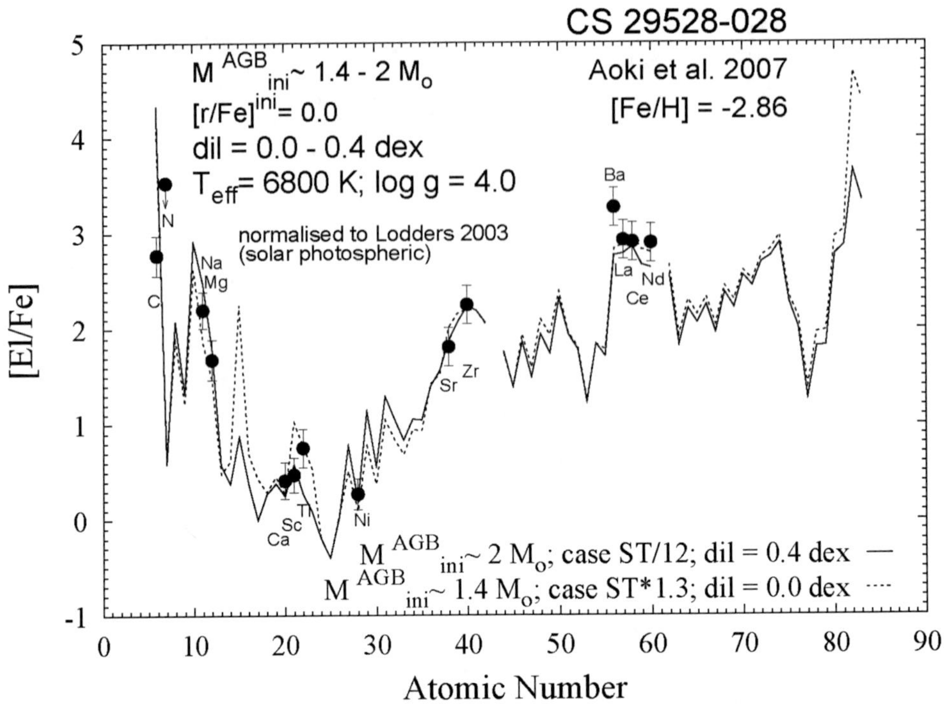

FIGURE 4. The main–sequence star CS 29528–028 by Aoki et al. [9] is interpreted with two AGB models: $M = 1.4$ M$_\odot$, ST*1.3 and no dilution factor (dotted line), and $M = 2$ M$_\odot$, ST/12, and dil = 0.4 dex (full line). The measured sodium, [Na/Fe] = 2.33, permits to exclude AGB initial masses lower than 1.4 M$_\odot$.

solution with M^{AGB}_{ini} = 1.3 M$_\odot$ and without dilution factor is preferable, in accord with a turnoff main–sequence star which lies before the first dredge–up episode, and which underwent a moderate thermohaline mixing effects.

The extremely metal–poor star **CS 22183–015** ([Fe/H] = −3.12), also called **HE 0058-0244**, was studied by Johnson and Bolte (JB02) [15], Tsangarides (T05) [16] and Cohen et al. (C06) [8]. Afterward, Aoki et al. (A07) [9] determined its Na abundance. The barium observed by JB02 and T05 is $\sim$ 0.5 dex higher with respect to the other elements of the hs peak, in contrast with our theoretical predictions ([Ba/Fe]$_{th}$ $\sim$ [La/Fe]$_{th}$). This huge discrepancy was not confirmed by C06, where all second peak elements are roughly 1.9 dex. This suggests that barium, in some cases, is overestimated, because of the strong and saturated lines. The enhancement in r–process elements needs an averaged initial enrichment of the interstellar cloud of [r/Fe]^{ini} = 1.5. As shown in Fig. 2, left panel, the only way to fit Na, Sr and Y, is to use an initial mass M^{AGB}_{ini} = 1.3 M$_\odot$, for which, an higher [hs/ls] value can be reached with respect to $M^{AGB}_{ini} \geq$ 1.4 M$_\odot$ (see Fig.2, right panel). No dilution factors are applied for M^{AGB}_{ini} = 1.3 M$_\odot$, according with a star which lies before the first dredge–up and a moderate thermohaline mixing effect.

Luck and Bond [17] found an s–process enhancement and a high [C/O] ratio in **HD 196944**. Subsequently, the star was observed with very high resolution spectroscopy by [18, 19, 20, 21], and revised by Van Eck et al. [22]. A recent Na measurement was made by Aoki et al. [9]. Van Eck et al. [22] found no clear evidence of the binarity for this star, contrary to the recent results by [16, 23], which classified this object as an extrinsic AGB. In Fig. 3, left and right panels, the theoretical interpretation with AGB stellar models of M_{ini}^{AGB} = 1.3 and 1.5 $M_{\odot}$ are shown. The low values of $T_{\rm eff}$ = 5250 K and log g = 1.8 [9], imply that the star lies on the giant phase, and, in general, an important dilution factor is necessary to explain the mixing of the material transferred from the AGB and the huge convective envelope of the observed star. Because of the high dilution applied to match the s–process elements, with a case M_{ini}^{AGB} = 1.3 $M_{\odot}$ Na is underestimated by the model. Instead, the solution with M_{ini}^{AGB} = 1.5 $M_{\odot}$, which underwent an higher number of thermal pulses, can reproduce Na and the neutron capture elements. A high dilution factor is adopted, dil = 1.9 dex. This is considered the best fit, since the case with higher initial mass (Fig. 3, right panel) overestimates the Na observed.

CS 29528–028 shows the higher overabundance in ls and hs elements among the observed stars, with an averaged [ls/Fe] = 2.03 dex and [hs/Fe] = 3.01 dex, contrary to the usual values which are 1 dex lower. Moreover, this star exhibit a large overabundance of Na and Mg. With [Na/Fe] = 2.33 and [Mg/Fe] = 1.69, no possible solutions are found for AGB models of initial mass M $\lesssim$ 1.3 $M_{\odot}$. For M_{ini}^{AGB} = 1.4 $M_{\odot}$, an interpretation is found only if a very efficient ^{13}C-pocket is used (ST*1.3), as shown in Fig. 4, dotted line. With the exception of Ba, all the s–process abundances are well fitted, and the lead prediction is very high, $[\mathrm{Pb/Fe}]_{th} \sim 4.8$. A similar prediction for ls and hs can be obtained with higher initial masses, for instance a M_{ini}^{AGB} = 2 $M_{\odot}$ and ST/12, with a dilution factor of about 0.4 dex (full line). A lower lead prediction is obtained in this case ($[\mathrm{Pb/Fe}]_{th} \sim 3.7$). Since this star is a main sequence star, close to the turnoff, and it had not seen the first dredge–up episode, a low dilution factor is in agreement with a moderate thermohaline mixing [24].

V453 Oph belongs to RV Tauri stars, variable objects with very high luminosity. Chemical abundances of this star were detected for the first time by Giridhar et al. [25, 26], which found a mild overabundance in Y and Ba. These s–enrichments were confirmed by the high resolution spectroscopy study of Deroo et al. [27], which found an overabundance of r–elements $[\mathrm{r/Fe}]^{ini}$ = 0.7 dex. Many lines are found for the neutron capture elements, with the exception of Ba, Eu and Dy, for which only 1 line is detected. The determination of lead is very difficult, and only a tentative of upper limit could be determined ([Pb/Fe] $<$ 2.0). Unfortunately, detailed radial velocity monitoring data, useful to exclude or confirm an extrinsic enrichment, are not available. Moreover, no dust–excess is observed, so the indirect evidence for binarity is lacking. The main peculiarity of this star is the deficiency of carbon, which appears underabundant with respect to solar ([C/Fe] = −0.26), hardly reconcilable with AGB stellar model predictions. For N, only an upper limit was obtained ([N/Fe] $\lesssim$ 0.7). To understand the nucleosynthesis and the evolution of this stars, we consider the results of the theoretical AGB models. Usually, an efficient cool bottom process might explain the overproduction of [N/C] observed in some stars (e.g. HE 1031−0020 and HE 1509−0806 by [8], CS 22960−053 and HE 0400−2030 by [9], CS 30322−023 by [28]), but, in this case, a negative value of C, together with a nitrogen lower than 0.7 dex is critical to interpret. In particular, with our AGB models, the carbon produced already at the first TDU episode is enhanced by 2 dex. By increasing the initial mass and using higher dilution factors, it is possible to decrease this value to [C/Fe] $\sim$ 1.2 dex, which is still very high. The r−process elements are in averaged higher than the s−elements, with a low s−enrichment if compared with CEMP−s stars ([r/Fe] $\sim$ 0.7 and [hs/Fe] $\sim$ 0.5). If we consider the solar system r−only curve calculated with the Galactic Chemical Evolution model by Travaglio et al. [29, 30], and we scale all the heavy elements abundances to the observed Eu, we obtain, for instance, $[\mathrm{La/Fe}]_r = 0.3$, $[\mathrm{Nd/Fe}]_r = 0.4$. Considering that also Na, Nd and Sm receive some contributions by the r−process, the observed abundances are not so far from an r−only prediction. We hypothesize that this stars formed by an interstellar cloud mildly enriched in r−elements, excluding the contribution from AGB stars. The theoretical results are 0.2 and 0.3 dex lower than the observed values of La and Nd, respectively. In this interpretation the spectroscopic uncertainty of the r−elements have to be considered, since only one line is detected for both Eu and Dy. Even lower estimates are found for Y and Zr, reconcilable with the abundance pattern of some well known r−rich stars, for which the multiplicity of the r−process site has been hypothesized, the weak and the strong r−processes [30, 31]. Travaglio et al. [30] computed the nucleosynthesis of Sr−Y−Zr, and they found that the GCE model and the r−process are not sufficient to explain the observed solar abundances, suggesting a contribution from an another primary process, the Light Element Primary Process (LEPP).

3. CONCLUSIONS

A sample of 74 very metal−poor, C− and s−rich rich stars were collected from the literature, and analyzed in detail with AGB stellar models of different masses, metallicities, s−process efficiencies, and initial r−process enrichment. Here we presented, as example, a discussion of two of these CEMP−s+r stars, as well as a CEMP−s giant. Moreover, a peculiar RV Tauri stars is treated, V453 Oph, which has been excluded from the sample since it is not s−process enhanced, and considered a mild r−rich star. The main results coming from the analysis of this large sample are presented here below.

First of all, by comparing Na observations in CEMP−s stars with theoretical models, we can obtain an indicator of the initial AGB mass. In addition, the Na overabundance give a constraint on the dilution factor, and, if the stars lie on the main−sequence (or before the first dredge−up episode), it can give an idea of the efficiency of the thermohaline effects [32]. In general, main−sequence or turnoff stars show a low Na value: low AGB initial mass are adopted, and the lack of dilution suggest that a moderate thermohaline mixing had occurred [24]. Instead, all the eight stars among the sample with high Na [6], with a maximum observed for CS 29528−028 ([Na/Fe] = 2.33, [9]), lie on the giant phase, and are all treated with important dilution factors, for which higher initial AGB masses are necessary.

Another indicator of the initial AGB mass is the ls peak: since the [ls/Fe] ratio is more sensitive to the initial mass of the AGB star, a low observed intrinsic indicator [hs/ls] may give a constraint on the initial mass, as well as it can be an indicator of the s−process efficiency. In particular, when very low Sr−Y−Zr, with respect to the hs elements, are detected, these abundances can be interpreted only with $M^{AGB}_{ini} \leq 1.4$ $M_\odot$ (e.g. the two stars analyzed here CS 22183−015, HE 0338−3945, as well as CS 22948−27, [33, 9]). In other cases, only Na can give a constraint of the initial mass, since the three s−process peaks are equally well matched by M^{AGB}_{ini} = 1.3, 1.5 or 2 $M_\odot$ (e.g. CS 29497−030, [34, 32]).

An important result is the interpretation of the growing number of CEMP−s+r stars detected. The scenario adopted here hypothesize that the molecular cloud, from which the binary system was formed, was pre−enriched in r−elements because of the pollution by various supernova generations. Different initial r−process enrichments are adopted to explain the observations: 6 stars need a very high enhancement ($[r/Fe]^{ini}$ = 2.0), while other 16 stars have $1.0 \leq [r/Fe]^{ini} \leq 1.8$. If the statistic of the CEMP−s+r stars is given among the 34 stars for which europium is measured, $\sim 47\%$ are CEMP−s+r.

ACKNOWLEDGMENTS

Thanks to the italian MIUR project.

REFERENCES

1. O. Straniero, I. Domínguez, S. Cristallo, and R. Gallino, *PASA* **20**, 389 (2003)
2. M. Busso, R. Gallino, and G.J. Wasserburg, *ARA&A* **37**, 239 (1999)
3. O. Straniero, A. Chieffi, and M. Limongi, *ApJ* **490**, 425 (1997)
4. R. Gallino, S. Bisterzo, L. Husti, F. Käppeler, S. Cristallo, and O. Straniero, *Nuclei in the Cosmos - IX, Proceedings of Science (PoS)*, 100 (2006)
5. R. Gallino, et al., *ApJ* **497**, 388 (1998)
6. S. Bisterzo, *UNPUBLISHED*, PhD Thesis (2008)
7. K. Jonsell, et al., *A&A* **451**, 651 (2006)
8. J.G. Cohen, et al., *AJ* **132**, 137 (2006)
9. W. Aoki, T. Beers, N. Christlieb, J.E. Norris, S.G. Ryan, and S. Tsangarides, *ApJ* **655**, 492 (2007)
10. P.S. Barklem, et al., *A&A* **493**, 129 (2005)
11. H.A.T. Vanhala and A.G.W. Cameron, *ApJ* **508**, 291 (1998)
12. A. Frebel, N. Christlieb, J.E. Norris, C. Thom, T.C. Beers, and J. Rhee, *ApJ* **660**, 117 (2007)
13. C. Sneden, G.W. Preston, and J.J. Cowan, *ApJ* **592**, 504 (2003)
14. N. Christlieb, et al., *A&A* **428**, 1027 (2004)
15. J.A. Johnson and M. Bolte, *ApJ* **579**, 87 (2002)
16. S.A. Tsangarides , *Open University (United Kingdom), DAI-C 66/04, PhD Thesis*, (2005)
17. R.E. Luck and H.E. Bond, *BAAS* **180**, 786 (1992)
18. L. Zacs, P.E. Nissen, and W.J. Schuster, *A&A* **337**, 216 (1998)
19. S. Van Eck, S. Goriely, A. Jorissen, and B. Plez, *Nature* **412**, 793 (2001)
20. W. Aoki, J.E. Norris, S.G. Ryan, T. C. Beers, and H. Ando, *Publ. Astron. Soc. Japan* **54**, 933 (2002)
21. W. Aoki, S.G. Ryan, J.E. Norris, T.C. Beers, H. Ando, and S. Tsangarides, *ApJ* **580**, 1149 (2002)
22. S. Van Eck, S. Goriely, A. Jorissen, and B. Plez, *A&A* **404**, 291 (2003)
23. S. Lucatello, S. Tsangarides, T.C. Beers, E. Carretta, R.G. Gratton, and S.G. Ryan, *ApJ* **625**, 825 (2005)
24. S. Vauclair, *ApJ* **605**, 874 (2004)
25. S. Giridhar, D.L. Lambert, and G. Gonzalez, *ApJ* **509**, 366 (1998)
26. S. Giridhar, D.L. Lambert, and G. Gonzalez, *ApJ* **531**, 521 (2000)
27. P. Deroo, M. Reyniers, H. van Winckel, S. Goriely, and L. Siess, *A&A* **438**, 987 (2005)
28. T. Masseron, et al., *A&A* **455**, 1059 (2006)

29. C. Travaglio, D. Galli, R. Gallino, M. Busso, F. Ferrini, and O. Straniero, *ApJ* **521**, 691 (1999)
30. C. Travaglio, R. Gallino, E. Arnone, J. Cowan, F. Jordan, and C. Sneden, *ApJ* **601**, 864 (2004)
31. Y.-Z. Qian and G.J. Wasserburg, *Physics Reports* **442**, 237 (2007)
32. S. Bisterzo, R. Gallino, O. Straniero, I.I. Ivans, G.W. Preston, and W. Aoki, *AIP Conference Proceedings, in press*, (2008)
33. B. Barbuy, M. Spite, F. Spite, V. Hill, R. Cayrel, B. Plez, and P. Petitjean, *A&A* **429**, 1031 (2005)
34. I.I. Ivans, C. Sneden, R. Gallino, J.J. Cowan, and G.W. Preston, *ApJ* **627**, 145 (2005)

Barium Stars: a Subclass of Extrinsic AGBs

L. Husti and R. Gallino

Dipartimento di Fisica Generale, Università di Torino, Via Pietro Giuria 1, 10125 Torino, Italy

Abstract. Barium stars are stars of Population I. They are considered extrinsic AGB stars as their enhancement in carbon and s-process elements is due to mass transfer in binary systems from the more massive companion, when it reaches the Thermal Pulsing Asymptotic Giant Branch. The accreted material from the AGB donor is mixed with the envelope of the secondary, resulting in a dilution of the s-element abundances. We have made an analysis of the s-element formation inside the AGB donor and the subsequent dilution inside the observed star, for a sample of $\sim$ 30 barium dwarfs and giants. We obtained theoretical fits for all the stars of our sample. We discuss the intrinsic aspects of these stars and comment on the dilution factors obtained from our fits.

Keywords: Stars: Abundances - Stars: Barium - Stars: AGB - Nucleosynthesis: s Elements
PACS: 97.10.Cv, 97.10.Tk, 97.20.Jg, 97.30.Hk, 98.62.Bj, 98.80.Ft

1. INTRODUCTION

Barium stars were first reported by Bidelman & Keenan [1] as a "newly discovered group of peculiar G- and K- type stars". The group consisted of five giant stars showing abnormally strong lines of ionized barium BaII (λ 4554 A) and SrII (λ 4077 A, λ 4215 A). The CH G band and the CN bands were also enhanced. Further abundance studies [2, 3, 4, 5] showed that the barium stars present excesses of ^{12}C and heavy s-process elements.

The production of ^{12}C, along with the the s-process species has been associated with the luminous TP-AGB phase of stellar evolution [6, 7].

The problem of barium stars arose when studies of luminosities [8, 9] found them to be characterized by much lower luminosities than the stars in TP-AGB phase. The discovery of some dwarf barium stars [10, 11, 12] emphasized this problem.

A promising solution to this mystery was provided by McClure et al. [13], who pointed out that probably barium stars belonged to binary systems. At about the same time, Böhm-Vitense [14] discovered that one of the well-known barium stars, ζ Cap, had a white dwarf companion. Further radial velocity studies [15, 16] have demonstrated that probably all barium stars are binaries with companions that have masses expected for white dwarfs. In some cases [17, 18] an ultraviolet excess was observed in barium stars, due to the white dwarf companion.

If barium stars have white dwarf companions, their chemical peculiarities are then directly related to the binarity by a mass transfer scenario. The companion star seen today as a white dwarf was initially the more massive star of the system, it evolved faster and became a TP-AGB, while the current barium star was still in an earlier evolutionary stage. The s-process enriched material was transferred onto the current barium star through mass loss mechanisms, most probably by stellar winds.

CP1001, *9th Torino Workshop on Evolution and Nucleosynthesis in AGB Stars and the 2nd Perugia Workshop on Nuclear Astrophysics*,
edited by R. Guandalini, S. Palmerini, and M. Busso

2. AGB MODEL CALCULATIONS FOR BARIUM STARS

The two neutron sources in low mass AGB stars are the $^{13}C(\alpha,n)^{16}O$ reaction [19, 20] and the $^{22}Ne(\alpha,n)^{25}Mg$ reaction [21].

Immediately after the occurrence of each thermal instability, third dredge up occurs, mixing the deep layers of the star with the envelope and allowing a small amount of hydrogen to reach the He intershell. At H reignition, these protons are captured by the large abundance of ^{12}C, giving rise to a ^{13}C pocket, via the reaction chain $^{12}C(p,\gamma)^{13}N(\beta^+\nu)^{13}C$, eventually followed by $^{13}C(p,\gamma)^{14}N$ in case some extra protons survive the capture on the abundant ^{12}C. Later on, the temperature in the He intershell reaches values of $\sim 0.8-0.9\times 10^8$ K. Thus ^{13}C is burned in radiative conditions, in the ^{13}C pocket, releasing at the most 10^7 n cm^{-3} which are used for the build up of s-process elements. The s-rich pocket is then engulfed by the next convective thermal instability.

The ^{13}C pocket is treated as free parameter. We consider a large range of ^{13}C pocket efficiencies [22]. The standard ST case corresponds to a mass pocket of 9.37×10^{-4} $M_\odot$ containing a total mass of ^{13}C atoms of 4.69×10^{-6} $M_\odot$ distributed in three subzones. Starting from the top of the pocket, the masses of the three subzones are 7.5×10^{-6} $M_\odot$, 5.3×10^{-4} $M_\odot$, and 4.0×10^{-4} $M_\odot$; each subzone contains a mass fraction $X_{\rm eff}(^{13}C)$ = 1.5×10^{-2}, 6.37×10^{-3}, and 3.0×10^{-3}, respectively. Here we consider the effective ^{13}C mass fractions, defined as $X(^{13}C) - 13/14\ X(^{14}N)$. In fact, during the subsequent release of neutrons by the reaction $^{13}C(\alpha,n)^{16}O$, any ^{14}N present in the zone would act as a major neutron poison via the resonant reaction $^{14}N(n,p)^{14}C$. For a general discussion of the choice of ^{13}C pocket and its profile we refer to [23]. The ST case was used by [24] and was shown to best reproduce the solar system main component with low mass AGB stellar models of half solar metallicity.

The H burning ashes contain ^{14}N from the original CNO nuclei. During each convective thermal instability in the He shell, all the ^{14}N nuclides are converted to ^{22}Ne by $^{14}N(\alpha,\gamma)^{18}F(\beta^+\nu)^{18}O$, followed by $^{18}O(\alpha,\gamma)^{22}Ne$. At the peak temperature reached at the base of the thermal pulse, the $^{22}Ne(\alpha,n)^{25}Mg$ reaction is partly activated, giving rise to a small neutron exposure. The contribution of this second neutron source to the total neutron irradiation is small, but important on the final abundances of *s*-branched nuclei, because of the high peak neutron density ($N_n = 10^{10}$n cm^{-3}).

The free neutrons provided by the two neutron sources are captured on iron seeds and previously *s*-processed material, giving rise to new s-elements. Third dredge up takes to the surface the fresh C, made by partial He burning in the thermal pulse, and the s-elements, thus the envelope of the AGB becomes s-process enhanced.

Barium stars are not intrinsic AGBs, but extrinsic ones, so their enhancement in *s*-process material is due to mass transfer from the intrinsic AGB companion. The *s*-enriched material transferred from the AGB star is mixed and diluted into the envelope of the observed star.

We consider the dilution factor $dil = \log(M^{\rm env}_{\rm obs}/M^{\rm AGB}_{\rm transf})$, where $M^{\rm env}_{\rm obs}$ is the mass of the convective envelope after mass transfer and $M^{\rm AGB}_{\rm transf}$ is the mass transfered from the AGB. This dilution will affect the observed abundance of each element in the envelope of the barium stars according to the formula:

$$\left[\frac{\mathrm{X}}{\mathrm{Fe}}\right]^{\mathrm{obs}} = \log\left(10^{\left[\frac{\mathrm{X}}{\mathrm{Fe}}\right]^{\mathrm{ini}}} \cdot (1-10^{-\mathrm{dil}}) + 10^{\left[\frac{\mathrm{X}}{\mathrm{Fe}}\right]^{\mathrm{AGB}}} \cdot 10^{-\mathrm{dil}}\right) \tag{1}$$

Dilution will lower the abundances of the elements, but will not change the shape of the curve beyond iron. The dilution factor *dil* is chosen in every case in order to have best fits for the observational data.

3. COMPARISON WITH OBSERVATIONS

We considered the sample of 26 barium stars from Allen & Barbuy [25] and the sample of classical and mild barium stars of Smiljanic et al. [26]. For both samples, the error bars of the elements plotted here are considered as the largest between the standard deviation of the value, and the error due to the uncertainties in the atmospheric parameters. In the sample of the stars from Allen & Barbuy [25], in some cases the discrepancy between the abundances derived from neutral and ionized lines of Sr and Zr being very large, we chose to plot the abundances of SrII and ZrII.

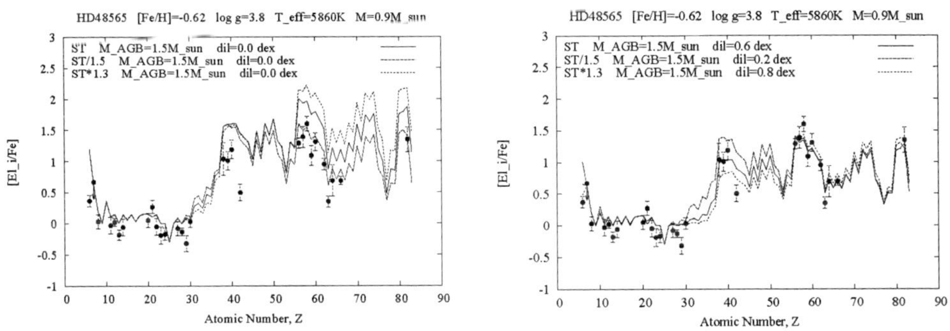

FIGURE 1. Star HD 48565. Choice of the ^{13}C-pocket. On top of each panel the observational parameters of the star are reported: metallicity, logarithm of the surface gravity, effective temperature and the estimated mass from isochrones. In the inset we report the parameters that determine the theoretical abundance: ^{13}C-pocket efficiency, initial mass of the AGB donor, and the dilution factor. In the left panel we plot the theoretical curves with no dilution. In the right panel, we consider the proper dilution factor to get an agreement between the observed values of the elements and the theoretical predictions in the heavy-s zone.

In order to obtain a fit, we search for the ^{13}C pocket efficiency that would give the right trend of the elements belonging to the three s-peaks: the light-s peak (Sr, Y, Zr), the heavy-s peak (Ba, La, Ce, Pr, Nd, Sm) and the Pb peak. Fig 1 shows how we obtain the best fit for a star. We chose for this purpose the dwarf star HD 48565 [25], and we plotted the ST, ST×1.3 and ST/1.5 cases (Fig. 1 left panel). Cases with lower ^{13}C pocket efficiency would give lower values than the observed ones for Pb and the heavy s-elements. Cases with higher ^{13}C pocket efficiency would have about the same values for the light s elements as case ST×1.3, but even higher values for Pb and heavy s-elements.

In Fig. 1 right panel we plot the same cases as in the left panel, but applying a proper dilution factor, in order to lower the theoretical curves to match the observed values in the heavy-s zone. We see that the ST case is in good agreement with the observed values of the elements, but we cannot fit the observations neither using a ST/1.5 or a ST $\times$ 1.3 ^{13}C pocket, nor any other. These later would fit well Pb and the elements of the Ba peak, but not those of the Sr peak. The discrepancy between the molibdenum abundances and ZrII abundances is very large in some stars, molibdenum abundances being in good agreement with ZrI abundances. We can either fit YII and ZrII, or ZrI and MoI. In the present work we opt for the first choice.

Applying the same procedure for each star, we usually found good fits using a unique choice of the ^{13}C pocket. For most stars, using the proper ^{13}C pocket, we obtained good fits with models of initial masses for the AGB donor of 1.5, 2 and 3 $M_\odot$. The only difference between the fits of various masses is the resulting dilution factor.

Subgiants and red giants have large convective envelopes and many of them have already suffered first dredge up, so we expect large dilution factors in their cases (*dil* $\sim$ 1 dex). Main sequence dwarfs have small convective envelopes and we expect smaller dilutions than in the case of giants. In order to estimate the mass of convective envelope of our sample stars, we used the Window To The Stars (WTTS) interface to Peter Eggleton's TWIN single and binary stellar evolution code [27].

In Fig. 2 (left panel) we present the giant star HD 116869, for which we obtain a dilution factor *dil* = 0.9 dex, in agreement with our expectations. In the right panel of the same figure we have the dwarf HD 123585, with a dilution factor *dil* = 0.25 dex, consistent with the mixing of the material transferred from the AGB donor over the small convective envelope characteristic of main sequence stars.

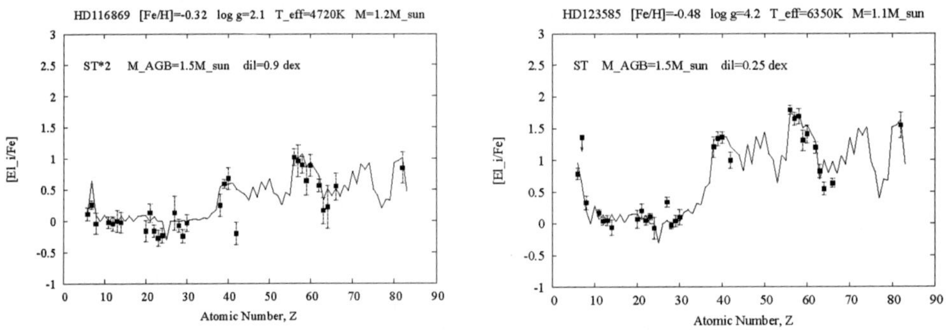

FIGURE 2. Giant barium star HD 116869 (left panel). Dwarf barium star HD 123585 (right panel).

The convective envelope of main sequence stars is known to increase with increasing metallicity and to decrease with increasing mass. Star HD 48565 has slightly lower metallicity and estimated mass than HD 123585. According to WTTS, this star would have a larger convective envelope than HD 123585, consistent with the dilution factor obtained from the fit. Similar results were found for the other barium dwarfs of the considered sample, when using the model with 1.5 $M_\odot$ initial mass AGB. In case of no thermohaline mixing, only the 1.5 $M_\odot$ model would give a dilution factor consistent with WTTS. If thermohaline mixing is present, we would have to exclude the 1.5 $M_\odot$

model and adopt a 2 or 3 $M_\odot$ model, for which the obtained dilution factors are larger of ~ 0.3 dex [28]. These models would be consistent with a moderate thermohaline mixing [29].

The ^{13}C pocket is determined by the relative abundances between the groups of elements corresponding to the three s-peaks. These relative abundances are not affected by dilution, since dilution would lower all of them by roughly the same amount. We consider [ls/Fe] the average of [Y/Fe] and [ZrII/Fe], [hs/Fe] as the average of [La/Fe], [Nd/Fe], [Sm/Fe]. Ba and Sr have often saturated lines and this affects the determination of their abundance, so they were not included in our [ls/Fe] and [hs/Fe] averages. In Fig. 3 we present the intrinsic indicators [hs/ls] and [Pb/hs] for the stars from Allen & Barbuy [25] and the theoretical intrinsic indicators for the models of 1.5 $M_\odot$. At these metallicities, the differences between [ls/Fe], [hs/Fe] and [Pb/Fe] are equal for models of various masses with the same metallicity and ^{13}C pocket efficiency, so the ^{13}C pocket indicated in Fig. 3 remains the same, even if we use a higher mass model to obtain the fit for a particular star.

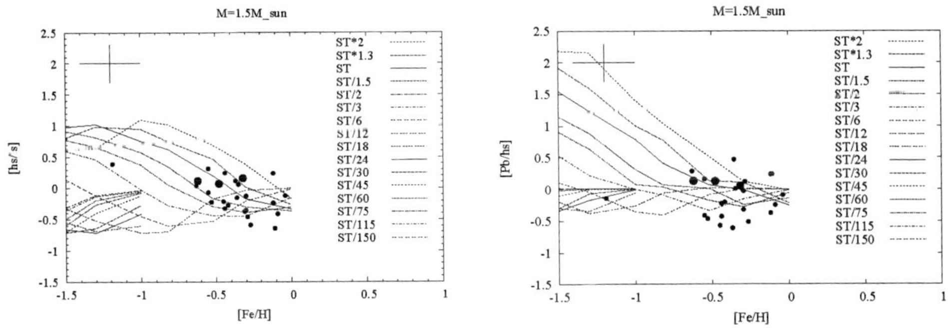

FIGURE 3. Intrinsic indicator [hs/ls] against metallicity (left panel). Intrinsic indicator [Pb/hs] against metallicity (right panel). Big circles stand for the stars HD 48565, HD 116869 and HD 123585, small circles for the other stars of Allen & Barbuy [25]

Comparing the plots of Fig. 3 left panel and right panel, we observe that at metallicities corresponding to barium stars, the production of light and heavy s-elements prevails over the production of lead. We notice that at [Fe/H] higher than -0.5, the predicted [hs/ls] presents a spread of ≤ 1 dex, depending on the considered ^{13}C-pocket efficiency (see Fig. 3 left panel). This spread increases when lowering the metallicity. At metallicities higher than [Fe/H] = -0.3, the predicted Pb never exceeds the heavy s-elemets, and at [Fe/H] between -0.3 to -0.5, lead production is higher than heavy s-element production only the in case of the highest considered ^{13}C-pocket efficiency, ST$\times$2 (see Fig. 3 right panel).

In both panels of Fig. 3 we see that the representative points of the stars previously discussed match within the error bars the points corresponding to the considered fits. The uncertainties are increased in this kind of plots because of the discrepancies sometimes present between the values of the elements belonging to the same s-peak.

The stars of Smiljanic et al. [26] do not have the abundance of Pb determined. In Fig. 4 we present the fits obtained for the classical barium stars HD 116713 (right panel) and the mild barium star HD 205011 (left panel).

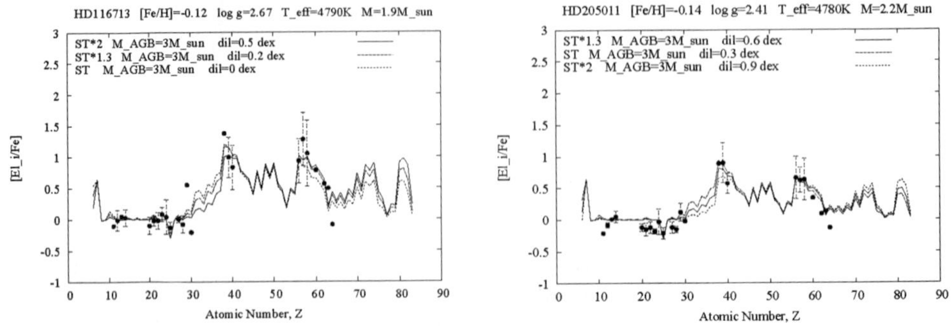

FIGURE 4. Classical barium giant HD 116713 (left panel). Mild barium giant HD 205011(right panel).

For star HD 116713, we obtain good fits using a 3 $M_\odot$ model, for all the three ^{13}C pockets plotted in the figure, due to the large error bars of the observational values of the light-s and heavy-s elements. We prefer the ST×2 case, as it gives both a good reproduction of the observed values of the elements and a higher dilution factor, suitable with the fact that HD 116713 is a giant. This fit allows us to predict a value for the lead abundance of $\sim$ 1 dex. The fact that dilution is smaller in case of HD 116713, than in case of HD 116869 may be explained by a smaller separation between the components and the higher mass of HD 116713, which would make more material to be transferred from the AGB to the observed star.

We also used a 3 $M_\odot$ model to fit the data of HD 205011. Having Pb not determined, we obtain again a triple solution (see Fig. 4 right panel). We would exclude the ST case, due to its low dilution, and prefer case ST×1.3 as it better reproduces the observed data. In this case we predict [Pb/Fe] $\sim$ 0.4 dex. The possible dilution factors for the mild barium giant HD 205011 are higher than for the classical barium giant HD 116713.

All stars of this sample are giants, of about the same mass and metallicity. For all these stars we have double or triple solutions, as described above. From our fits, we usually get a range of higher dilution factors for the mild barium stars than for the classical giants, supporting the suggestion of [30], that the separation between the components in the binary system of the mild barium stars is wider than that of classical barium stars.

4. CONCLUSIONS

We have studied a sample of $\sim$ 30 barium stars and found theoretical fits for the spectroscopic data using ^{13}C-pocket efficiencies from ST×2 down to ST/3. Good fits were found using models of 1.5, 2 and 3 $M_\odot$, with various dilution factors. Generally, we obtain larger dilution factors for the giant stars than for the dwarfs. Considering the expected dilution factors for barium dwarfs, the fits we obtain with models of 1.5 $M_\odot$ give a dilution factor that can be explained by the mixing of the transferred material within the convective envelope of the barium star. The dilutions obtained with models of 2 and 3 $M_\odot$ can account for a moderate thermohaline mixing. The intrinsic indicator [hs/ls] shows that the representative points of the stars spread in a range of ~ 0.3 dex around the ST case, while in case of [Pb/hs] the spread is larger. We predict the lead abundance for the stars that do not have an observational value of it. The predicted value is about ~ 1 dex in case of the classical barium giants, and of less than 0.5 dex in case of mild barium stars. We obtain a range of higher dilution factors for the mild barium giants than for the classical ones. This fact supports the idea of a wider separation between the components of the binary system in case of mild barium stars.

ACKNOWLEDGMENTS

L.H. thanks Sara Bisterzo, Dinah M. Allen and Robert Izzard for useful discussions concerning respectively stellar nucleosynthesis, the uncertainties of the observed elements reported in the Allen & Barbuy [25] article and mass transfer in binary systems.

REFERENCES

1. W.P. Bidelman and P.C. Keenan, *ApJ* **114**, 473 (1951)
2. E.M. Burbidge and G.R. Burbidge, *ApJ* **126**, 357 (1957)
3. B. Warner, *MNRAS* **129**, 263 (1965)
4. V.V. Smith, *A&A* **132**, 326 (1984)
5. N. Kovacs, *A&A* **150**, 232 (1985)
6. I. Iben Jr. and J.W. Truran, *ApJ* **220**, 980 (1978)
7. I. Iben Jr. and A. Renzini, *ARAA* **21**, 271 (1983)
8. F.J. MacConnel, P.L. Frye, and A.R. Upgren, *AJ* **77**, 384 (1972)
9. R.D. McClure, *IAUS* **54**, 225 (1973)
10. W.P. Bidelman, *AJ* **86**, 553 (1981)
11. J. Tomkin, et al., *A&A* **219**, L15 (1989)
12. P. North, et al., *A&A* **468**, 679 (1994)
13. R.D. McClure, J.M. Fletcher, and J.M. Nemec, *ApJ* **238**, L35 (1980)
14. E. Böhm-Vitense, *ApJ* **239**, L79 (1980)
15. R.D. McClure, *ApJ* **268**, 264 (1983)
16. R.D. McClure and A.W. Woodsworth, *ApJ* **352**, 709 (1990)
17. E. Böhm-Vitense and H.R. Johnson, *ApJ* **293**, 288 (1985)
18. E. Böhm-Vitense, *ApJ* **533**, 969 (2000)
19. A.G.W. Cameron, *Phys.Rep.* **93**, 932 (1954)
20. A.G.W. Cameron, *Chalk River Rep.*, CRL-41 (1957)
21. A.G.W. Cameron, *AJ* **65**, 485 (1960)
22. M. Busso, R. Gallino, and G.J. Wasserburg, *ARAA* **37**, 239 (1999)
23. R. Gallino, et al., *ApJ* **497**, 388 (1998)
24. C. Arlandini, et al., *ApJ* **525**, 886 (1999)
25. D.M. Allen and B. Barbuy, *A&A* **454**, 895 (2006)
26. R. Smiljanic, G.F. Porto de Mello, and L. da Silva, *A&A* **468**, 679 (2007)
27. R.G. Izzard and E. Glebbeek, *New Astronomy* **12**, 161 (2006)
28. L. Husti, *AIP*, in press (2007)
29. S. Vauclair, *ApJ* **605**, 874 (2004)
30. E. Böhm-Vitense, J. Nemec, and C. Proffit, *ApJ* **278**, 726 (1984)

Rubidium and Zirconium Production in Massive AGB Stars

M.A. van Raai*, M. Lugaro*, A.I. Karakas† and D.A. García-Hernández**

*Sterrekundig Instituut, University of Utrecht, Postbus 80000, 3508 TA Utrecht, The Netherlands
†Research School of Astronomy and Astrophysics, Mt. Stromlo Observatory, Cotter Rd., Weston, ACT Australia
**Instituto de Astrofisica de Canarias, La Laguna (Tenerife), España

Abstract. A recent survey of a large sample of massive Galactic asymptotic giant branch (AGB) stars shows that significant overabundances of rubidium (up to 100 times solar), but merely solar zirconium, are present in these stars. These observations can set constraints on our theoretical notion of the *slow* neutron capture process (the *s* process) in AGB stars, as well as on the rates of the neutron capture reactions involved in the production of Rb and Zr. We use the Monash nucleosynthesis code with a recently extended network to try to reproduce these observations. We present results for AGB stars of masses 5, 6, and 6.5 $M_\odot$ and solar metallicity. We also show results for different available choices of the neutron capture rates, as well as for the possible inclusion of a partial mixing zone (PMZ), leading to the activation of the ^{13}C neutron source. We find increasing Rb overabundances with increasing stellar mass, as observed, but we are far from matching the highest observed Rb enhancements. Inclusion of a PMZ increases the Rb abundance, but also produces an overabundance of Zr, contrary to what is observed. Only if the third dredge up efficiency remains as high as before the onset of the superwind phase during the final few pulses of a massive AGB star, can we match the highest [Rb/Fe] ratios observed by García-Hernández et al. [1]. A better understanding of the third dredge up efficiency with decreasing envelope mass for massive AGB stars is essential for further investigation of this issue.

Keywords: AGB Stars, Nucleosynthesis
PACS: 26.20.Kn

1. INTRODUCTION

Stars with an initial mass in the range $\approx$ 0.8 to 8 $M_\odot$ will evolve through the thermally-pulsing asymptotic giant branch (TP-AGB) before eventually ending their lives as white dwarfs. Figure 1 shows a schematic representation of the structure of an AGB star. The H-burning shell is responsible for the nuclear energy production most of the time, but every 10^{4-5} years or so, the He shell ignites, driving convection throughout most of the He intershell, this is known as a thermal pulse (TP). When a TP occurs the H-burning shell is temporarily shut off. After a TP has occurred, at which time the H-burning shell still lies dormant, the convective envelope extends inward and may reach the He-intershell; this is known as the third dredge-up (TDU) and can occur periodically after each TP. The TDU mixes nuclear burning products from the He and H-burning shells to the stellar surface [2, 3]. In the He intershell of AGB stars neutrons can be released by the $^{13}C(\alpha,n)^{16}O$ reaction during the interpulse periods, and by the $^{22}Ne(\alpha,n)^{25}Mg$ reaction in the convective TPs. Due to the availability of free neutrons, *slow* neutron captures (the *s* process) can occur leading to the production of elements heavier than Fe

CP1001, *9th Torino Workshop on Evolution and Nucleosynthesis in AGB Stars and the 2nd Perugia Workshop on Nuclear Astrophysics*, edited by R. Guandalini, S. Palmerini, and M. Busso

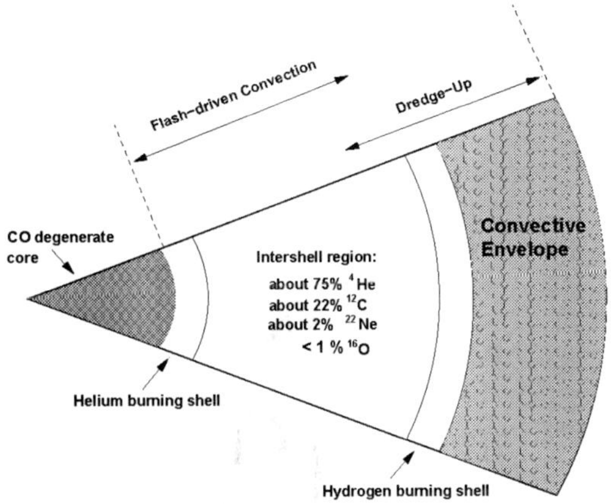

FIGURE 1. Schematic representation of the structure of an AGB star.

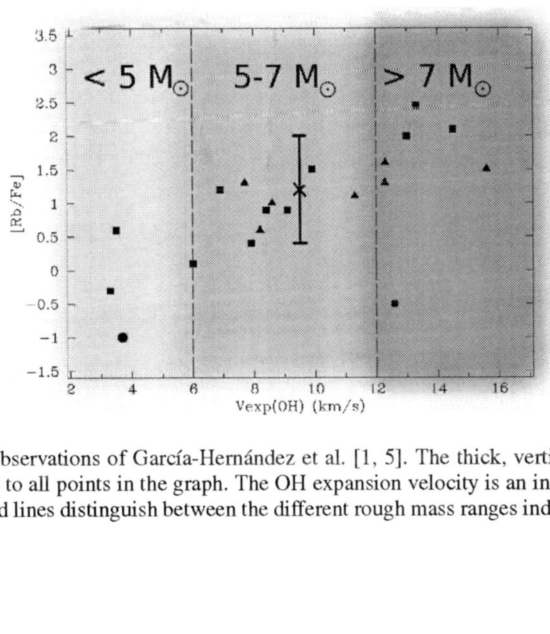

FIGURE 2. The observations of García-Hernández et al. [1, 5]. The thick, vertical bar represents the error bar and applies to all points in the graph. The OH expansion velocity is an indicator of stellar mass (see text). The dashed lines distinguish between the different rough mass ranges indicated at the top of the graph.

(see, e.g. [4]).

A recent survey of massive Galactic AGB stars shows significant enhancements in rubidium, but not in zirconium [1, 5]. Figure 2 shows the observations of García-Hernández et al. [1, 5].

The mass, which can be estimated from the observed Mira pulsation period if the distance to the star is known (see, e.g. [6]), in general, shows a correlation with the OH expansion velocity (v_{exp}(OH)) (note that for the observations of [1, 5] the distance is unknown). Independently, for this sample, the correspondence of the v_{exp}(OH) with mass is obtained using the location of the observed stars with respect to the Galactic plane.

TABLE 1. Ranges of observed Rb and Zr abundances in AGB stars of different initial masses.

	Low Mass (1-3 $M_{\odot}$)	High Mass (5-7 $M_{\odot}$)
[Rb/Fe]	-0.3 to 0.6	0.6 to 1.6
[Zr/Fe]	up to 1.0	< 0.5
[Rb/Zr]	< 0	0.1 to 2.1
	Abia et al. [7], Lambert et al. [8]	García-Hernández et al. [1]

Furthermore, lithium is not detected for any of the stars in this sample with v_{exp}(OH) < 6 km s^{-1}, whereas Li is detected for some of the other stars in the sample; the detection of Li is a signature of hot bottom burning, which only occurs in stars of at least $\approx$ 4-5 $M_{\odot}$. See §5.2 and §5.3 of [5] for a full discussion on the estimation of mass from the OH expansion velocity. From Fig.2, hence, it is estimated that the [Rb/Fe] ratio increases with the initial stellar mass.

The aim of this paper is to see if we can match these observations with our theoretical models. We focus here on the v_{exp}(OH) range 6-12 Km/s, which may correspond roughly to a mass range of $\approx$ 5-7 $M_{\odot}$. Stars with higher v_{exp}(OH) may have stellar masses > 7 $M_{\odot}$, for which we do not have models available yet.

In Table 1 we compare the Rb and Zr abundances observed in low-mass and high-mass AGB stars. In the low-mass case the Rb abundance is lower than in the high-mass case, while the opposite is true for Zr. A further check will be to see if the models can reproduce these differences.

2. MASSES AND MODELS

We calculate the nucleosynthesis with a detailed postprocessing code for which the stellar structure inputs were calculated beforehand. The stellar structure models are from [3] and were calculated with the Monash version of the Mount Stromlo Stellar Structure Program; see [9] and references therein for details. The code calculates the abundances due to convective mixing and nuclear reaction rates for a large number of species [10]. We use a nuclear network of 166 species from neutrons and protons to S and from Fe to Nb. The missing species are accounted for by means of artificial neutron sinks. The bulk of our 1285 reaction rates are from reaclib [11], with updated tables as described by [12]. We further updated the neutron capture rates to those of Bao et al. [13].

Since the observations are of Galactic-disk AGB stars, we focus on AGB models of solar metallicity. The initial abundances we use were taken from Anders and Grevesse [14] for Z = 0.02. We also calculated one model using the Asplund et al. [15] and Lodders [16] solar abundances table, which prescribe a solar metallicity of Z = 0.012. We calculated models with masses 4, 5, 6, and 6.5 $M_{\odot}$. As a consistency check we also calculated models of 3 $M_{\odot}$ to see if we could match the data for low-mass stars by [7, 8] (see Table 1).

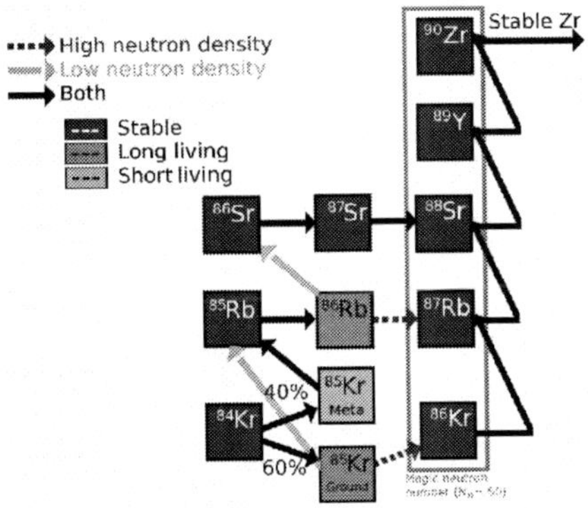

FIGURE 3. Section of the chart of the nuclides from Kr to Zr. Stable isotopes are depicted with their designation in white on a dark background, unstable isotopes are depicted with their designation in black on a light background. The light arrows show the neutron capture flux for low neutron densities, the darker, dashed arrows for high neutron densities, and the black, solid arrows for both high and low neutron densities. Isotopes with magic neutron number N = 50 are surrounded by the large open rectangle.

We artificially included a partial mixing zone (PMZ) in the top layers of the intershell in the way described in detail by [12]. There, protons combine with ^{12}C to make enough ^{13}C to efficiently activate the ^{13}C(α,n)^{16}O reaction. We varied the size of the PMZ from 0 to 2×10^{-3} $M_\odot$ for the 3 $M_\odot$ model and calculated all higher mass models with either no PMZ, or a PMZ of size 1×10^{-4} $M_\odot$. This is because the mass of the intershell is about one order of magnitude smaller in the higher than in the lower mass models, so we scale the mass of the PMZ accordingly.

3. THE PRODUCTION OF RUBIDIUM VIA THE *S* PROCESS

Figure 3 shows a section of the chart of the nuclides from Kr to Zr. When ^{84}Kr captures a neutron, 50% of the flux proceeds to the ground state of ^{85}Kr ($\tau_{1/2}$ = 3934.4 days) and the other 50% proceeds to the metastable state of ^{85}Kr ($\tau_{1/2}$ = 4.480 hours). Of the metastable ^{85}Kr, 20% decays to its ground state, while the remaining 80% decays into ^{85}Rb. Thus, effectively, 40% of ^{84}Kr+n results in the production of ^{85}Rb, whereas the other 60% results in the production of the ^{85}Kr ground state. The long decay time of the ^{85}Kr ground state allows it to capture another neutron, thus producing ^{86}Kr instead of ^{85}Rb, provided the neutron density is higher than $\approx 5\times10^{8}$ N$_n$ cm^{-3}. ^{86}Kr can also capture a neutron producing ^{87}Kr, which quickly decays into ^{87}Rb. The other unstable isotope in Fig. 3 is ^{86}Rb ($\tau_{1/2}$ = 18.63 days). This branching point produces ^{87}Rb directly, with $\approx$ 50 % of the flux going to ^{87}Rb if the neutron density is $\gtrsim 10^{10}$ N$_n$ cm^{-3}.

The maximum neutron density produced by the ^{22}Ne(α,n)^{25}Mg reaction in the thermal pulse of a 6.5 $M_\odot$ Z = 0.02 model is about 10^{12-13} N$_n$ cm^{-3}, two orders of magni-

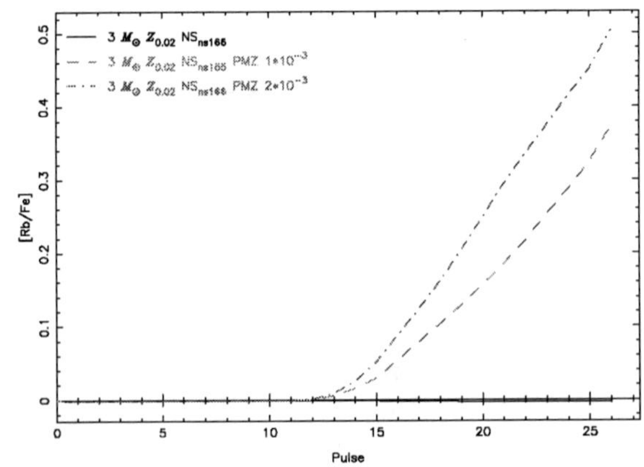

FIGURE 4. Model results for [Rb/Fe] as a function of the pulse number for a 3 $M_\odot$ $Z = 0.02$ star with and without a PMZ. The observed [Rb/Fe] ranges from -0.3 to 0.6, see Table 1.

tude higher than in a 3 $M_\odot$ $Z = 0.02$ model. Thus, due to the activation of the branchings described above, a high-mass AGB star will produce a higher Rb/Zr ratio than a low-mass AGB star. [1]

Traditionally, ^{85}Kr is quoted as the main branching point leading to the production of Rb [7], but we find that the opening of the ^{85}Kr channel actually results in the production of less elemental Rb, than with it switched off. In fact, if we artificially remove the ^{85}Kr channel from our network, we produce slightly less ^{87}Rb, but we retain a higher abundance of ^{85}Rb, resulting in a higher total Rb abundance. On the other hand, we confirmed that shutting off the ^{86}Rb channel does result in a lower Rb/Zr ratio.

4. RESULTS

Results are plotted in Fig. 4, 5, and 6 all in the following format: pulse number on the horizontal axis and the logarithm of a chosen elemental ratio against solar, i.e., [X/Y], on the vertical axis. For convenience, the observed range of values from Table 1 is reported in the respective captions.

Figure 4 shows that the observed range of the [Rb/Fe] ratio is well matched for stars of 3 $M_\odot$. Since high Zr (as well as other *s*-process elements such as Ba and Nd) are observed to be highly enhanced in these stars, the inclusion of a PMZ for this type of AGB star is necessary and is also in agreement with the constraints coming from the Rb abundances (see, e.g., [7, 17]).

[1] A link to the zip file of two movies showing the neutron density and the convective regions in and around the intershell of our AGB star models during the final pulses before the peak of the AGB can be found here: *http://www.fisica.unipg.it/ torino/i1.html* (or, direct link: *http://www.fisica.unipg.it/ torino/vanRaai.zip*). The red cross-hatched regions in these movies indicate convection. A plot of the maximum neutron density at each time step in the aforementioned movies can be found in the same zip file.

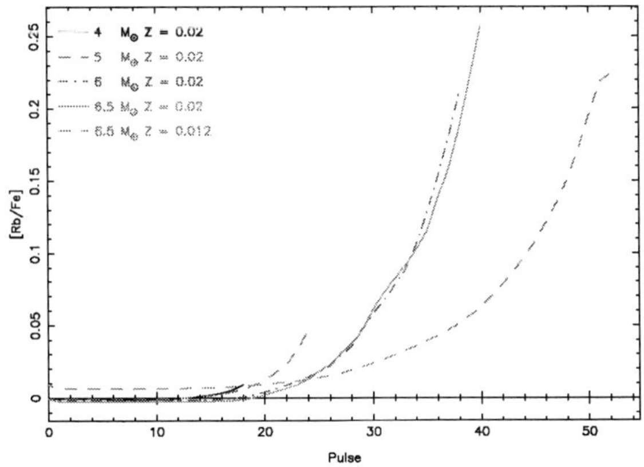

FIGURE 5. Model results for the [Rb/Fe] against pulse number for high mass stars, without the inclusion of a PMZ. The observed [Rb/Fe] ranges from 0.6 to 1.6 ($\pm$ 0.8, see Table 1 and Fig. 1).

Moving on to the massive AGB stars, first of all we note that, since the observed stars show very high mass loss rates (up to several times $10^{-5} M_\odot$) and are strong IR sources, which means that they are surrounded by dust, they are most likely located at the tip of the AGB. Hence, the final predicted values are those of more interest for comparison to the observations. From Fig. 5 we note that the theoretical [Rb/Fe] increases with the stellar mass, as observed, however, the predicted final values are well below the observed range. They are still within the error of the lower end of the observational range, however, values of at least [Rb/Fe] $\approx$ 0.8 are needed to be within the 0.8 dex error of the highest observed value of 1.6.

The [Rb/Zr] is $\approx$ 0.15, 0.04, and 0 for our 6 and 6.5 $M_\odot$, 5 $M_\odot$, and 4 $M_\odot$ models, respectively. Hence, the [Rb/Zr] value is within the range of the observations, but it fails to cover the majority of them. On the other hand, the [Zr/Fe] of our models reaches 0.1 at most, which is well below the 0.5 upper limit value derived from the observations. Including a PMZ, the final [Rb/Fe] value for our 6.5 $M_\odot$ model reaches 0.6, however, the inclusion of the PMZ also increases the [Zr/Fe] value to 0.6, which is no longer within the observed range.

4.1. Nuclear Uncertainties

If we vary the neutron-capture cross-sections of the involved isotopes within the uncertainties from [13], the [Rb/Fe] ratio increases, at most, by 15%. This occurs, e.g, in the 6.5 $M_\odot$ $Z = 0.02$ model, when we increase the neutron capture cross sections of both ^{86}Kr and ^{86}Rb by a factor of two and reduce the ^{85}Kr cross section by a factor of two. Changing the rate of the neutron source responsible for the Rb production, i.e. the ^{22}Ne(α,n)^{25}Mg rate, between its lower and upper limit as described in [18], we obtain values for [Rb/Fe] $\approx$ 0.21$\pm$0.14, for a 6 $M_\odot$ $Z = 0.02$ star.

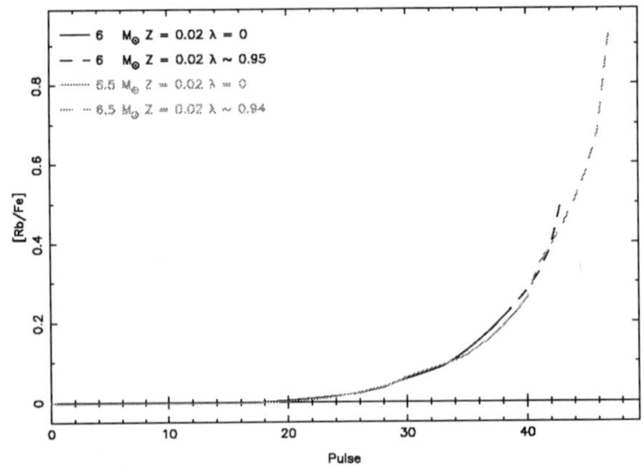

FIGURE 6. Results of synthetic extended evolution (see text). The solid part of the lines are the same as in Fig. 5 and are equivalent to a TDU efficiency of 0 during the extended evolution. The dashed lines are the result of the synthetic extension when the TDU efficiency is chosen to remain constant.

4.2. Stellar Modelling Uncertainties

The stellar structure code experiences convergence difficulties after the onset of the superwind phase; the final envelope mass is such that further TPs and TDUs may occur. We take from [19, 3] the number of remaining pulses, as calculated from the envelope mass and the mass loss rate. For the 6 $M_{\odot}$ and 6.5 $M_{\odot}$ $Z = 0.02$ models we obtain 5 and 7 more pulses, respectively. Then, we take the same intershell abundances from the last computed pulse and augment surface abundances with the extra pulses. The main uncertainty here is the TDU efficiency with decreasing size of the envelope. The efficiency could cover all possible ranges from zero [20] to staying at the efficiency of the last pulse computed in detail [21]. The resulting abundances are plotted in Fig. 6. The black line (for the 6 $M_{\odot}$ model) and grey line (for the 6.5 $M_{\odot}$ model) are the same results presented in Fig. 5 - i.e., they are equivalent to assuming no dredge up at all during the superwind phase - while the dashed lines show the result of extending the models using the same dredge up efficiency as during the last computed pulse before the start of the superwind phase - i.e., the maximum possible dredge up efficiency. In the latter case we obtain a much better match to the observations.

5. CONCLUSIONS

The [Rb/Zr] ratio is a good indicator of the neutron source at work in AGB stars and therefore of the mass of the star. The observations of Rb and Zr in massive AGBs from García-Hernández et al. [1] are reproduced by our models with regards to the increase of the [Rb/Fe] ratio with the stellar mass. However, we do not produce high enough [Rb/Fe] ratios to match the whole range of the observations. Introducing a partial mixing zone mitigates this problem, but results in the production of too high Zr to match the observations. Therefore, a partial mixing zone is undesirable in this context. We find that ^{86}Rb is the main branching point, responsible for the Rb enhancement, rather than ^{85}Kr. Finally, if the third dredge up efficiency remains as high as before the onset of the superwind phase during the final few pulses of a massive AGB star, we can match the [Rb/Fe] ratios observed by García-Hernández et al. [1]. A better understanding of the third dredge up efficiency with decreasing envelope mass for massive AGB stars is essential for further investigation of this issue.

REFERENCES

1. D.A. García-Hernández, P. García-Lario, B. Plez, F. D'Antona, A. Manchado, and J.M. Trigo-Rodriguez, *Science* **314**, 1751 (2006), astro-ph/0611319
2. F. Herwig, *ApJ* **605**, 425 (2004)
3. A. Karakas and J.C. Lattanzio, *Publications of the Astronomical Society of Australia* **24**, 103 (2007), arXiv:0708.4385
4. R. Gallino, C. Arlandini, M. Busso, M. Lugaro, C. Travaglio, O. Straniero, A. Chieffi, and M. Limongi, *ApJ* **497**, 388 (1998)
5. D.A. García-Hernández, P. García-Lario, B. Plez, A. Manchado, F. D'Antona, J. Lub, and H. Habing, *A&A* **462**, 711 (2007), astro-ph/0609106
6. P.R. Wood, M.S. Bessell, and M.W. Fox, *ApJ* **272**, 99 (1983)
7. C. Abia, M. Busso, R. Gallino, I. Domínguez, O. Straniero, and J. Isern, *ApJ* **559**, 1117 (2001), arXiv:astro-ph/0105486
8. D.L. Lambert, V.V. Smith, M. Busso, R. Gallino, and O. Straniero, *ApJ* **450**, 302 (1995)
9. C.A. Frost and J.C. Lattanzio, *ApJ* **473**, 383 (1996)
10. R.C. Cannon, *MNRAS* **263**, 817 (1993)
11. F.K. Thielemann, J.W. Truran, and M. Arnould, "Thermonuclear reaction rates from statistical model calculations," in *Advances in Nuclear Astrophysics*, edited by E. Vangioni-Flam, J. Audouze, M. Casse, J.-P. Chieze, and J. Tran Thanh van, 1986, pp. 525–540
12. M. Lugaro, C. Ugalde, A.I. Karakas, J. Görres, M. Wiescher, J.C. Lattanzio, and R.C. Cannon, *ApJ* **615**, 934 (2004)
13. Z.Y. Bao, H. Beer, F. Käppeler, F. Voss, K. Wisshak, and T. Rauscher, *Atomic Data and Nuclear Data Tables* **76**, 70 (2000)
14. E. Anders and N. Grevesse, *Geochim. Cosmochim. Acta* **53**, 197 (1989)
15. M. Asplund, N. Grevesse, and A.J. Sauval, "The Solar Chemical Composition," in *ASP Conf. Ser. 336: Cosmic Abundances as Records of Stellar Evolution and Nucleosynthesis*, edited by T. G. Barnes, III, and F. N. Bash, p. 25 (2005)
16. K. Lodders, *ApJ* **591**, 1220 (2003)
17. M. Busso, R. Gallino, D.L. Lambert, C. Travaglio, and V.V. Smith, *ApJ* **557**, 802 (2001)
18. A.I. Karakas, M. Lugaro, M. Wiescher, J. Goerres, and C. Ugalde, *ApJ* **643**, 471 (2006)
19. A.I. Karakas, and J.C. Lattanzio, *PASA* **20**, 393 (2003)
20. O. Straniero, A. Chieffi, M. Limongi, M. Busso, R. Gallino, and C. Arlandini, *ApJ* **478**, 332 (1997)
21. R. J. Stancliffe and C. S. Jeffery, *MNRAS* **375**, 1280 (2007), astro-ph/0612005

Abundances of Heavy Elements in PNe and AGB Model Predictions

M. Pignatari[*,†], R. Gallino[**], S. Cristallo[‡] and O. Straniero[‡]

[*]*Keele Astrophysics Group, Keele University, ST5 5BG UK*
[†]*Joint Institute for Nuclear Astrophysics, University of Notre Dame, IN 46556, USA*
[**]*Dipartimento di Fisica Generale, Universita' di Torino, Via Pietro Giuria 1, 10125, Torino, Italy*
[‡]*Osservatorio Astronomico di Collurania, 64100 Teramo, Italy*

Abstract. During the AGB phase of low and intermediate mass stars heavy $s-$process elements are produced and dredged−up in the envelope. In the following Planetary Nebula phase part of the AGB envelope will form the nebula around the hot central star, carrying the $s-$process signature unchanged. Recently, heavy elements have been observed is many Planetary Nebulae, providing a new tool to analyze the $s-$process nucleosynthesis working in the AGB stars. In this paper we present $s-$process predictions for a large spread of stellar masses, metallicities and ^{13}C−pocket efficiencies. Theoretical results are compared with the most recent observations in PNe, finding a general good agreement.

Keywords: Stars: Evolution; Stars: Nucleosynthesis; Planetary Nebulae
PACS: 97.10.Cv; 98.38.Ly

1. INTRODUCTION

After the asymptotic giant branch phase (AGB), low and intermediate mass star ($M \leq 8\ M_\odot$) follow their evolution in the HR diagram, becoming post−AGB stars and then planetary nebulae (PNe), when the surface temperature of the central star is high enough to ionize the interstellar material around. Finally, they will end their evolution cooling down as white dwarfs (WD)(for a detailed discussion, see [1, 2, 3]). PNe have been classified by their chemical composition and by their morphology:

- According to [4] or [5], PNe may be classified as He, N rich (Type I) and not N enriched (non−Type I).
- According to [6] and reference therein, we classify PNe in bipolar, round, elliptical, and irregular;

N enrichment should be related to the Hot Bottom Burning (HBB, see for details [7, 8]), which indicates an intermediate mass AGB progenitor (IMS−AGB) for Type I PNe in agreement with nebula masses analysis (e.g.[9]). On the other hand, PNe without N enrichment should have a low mass AGB progenitor (lms−AGB). Concerning the morphology, bipolar PNe could be formed in a binary system (e.g.[10, 11]). However, the binarity fraction over the bipolar PNe sample is still uncertain. On the other hand, there is a quite clear correlation between Type I Pne and bipolar PNe (e.g.[12, 13]), also if few PNe are known to be bipolar but not Type I (e.g. NGC 6445[6]). According to [13], elliptical and round PNe are not N enriched (non−Type I), pointing to a lms−AGB progenitor. [14] proposed that most of the elliptical PNe are formed in a binary system,

CP1001, *9th Torino Workshop on Evolution and Nucleosynthesis in AGB Stars and the 2nd Perugia Workshop on Nuclear Astrophysics*,
edited by R. Guandalini, S. Palmerini, and M. Busso

whereas round PNe have been formed by a progenitor without a companion star. However, further investigations are required to confirm this scenario.

In this work we will focus on the heavy element observations in PNe. For an updated analysis of light element observations in PNe we refer to [13, 15, 16]. The first detection of heavy $s-$process elements in PNe is presented by [17], with the detection of emission lines of Kr, Xe, Ba and many other elements indicating an $s-$process enrichment in NGC 7027. The s elements have been produced in the previous AGB phase, and the fraction of the AGB envelope that is part of the nebula still show the $s-$process signature. Following works confirm this first important result in many other PNe. According to [6], about 50 % of 78 PNe shows an $s-$process enrichment. Such enrichment increases with increasing the C/O, in agreement with AGB models predictions (e.g.[18]). On average, bipolar/Type I PNe show lower $s-$process abundances and elliptical/non−Type I PNe higher $s-$process abundances (e.g.[6]), confirming IMS−AGB and lms−AGB progenitors, respectively.

In the next section we present AGB model calculations for a large range of initial masses and metallicities, and we compare our results with recent PNe observations of heavy elements.

2. AGB MODEL PREDICTIONS

In this section we briefly present updated $s-$process calculations by the Torino group AGB models. We refer to [18, 19] and [20] for the ^{13}C−pocket definition and for a detailed description of the AGB models and of the s process.

[21] proposed to use [Xe/Kr] to evaluate the intrinsic index of the $s-$process [hs/ls] [22, 23], where Xe is used instead of hs typical elements (Ba, Nd, La, Sm) and Kr instead of ls typical elements (Sr, Y, Zr). The reason is that ls elements and hs elements are more difficult to observe in PNe than Kr and Xe, respectively. Kr and Xe emission lines are observed with reasonable uncertainty, and they have similar ionizations potentials [21, 24]. According to [25], low mass AGB stars and intermediate mass AGB stars produced 17 % and 12 % of the solar Kr, respectively, with a total contribution of 29 %. Low mass AGB stars provide the dominant $s-$process contribution to the solar Xe (about 15 %). Despite Xe is mostly produced by the r−process ([26] and references therein), a non−marginal contribution from AGB stars is expected, and theoretical [Xe/Kr] predicted in AGB star envelopes from $s-$process calculations can be compared with observations in PNe.

In Fig. 1 [hs/ls] ratio is reported for M = 1.5 $M_\odot$ within a large range of ^{13}C−pockets and metallicities (Left Panel). For the same cases, [Xe/Kr] is reported in the Right Panel. Here the average [Xe/Kr] observed over 5 PNe by [21] is also reported ([Xe/Kr] = +0.09). We assume an uncertainty of 0.3 dex for the average [Xe/Kr] (indicated by the top and the bottom borders of the box), in agreement with observations. According to [27], for the observed PNe we consider a range of initial metallicity between solar and 1/3 of solar (right and left borders of the box, respectively). Notice that over the 5 PNe considered three of them are clearly $s-$process enriched within the reported uncertainty of [Kr/Ar] and [Xe/Ar] (IC 4191, NGC 7027, IC 418). The average [Xe/Kr] considering only the s−process rich PNe is + 0.07, in agreement with the ratio reported in Fig. 1. The observed [Xe/Kr] is reproduced by the AGB models predictions, within the spread of ^{13}C−pocket efficiency considered.

In Fig. 2, 3 we report the [hs/ls] and the [Xe/Kr] ratios for M = 2, 3 $M_\odot$, respectively. In Fig. 2 also recent FRANEC AGB model predictions by Cristallo & Straniero (see contribution in this volume) have been included (grey pentagons): M = 2 $M_\odot$ and [Fe/H] = −0.14, −0.67 (according to the solar [Fe/H] = 0 given by [28]). Those models are both in agreement with Torino group predictions, in particular with a ^{13}C−pocket = u1p3 (ST*1.3) – u2 (ST*2). The M = 2 $M_\odot$ and [Fe/H] = −0.14 model is within the metallicity range of PNe and the predicted [Xe/Kr] well reproduces the observations.

Finally, in Fig. 4 [hs/ls] and the [Xe/Kr] ratios for M = 5 $M_\odot$ models are reported, with ^{13}C−pocket efficiency spreading over a factor of two. In this case, the observations are not reproduced. Only the cases at solar metallicity barely agree with the lower limit of the observed [Xe/Kr]. The reason is that intermediate mass AGB stars give a significant contribution to Kr and not to Xe, causing a final low [Xe/Kr]. This argument continues to be true for [hs/ls].

We concluded that, according to their [Xe/Kr] ratio, all the 5 PNe reported in [21] should have low mass AGB progenitors. For the two PNe showing no clear s−process enhancements (NGC 2440 and IC 2501), a 1.5 $M_\odot$ model with low ^{13}C−pocket efficiency could reproduce the observed [Kr/Ar] and [Xe/Ar]. However, for them we cannot rule out the possibility of an intermediate mass AGB progenitor within the present uncertainty.

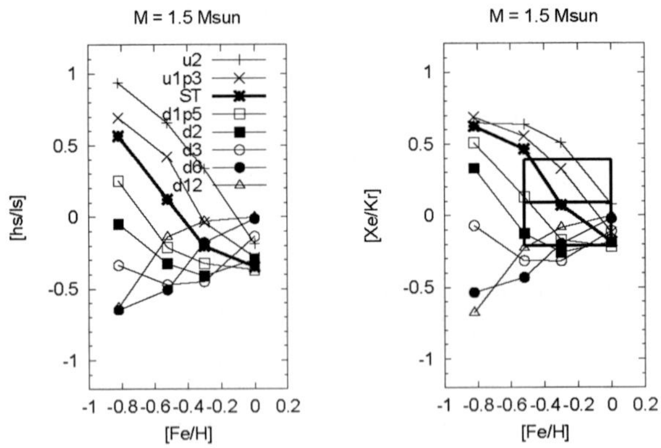

FIGURE 1. Left Panel: The [hs/ls] ratio is reported for M = 1.5 $M_\odot$ in the spread of ^{13}C−pocket efficiency d12 (ST/12) – u2 (ST*2) at different metallicities ([Fe/H] = 0, −0.30, −0.52, −0.82). Right Panel: The [Xe/Kr] ratio is reported for the same mass, ^{13}C−pocket and metallicity ranges of Left Panel. The thick box reports the average observed ratio in PNe according to [21], assuming 0.3 dex of uncertainty (see the text for more details).

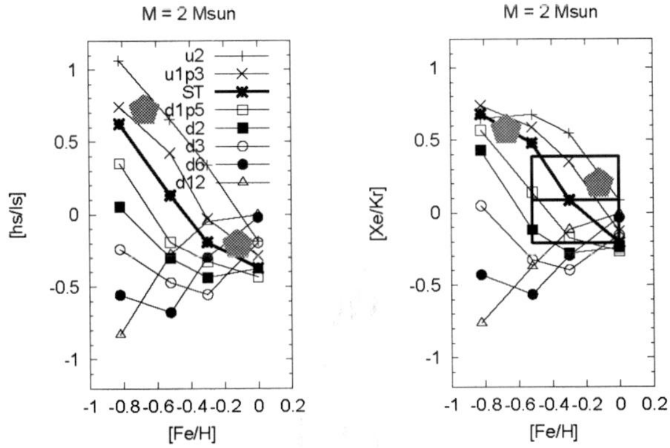

FIGURE 2. The [hs/ls] and [Xe/Kr] ratios are reported as in Fig.1 for M = 2 $M_{\odot}$ in the same spread of ^{13}C−pocket efficiency and metallicity. As a comparison, [hs/ls] and [Xe/Kr] are also presented for M = 2 $M_{\odot}$ and [Fe/H] = −0.15, −0.67 AGB models by Cristallo & Straniero contribution in this volume (grey pentagons).

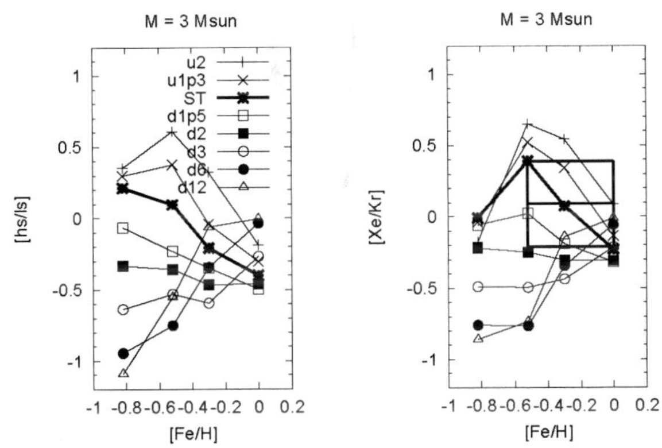

FIGURE 3. The [hs/ls] and [Xe/Kr] ratios are reported as in Fig. 1 for M = 3 $M_{\odot}$ in the same spread of ^{13}C−pocket efficiency and metallicity.

ACKNOWLEDGMENTS

This research was supported by a Marie Curie International Reintegration Grant within the 6th European Community Framework Programme, grant MIRG-CT-2006-046520, and by NSF grants PHY 02-16783 (Joint Institute for Nuclear Astrophysics). We acknowledge the support of the Italian MIUR-PRIN06 Project 'Late phases of Stellar Evolution: Nucleosynthesis in Supernovae, AGB stars, Planetary Nebulae'.

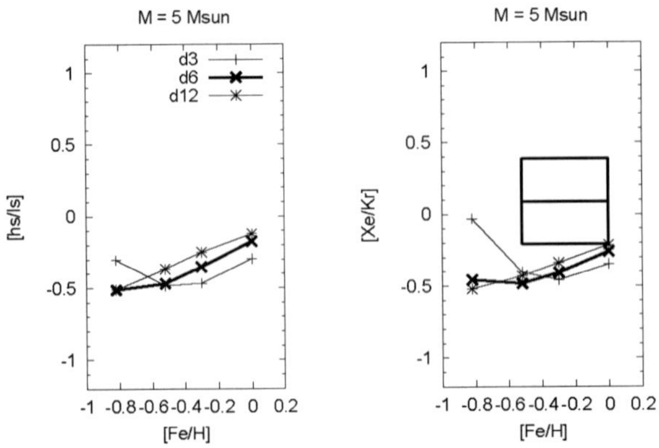

FIGURE 4. The [hs/ls] and [Xe/Kr] ratios are reported as in Fig. 1 for M = 5 $M_\odot$ in the same spread of metallicity and in the spread of ^{13}C−pocket efficiency d12 (d6/2) – d3 (d6*2).

REFERENCES

1. H. van Winckel, *ARA&A* **41**, 391 (2003)
2. T. Bloecker, *A&A* **299**, 755 (1995)
3. K. Werner and F. Herwig, *PASP* **118**, 183 (2006), arXiv:astro-ph/0512320
4. M. Peimbert, *ŞChemical abundances in planetary nebulae,Ť in Planetary Nebulae, edited by Y. Terzian, vol. 76 of IAU Symposium*, p. 215 (1978)
5. R.L. Kingsburgh and M.J. Barlow, *MNRAS* **271**, 257 (1994)
6. N.C. Sterling and H.L. Dinerstein, *ApJS* **174**, 158 (2008)
7. A.I. Boothroyd, I.-J. Sackmann, and S.C. Ahern, *ApJ* **416**, 762 (1993)
8. J. Lattanzio and M. Forestini, *ŞNucleosynthesis in AGB Stars,Ť in Asymptotic Giant Branch Stars, edited by T. Le Bertre, A. Lebre, and C. Waelkens, vol. 191 of IAU Symposium*, p. 31 (1999)
9. S. Torres-Peimbert and M. Peimbert, *ŞPlanetary Nebulae of type I revisited (Invited Review),Ť in Planetary Nebulae, edited by H. J. Habing, and H.J.G.L.M. Lamers,vol. 180 of IAU Symposium*, p. 175 (1997)
10. O. De Marco and N. Soker, *PASP* **114**, 602 (2002) arXiv:astro-ph/0204230
11. B. Balick and A. Frank, *ARA&A* **40**, 439 (2002)
12. M. Peimbert and S. Torres-Peimbert, *ŞType I planetary nebulae,Ť in Planetary Nebulae, edited by D. R. Flower, vol. 103 of IAU Symposium*, p. 233 (1983)
13. L. Stanghellini, M.A. Guerrero, K. Cunha, A. Manchado, and E. Villaver, *ApJ* **651**, 898 (2006), arXiv:astro-ph/0607480
14. O. De Marco, H.E. Bond, D. Harmer, and A.J. Fleming, *ApJ* **602**, L93 (2004) arXiv: astro-ph/0312410
15. M. Perinotto, L. Morbidelli, and A. Scatarzi, *MNRAS* **349**, 793 (2004)
16. R.B.C. Henry, K.B. Kwitter, and B. Balick, *AJ* **127**, 2284 (2004), arXiv:astro-ph/ 0401156.
17. D. Pequignot and J.P. Baluteau, *A&A* **289**, 659 (1994)
18. R. Gallino, C. Arlandini, M. Busso, M. Lugaro, C. Travaglio, O. Straniero, A. Chieffi, and M. Limongi, *ApJ* **497**, 388 (1998)
19. O. Straniero, A. Chieffi, M. Limongi, M. Busso, R. Gallino, and C. Arlandini, *ApJ* **478**, 332 (1997)
20. M. Lugaro, F. Herwig, J.C. Lattanzio, R. Gallino, and O. Straniero, *ApJ* **586**, 1305 (2003), arXiv:astro-ph/0212364
21. B. Sharpee, Y. Zhang, R. Williams, E. Pellegrini, K. Cavagnolo, J.A. Baldwin, M. Phillips, and X.-W. Liu, *ApJ* **659**, 1265 (2007), arXiv:astro-ph/0612101

22. R.E. Luck and H.E. Bond, *ApJS* **77**, 515 (1991)
23. M. Busso, R. Gallino, and G.J. Wasserburg, *ARA&A* **37**, 239 (1999)
24. H.L. Dinerstein, *ApJ* **550**, L223 (2001)
25. C. Travaglio, R. Gallino, E. Arnone, J. Cowan, F. Jordan, and C. Sneden, *ApJ* **601**, 864 (2004), arXiv:astro-ph/0310189
26. K.-L. Kratz, K. Farouqi, B. Pfeiffer, J.W. Truran, C. Sneden, and J.J. Cowan, *ApJ* **662**, 39 (2007), arXiv:astro-ph/0703091
27. N.C. Sterling and H. L. Dinerstein, *ŞAn Infrared Survey of Neutron-Capture Elements in Planetary Nebulae,Ť in Revista Mexicana de Astronomia y Astrofisica Conference Series, edited by S. Torres-Peimbert and G. MacAlpine, vol. 23 of Revista Mexicana de Astronomia y Astrofisica Conference Series*, p. 1 (2005)
28. E. Anders and N. Grevesse, *Geochim. Cosmochim. Acta* **53**, 197 (1989)

SESSION V:

SPECTROSCOPIC, INTERFEROMETRIC, AND PHOTOMETRIC CONSTRAINTS

Nitrogen Production in Intermediate-Mass Stars by Hot-Bottom Burning

P. R. Wood*, J. A. McSaveney*, M. Scholz†,**, J. C. Lattanzio‡, K. H. Hinkle§ and A. I. Karakas*

*Mount Stromlo Observatory, Australian National University
†Institut f. Theoretische Astrophysik d. Univ. Heidelberg
**School of Physics, University of Sydney
‡Department of Mathematical Sciences, Monash University
§National Optical Astronomy Observatory, Tucson

Abstract. The PHOENIX spectrograph has been used on Gemini South in order to determine C, N and O abundances for intermediate mass AGB stars in the Magellanic Clouds. Although seven stars were observed, the spectra of three stars were unusable because they were too cool, causing the molecular features to be too strong for modelling. In another two stars, the spectra were contaminated by emission from shock waves. Pulsation modelling showed that spectra should only be taken near minimum visual light to avoid this problem. The two stars analysed for abundances showed strong enhancements of N and reductions of C, consistent with the expectation of hot-bottom burning. The observed N enhancements and C reductions suggest that envelope convection goes deeper than predicted by evolutionary models using the mixing-length theory of convection. These results provide the first observational in-situ confirmation of primary nitrogen production by hot-bottom burning in intermediate mass AGB stars.

Keywords: Stars: Abundances, Stars: AGB and Post-AGB, Stars: Evolution
PACS: 97.10.Cv; 97.10.Tk; 97.30.Jm

1. INTRODUCTION

The phenomenon of hot-bottom burning (HBB) was first described by [1]. During HBB, the envelope convection region in an intermediate-mass ($\sim$3-7 $M_\odot$) AGB star extends inwards far enough that some CN cycling of all the envelope material occurs, converting envelope ^{12}C into ^{14}N. In these AGB stars, the ^{12}C in the envelope is made up from both the original C present in the star at birth along with ^{12}C dredged up by the third dredge-up. The dredged-up ^{12}C is primary carbon since it is the result of interior conversion of H to C via H-burning and the triple-alpha process. The ^{14}N produced from this ^{12}C by HBB is thus primary nitrogen. Intermediate-mass AGB stars are one of the main candidates for the production of nitrogen in the universe (rapidly rotating massive stars are the other main possible source of primary nitrogen).

At this stage, there are many theoretical predictions of the production of nitrogen by HBB (eg. [2, 3, 4]). However, there have been no definitive observational confirmations or calibrations of the HBB process. The main reason for this is that by the time the HBB process is in operation, the intermediate-mass AGB stars are very cool, large amplitude, long-period variables. Abundance analysis of these stars is very difficult.

Intermediate-mass AGB stars are rare and, in the Galaxy, it is difficult to determine

CP1001, 9th Torino Workshop on Evolution and Nucleosynthesis in AGB Stars and the 2nd Perugia Workshop on Nuclear Astrophysics, edited by R. Guandalini, S. Palmerini, and M. Busso

their distances and hence their masses and luminosities: the luminosity and mass are required for meaningful comparison with theoretical evolutionary models. The Magellanic Clouds do, however, provide well-known examples of relatively nearby intermediate-mass AGB stars at a known distance - see [5]. Some indications of the presence of HBB have been noted in these stars. [6] estimated that the CN bands in the optical spectra of some of the intermediate-mass AGB stars appeared stronger than expected if third dredge-up alone was occurring in these stars. [7] (see also [8, 9]) found that intermediate-mass AGB stars in the LMC showed enhanced Li line strengths, which could be explained by the HBB process, or more specifically by the beryllium transport mechanism. In this mechanism, ^{7}Be produced by the proton-proton chains is transported up from the hot burning region where it is produced before it absorbs an electron to produce ^{7}Li: the ^{7}Li would be quickly burnt if the Be^{7} stayed in the hot region where it was produced. Finally, [10] showed that a few Galactic S stars had C and N abundances that suggested that HBB had occurred, although most Galactic and LMC M and S stars showed C and N abundances in their envelopes that were consistent with CN cycling of their initial C and N i.e. no HBB was required to explain the observed abundances. Since the Galactic S stars with enhanced N were field stars, it was not possible to accurately determine masses or luminosities, and there is no particular reasons that HBB should be seen only in S stars.

With the advent of high dispersion infrared spectrographs on 8-m telescopes in the southern hemisphere (PHOENIX on Gemini South and CRIRES on the VLT), it has become possible to obtain high dispersion near-IR spectra for intermediate-mass AGB stars in the Magellanic Clouds. Coupled with appropriate model atmospheres that take into account the pulsation motions of the star, it is possible to derive C, N, O and metal abundances. The known distance to the LMC and SMC, combined with near infrared photometry, also allows calculation of the luminosity, and pulsation theory allows calculation of the stellar mass. This provides all the information needed for comparing observed CNO abundances with AGB models that include the effects of third dredge-up and HBB. Here, some results reported in detail by [11] are discussed.

2. LESSONS ON CANDIDATE STAR SELECTION

Clearly, the later in its life an AGB star is observed, the more third dredge-up of ^{12}C will have occurred, and the more HBB will have converted the dredged-up and initial ^{12}C into ^{14}N. Thus, the ideal targets for the study of primary nitrogen production in intermediate-mass AGB stars should be AGB stars with high mass loss rates and very long pulsation periods: these stars are near the end of the AGB and should currently be returning copious amounts of primary nitrogen to the interstellar medium. [11] chose two IRAS sources (IRAS 04516-6902 and IRAS 04509-6922) and a very red, luminous and long period intermediate-mass AGB star (HV 12149) as examples of such stars. Sections of their spectra are shown in Fig. 1. Also shown in Fig. 1 are the spectra of the four other, less evolved intermediate-mass AGB stars. As can be seen in Fig. 1, the spectra of the 3 very evolved stars are dominated by molecular absorption: no attempt was made to model them or derive abundances.

In principle, it should have been possible to model the spectra of the other four stars.

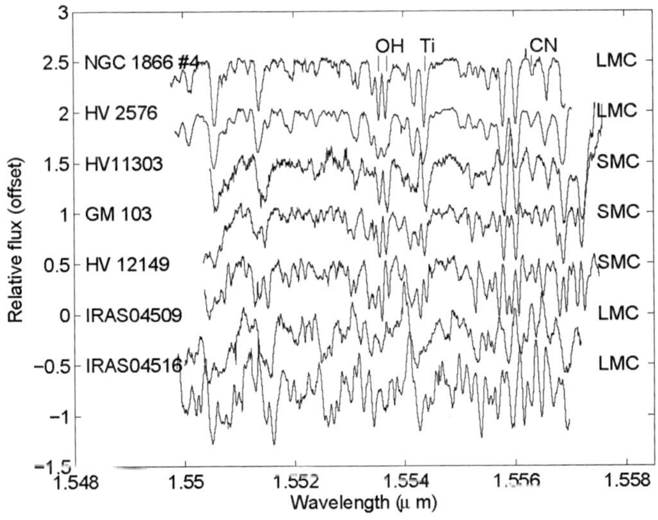

FIGURE 1. Spectra in the region of 1.554 μm of the intermediate-mass AGB stars observed in the Magellanic Clouds. From [11].

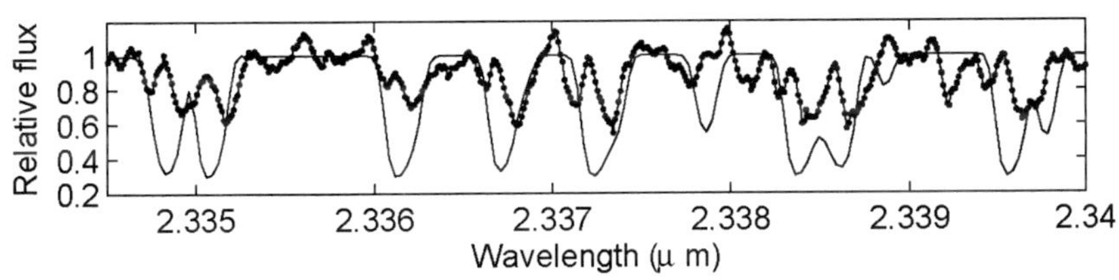

FIGURE 2. The spectrum of GM 103 near 2.337 μm with an LTE spectrum (thin line) overlaid. Each absorption line is filled in by an emission component. From [11].

However, after pulsation models of the stars GM 103 and HV 11303 were made, it was found that there were large amplitude shock waves present deep in the photosphere. Model spectra made in LTE, without any contribution from the immediate post-shock zone, showed that many of the lines in the star have a strong emission component filling up the absorption line produced in other layers of the atmosphere (see Fig. 2). From an analysis of the pulsation models, [11] estimated that the only time these large-amplitude variables could be observed without emission contamination is near minimum light, at visual phases ~0.4–0.6.

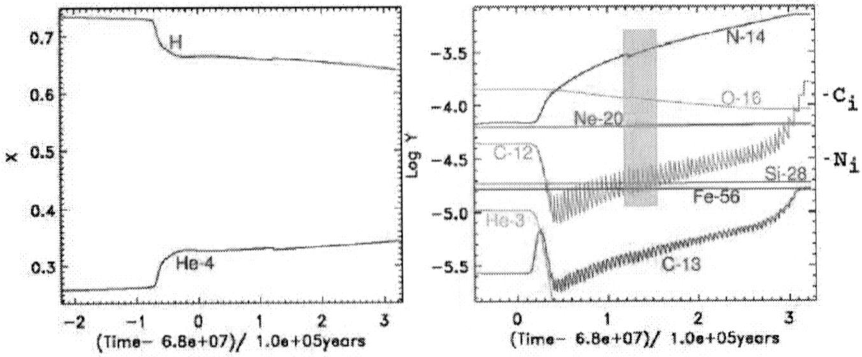

FIGURE 3. The evolution on the AGB of various surface abundances for a star with $M = 6\,M_\odot$ and metal abundance Z = 0.008. The initial, main-sequence abundances of carbon and nitrogen, C_i and N_i respectively, are marked on the right of the figure. The shaded region is the interval where the model abundances are close to those of HV 2576 (N abundance up by 10, and C abundance down by 5). The model calculations are from [4].

3. COMPARISON OF OBSERVED C AND N ABUNDANCES WITH MODELS

For the two LMC stars (HV 2576 and NGC 1866 #4) whose abundances were determined by McSaveney et al, the metallicity (Fe abundance) was similar to the half-solar value typical of the LMC (eg. [12, 10]). However, in each case the ^{14}N abundance was enhanced over the typical initial LMC value by a factor of $\sim$10 and the ^{12}C abundance was decreased from the typical initial LMC value by a factor of $\sim$5. The mass derived from pulsation theory for HV 2576 was 6 $M_\odot$ and the mass derived from pulsation theory for NGC 1866 #4 was 4 $M_\odot$. Since all the parameters of these two stars are known, their observed C and N abundances can be directly compared to those in theoretical models. Here, we compare to the models of Karakas [4].

Figure 3 shows the variation of H, He, ^{12}C and ^{14}N abundances during the AGB evolution of a 6 $M_\odot$ star with typical LMC abundances. This model is directly comparable with the star HV 2576. In Fig. 3, the second dredge-up occurs near the time -0.5×10^5 years and then HBB and helium shell flashes start around a time of 0.3×10^5 years. The initial HBB processing of the post-second dredge-up envelope material around this time reduces the ^{12}C abundance to $\sim$0.1 times the initial value and increases the ^{14}N abundance to $\sim$4 times the initial value. These values are about half the values observed in HV 2576. In order to increase both the ^{12}C and ^{14}N abundances to the levels observed in HV 2576, a large number of helium shell flashes, accompanied by hot-bottom burning between them, is required. In Fig. 3, the observed abundances are reached about 1.1×10^5 years after the onset of helium shell flashing. HV 2576 thus provides the first observational measurement of the production of primary nitrogen in an intermediate-mass AGB star.

Observationally, the carbon and nitrogen abundances of NGC 1866 #4 are similar to

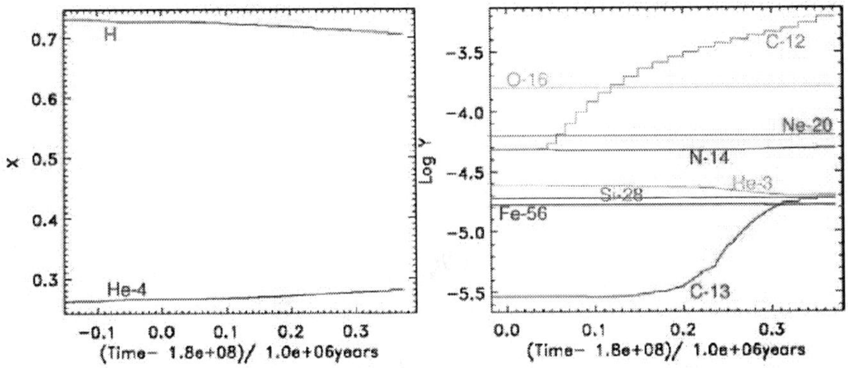

FIGURE 4. The same as Fig. 3 but for a model with $M = 4\,M_{\odot}$.

those of HV 2576 (^{14}N abundance enhanced by $\sim$10 and the ^{12}C abundance decreased by $\sim$5). However, this star has a lower mass of $\sim 4\,M_{\odot}$ according to pulsation theory, which is also consistent with the cluster age. The evolution of the 4 $M_{\odot}$ AGB models of [4], with which NGC 1866 #4 should be compared, is shown in Fig. 4. It is immediately clear that there is no hot-bottom burning occurring in this model sequence, just a monotonic increase in the ^{12}C abundance with time as a result of third dredge-up events. The Karakas 4 $M_{\odot}$ AGB models are typical of evolutionary calculations when standard mixing-length theory is used to treat convection in the envelope. However, other theories of convection that have been used can cause more extensive envelope convection. For example, the AGB models of [13, 14] which use a non-mixing-length theory of convection that provides a downward extension of the envelope convection zone compared to standard mixing-length theory, do show HBB in operation at 4 $M_{\odot}$.

4. SUMMARY

High dispersion infrared spectra of two intermediate-mass AGB stars in the LMC provide the first direct observational evidence for the production of primary nitrogen in such stars. Attempts to determine abundances in a number of other intermediate-mass AGB stars show that the coolest stars need to be avoided, and observations should be performed near minimum visible light. It is shown that the convection zones in AGB models using standard mixing-length theory are too shallow, and that a convection theory that extends the convection zone in depth is required in order to study hot-bottom burning and related nucleosynthesis in AGB stars.

REFERENCES

1. J.M. Scalo, K.H. Despain, and R.K. Ulrich, *ApJ* **196**, 805 (1975)
2. I. Iben and J.W. Truran, *ApJ* **220**, 980 (1978)

3. L.M. Dray, C.A. Tout, A.I. Karakas, and J.C. Lattanzio, *MNRAS* **338**, 973 (2003)
4. A.I. Karakas, *PhD thesis, Monash University*, (2003) `http://www.cspa.monash.edu.au/cspa-pub.html`
5. P.R. Wood, M.S. Bessell, and M.W. Fox, *ApJ* **272**, 99 (1983)
6. J.M. Brett, *MNRAS* **249**, 538 (1991)
7. V.V. Smith and D.L. Lambert, *ApJL* **361**, 69 (1990)
8. B. Plez, V.V. Smith, and D.L. Lambert, *ApJ* **418**, 812 (1193)
9. V.V Smith, B. Plez, D.L. Lambert, and D.A. Lubowich, *ApJ* **441**, 735 (1995)
10. V.V. Smith, K.H. Hinkle, K. Cunha, et al., *AJ* **124**, 3241 (2002)
11. J.A. McSaveney, P.R. Wood, M. Scholz M., J.C. Lattanzio, and K.H. Hinkle, *MNRAS* **378**, 1089 (2007)
12. V. Hill, P. François, M. Spite, F. Primas, and F. Spite, *A&A* **364**, L19 (2000)
13. P. Ventura, F. D'Antona, and I. Mazzitelli, *A&A* **363**, 605 (2000)
14. P. Ventura, F. D'Antona, and I. Mazzitelli, *A&A* **393**, 215 (2002)

The Lithium-, *r*- and *s*-Enhanced Metal-Poor Giant HK-II 17435-00532

Ian U. Roederer*, Anna Frebel*,†, Matthew Shetrone*,†, Carlos Allende Prieto*, Jaehyon Rhee**, Roberto Gallino‡,§, Sara Bisterzo‡, Christopher Sneden*, Timothy C. Beers¶ and John J. Cowan‖

**Dept. of Astronomy, University of Texas at Austin, 1 University Station, C1400, Austin, TX 78712*
†McDonald Observatory, University of Texas, Fort Davis, TX
***Dept. of Physics, Purdue University, 525 Northwestern Avenue, West Lafayette, IN 47907*
‡Dipartimento di Fisica Generale, Università di Torino, Torino, Italy
§Centre for Stellar and Planetary Astrophysics, Monash University, PO BOX 28M, Clayton, VIC 3800 Australia
¶Dept. of Physics and Astronomy, Center for the Study of Cosmic Evolution, and Joint Institute for Nuclear Astrophysics, Michigan State University, East Lansing, MI
‖Homer L. Dodge Dept. of Physics and Astronomy, University of Oklahoma, Norman, OK

Abstract. We present the first detailed abundance analysis of the metal-poor giant HK-II 17435-00532. This star was observed as part of the University of Texas Long-Term *Chemical Abundances of Stars in the Halo* (*CASH*) Project. A spectrum was obtained with the High Resolution Spectrograph (HRS) on the Hobby-Eberly Telescope with a resolving power of $R \sim 15,000$. Our analysis reveals that this star may be located on the red giant branch, red horizontal branch, or early asymptotic giant branch. We find that this metal-poor ([Fe/H] $= -2.2$) star has an unusually high lithium abundance ($\log \varepsilon\,(\mathrm{Li}) = +2.1$), mild carbon ([C/Fe] $= +0.7$) and sodium ([Na/Fe] $= +0.6$) enhancement, as well as enhancement of both *s*-process ([Ba/Fe] $= +0.8$) and *r*-process ([Eu/Fe] $= +0.5$) material. The high Li abundance can be explained by self-enrichment through extra mixing mechanisms that connect the convective envelope with the outer regions of the H-burning shell. If so, HK-II 17435-00532 is the most metal-poor star in which this short-lived phase of Li enrichment has been observed. The *r*- and *s*-process material was not produced in this star but was either present in the gas from which HK-II 17435-00532 formed or was transferred to it from a more massive binary companion. Despite the current non-detection of radial velocity variations (over a time span of ~ 180 days), it is possible that HK-II 17435-00532 is in a long-period binary system, similar to other stars with both *r* and *s* enrichment.

Keywords: Nuclear Reactions, Nucleosynthesis, Abundances — Stars: Abundances — Stars: Individual (HK-II 17435-00532) — Stars: Population II
PACS: 97.10.Cv, 97.10.Ex, 97.10.Tk, 97.20.Tr

1. INTRODUCTION

The story of early Galactic nucleosynthesis is written in the chemical compositions of very metal-poor halo stars. Li, the only metal produced during the Big Bang, is often observed in unevolved metal-poor stars. Information on the primordial Li abundance can be inferred from, e.g., the Spite Plateau Li abundance that most metal-poor main-sequence stars tend to possess, $\log \varepsilon\,(\mathrm{Li}) = 2.21 \pm 0.09$ [1, 2], although the stellar result does not agree with the WMAP estimate of the primordial Li abundance, $\log \varepsilon\,(\mathrm{Li}) = 2.64 \pm 0.03$ [3]. Some evolved stars, however, appear to have *overabundances* of Li—

CP1001, *9th Torino Workshop on Evolution and Nucleosynthesis in AGB Stars and the 2nd Perugia Workshop on Nuclear Astrophysics*, edited by R. Guandalini, S. Palmerini, and M. Busso

in contrast to the sparse galactic production mechanisms of Li through cosmic ray spallation, and in spite of the fact that Li is completely diluted and destroyed in the stellar atmosphere by the time the star reaches the giant branch. Some reactions enable a star to self-enrich its atmosphere with Li, but only for a very short period of time before the freshly produced Li burns again (e.g., the ^{7}Be transport mechanism [4]). In this work we present HK-II 17435-00532, a metal-poor object that falls into the same category of evolved Li-rich stars. The neutron (n)-capture species are often used to diagnose the chemical enrichment history of a star, and we find that HK-II 17435-00532 possesses enhancements of material from both the slow (s) and rapid (r) n-capture processes.

2. OBSERVATIONS AND ANALYSIS

The HK-II survey [5] originated as an extension of the original HK objective-prism survey [6, 7]. The HK-II survey was designed to discover a large sample of very metal-poor red giant stars with [Fe/H] ≤ -2.0 by using digitized objective-prism spectra and 2MASS JHK colors. We have recently started the *Chemical Abundances of Stars in the Halo* (*CASH*) Project with the Hobby-Eberly Telescope (HET) located at McDonald Observatory (Frebel et al., *in prep*). This project aims to characterize the chemical composition of the Galactic halo by means of abundance analyses of metal-poor stars. The HK-II survey is one source of targets for our HET-*CASH* Project, and the spectrum of HK-II 17435-00532 was taken as part of this project.

HK-II 17435-00532 was observed on 2007 February 09 and 28 with the high resolution spectrograph (HRS; [8]) at the Hobby-Eberly Telescope at McDonald Observatory as part of normal queue mode scheduled observing. Our spectra have $R \sim 15{,}000$ and cover the range from 4120 to 7850 Å. One 10 min exposure was taken on each night. Data reduction was performed both with standard IRAF routines and a data pipeline structured around the REDUCE software package [9]. The final S/N values per pixel range from ~ 50 at 4480 Å to ~ 130 at 5730 Å to ~ 160 at 6700 Å.

To check for radial velocity variations, two supplemental exposures were taken on 2007 July 15 and 2007 August 05 with the Cross-Dispersed Echelle Spectrometer (2dCoude; [10]) on the Harlan J. Smith 2.7 m Telescope at $R \sim 60{,}000$. The weighted average velocity is 38.9 km s^{-1} with an error of the mean of 0.3 km s^{-1}. All of the velocity measurements are within 1 σ of the average velocity. We find no evidence from these radial velocity measurements that HK-II 17435-00532 is in a binary system. If it were in a binary, it would have a very long period and/or an amplitude smaller than ~ 1.0 km s^{-1}, a result of a face-on orbit with very large $\sin i$.

We derive the effective temperature of HK-II 17435-00532 from recent color-$T_{\rm eff}$ calibrations [11]. The surface gravity and microturbulence were derived by the usual spectroscopic methods. We find $T_{\rm eff}/\log(g)/v_t/$[Fe/H] $= 5200\,{\rm K}/2.15/2.0\,{\rm km\,s^{-1}}/-2.23$. We perform our analyses using the LTE spectrum analysis code MOOG [12].

In Table 1 we list our derived LTE abundances, along with non-LTE corrections when available (Li: [13]; O: [14], [15]; Na: [16]; Mg: [17], [18]; K: [19]), for 23 elements in HK-II 17435-00532. We reference the elemental abundance ratios to the solar photospheric abundances [20].

We derive the Li abundance from a synthesis of the Li I 6707 Å doublet. We synthesize

TABLE 1. Abundance summary for HK-II 17435-00532

Species	LTE log ε	NLTE log ε	LTE [X/Fe]	σ	No. lines	log $\varepsilon_{\odot}$	Notes
Fe I	5.27		−2.23	0.23	86	7.50	EW
Fe II	5.27		−2.23	0.17	6	7.50	EW
Li I	2.06	2.16		0.16	1	1.10	synth
C	6.97		+0.68	0.3		8.52	synth
O I	7.56	∼7.4	+1.06	0.27	3	8.73	EW
Na I	4.79	∼4.7	+0.69	0.16	2	6.33	EW
Mg I	5.77	∼5.9	+0.42	0.25	4	7.58	EW
K I	3.33	∼2.9	+0.44	0.17	1	5.12	EW
Ca I	4.56		+0.43	0.19	13	6.36	EW
Sc II	1.14		+0.20	0.21	3	3.17	EW
Ti I	3.01		+0.22	0.22	9	5.02	EW
Ti II	3.12		+0.33	0.20	9	5.02	EW
Cr I	3.29		−0.15	0.16	4	5.67	EW
Mn I	2.85		−0.31	0.24	1	5.39	EW
Ni I	4.13		+0.11	0.18	3	6.25	EW
Sr II	1.05		+0.36	0.26	1	2.92	synth
Y II	0.18		+0.17	0.19	4	2.24	synth
Zr II	1.14		+0.77	0.27	2	2.60	synth
Ba II	0.76		+0.86	0.29	4	2.13	synth
La II	−0.28		+0.78	0.21	6	1.17	synth
Ce II	0.52		+1.17	0.20	4	1.58	synth
Pr II	−0.42		+1.10	0.20	3	0.71	synth
Nd II	0.31		+1.04	0.19	12	1.50	synth
Sm II	−0.49		+0.73	0.22	3	1.01	synth
Eu II	−1.24		+0.48	0.20	2	0.51	synth

portions of the CH G-band between 4290 and 4330 Å to determine the C abundance. We cannot estimate the $^{12}C/^{13}C$ ratio from the CH G-band at our spectral resolution and S/N levels. We measure equivalent widths for the O I triplet at 7771, 7774, and 7775Å. The Na abundance is derived from equivalent width measurements of the Na I lines at 5682 and 5688 Å. We measure abundances for 10 n-capture elements (Sr, Y, Zr, Ba, La, Ce, Pr, Nd, Sm and Eu) in HK-II 17435-00532 by generating synthetic spectra for each line. Hyperfine splitting structure was accounted for when necessary. We can only measure an upper limit for Tc, and our spectrum does not cover the Pb I line at 4057 Å.

3. EVOLUTIONARY STATUS

The location of HK-II 17435-00532 on the $T_{\rm eff}$-log(g) diagram is shown in Figure 1. HK-II 17435-00532 coincidentally appears to lie on the $M = 0.80 M_{\odot}$ horizontal branch (HB) track; however, due to the large uncertainties in our determination of the temperature and surface gravity of this star, we can not draw any firm conclusions from this. This star could possibly be ascending the red giant branch (RGB) for the first time or ascending the early asymptotic giant branch (AGB) from the HB. From fundamental relations, we calculate $\log(L/L_{\odot}) = 2.01 \pm 0.50$ for HK-II 17435-00532.

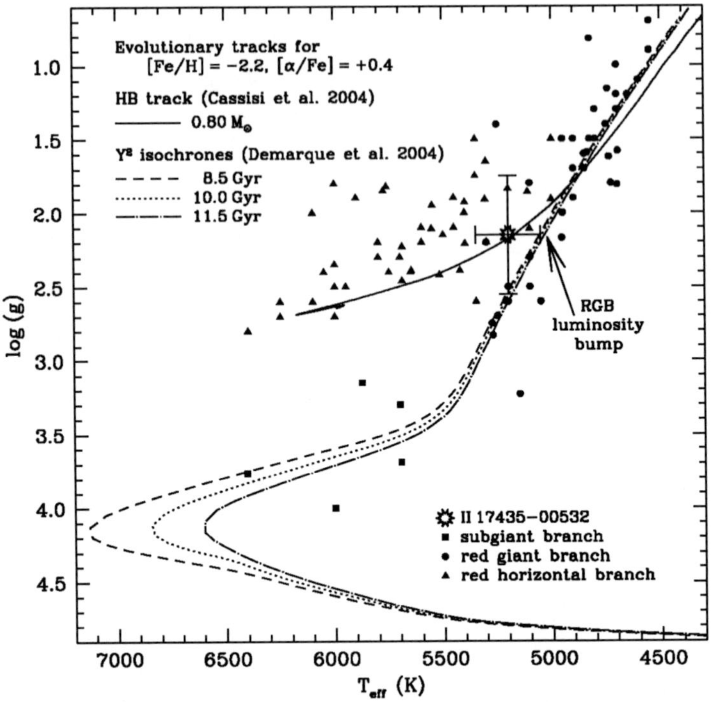

FIGURE 1. Spectroscopic gravities are shown as a function of effective temperature for HK-II 17435-00532 and a sample of other evolved metal-poor stars from previous studies [21, 22, 23]. Three sets of Y^2 isochrones [24] are displayed, along with a synthetic HB track [25]. The position of the RGB luminosity bump is indicated.

4. RESULTS

Our derived Li abundance in HK-II 17435-00532, $\log \varepsilon\,(\mathrm{Li}) = 2.1$, is consistent with the Li abundance of unevolved metal-poor stars on the Spite Plateau, but this star is far too cool to lie on the Spite Plateau. High levels of Li have been observed in other evolved stars, too, but most of these stars have metallicities close to solar. The surface ^{7}Li abundance in dwarf stars is depleted during main sequence core H-burning, and is has been shown [26] that Li abundances in metal-poor subgiants continue to decline by an additional factor of $\sim 10^1$–10^2 relative to the standard evolutionary model predictions as these stars ascend the RGB and AGB. The derived ^{7}Li abundance in HK-II 17435-00532 cannot be its zero-age main sequence ^{7}Li abundance. There are several reasons to suggest that the Li in HK-II 17435-00532 was not transferred from an undetected companion. Even if HK-II 17435-00532 is the secondary star of a binary system where the now-unseen and presumably more massive primary swelled in size during its TP-AGB phase and transferred mass—including freshly-synthesized Li—to the secondary, such a scenario would require extremely efficient mass transfer to enrich HK-II 17435-00532 to $\log \varepsilon\,(\mathrm{Li}) \sim 2.1$. Measurements of C and s-process abundances in other Li-rich subgiants do not suggest that the Li-enrichment has come from mass transfer from a companion that passed through the AGB phase, implying that Li enrichment by a factor of ~ 10 via mass transfer is rare if not impossible. We conclude that the Li should have originated within this star.

C and O are overabundant in HK-II 17435-00532, with [C/Fe] $= +0.7$ and [O/Fe] $= +1.1$. While the enhanced C may signal some enrichment has occurred, the O abundance in HK-II 17435-00532 appears typical for metal-poor stars and does not suggest that any extra enrichment of O has occurred [22]. A comparison of HK-II 17435-00532 with the ten most metal-rich stars ($-2.8 \leq$ [Fe/H] ≤ -2.0) in the sample of [27] suggests that all α- and Fe-peak abundances are consistent with "typical" metal-poor stars. The exception is Na, which is clearly enhanced in HK-II 17435-00532, [Na/Fe] $= +0.6$.

HK-II 17435-00532 exhibits overabundances of all n-capture elements relative to Fe when compared with the Solar System abundance ratios. Both Ba and Eu exhibit significant overabundances in HK-II 17435-00532, [Ba/Fe] $= +0.86$ and [Eu/Fe] $= +0.48$. These ratios suggest that HK-II 17435-00532 is enriched to some degree in both s- and r-process material, although neither the scaled-solar s- nor r-process abundances alone provide a satisfactory fit to the abundance pattern in HK-II 17435-00532. [28] analyzed a number of metal-poor stars exhibiting the $r+s$ enrichment pattern ([Ba/Fe] $> +1.0$, [Eu/Fe] $> +1.0$). Even though HK-II 17435-00532 has less overall n-capture enrichment, the n-capture abundance pattern between this star and the $r+s$ stars in their sample is very similar. Given the metallicity of HK-II 17435-00532, we can safely surmise that this star is very old (~ 10 Gyr) and has a low mass ($\sim 0.8 M_{\odot}$). The s and r nucleosynthesis reactions are not expected to operate in stars of such low mass, so they must have been present in the material from which this star formed or were transferred to it from an undetected binary companion.

5. INTERPRETATION AND DISCUSSION

We first consider the case of Li self-enrichment on the RGB. We investigate the possibility that the Li-enrichment has occurred because an extra mixing mechanism may be in operation, connecting the outer convective envelope with the H-burning shell, thus enabling fresh Li production via the Cameron-Fowler mechanism. To explain the Li enrichment observed in some evolved stars, it has been proposed [29, 30] that an extra, yet unspecified, mixing process may occur in low- and intermediate-mass stars (i.e., $1 < M < 5 M_{\odot}$) after the molecular weight discontinuity has been erased at the RGB luminosity bump. 3D simulations of the He-flash [31] also suggest that this extra mixing may be present outside the H-burning shell. This hypothesis explains the low- and intermediate-mass Li-enriched stars found at the RGB luminosity bump. As stars evolve further up the RGB and away from the luminosity bump, the outer convective envelope deepens further, carrying the ^{7}Li to these hotter regions where it is destroyed. The Li rich phase is fleeting; from the calculations for Li-enrichment in low-mass, metal-poor stars along the RGB [32], we estimate that the Li-rich phase will not last more than 3–4 Myr.

Thermohaline mixing has been proposed to describe the extra mixing on physical grounds [33, 34, 35]. Another possibility, the Li flash [36], could also explain Li enrichment at the RGB luminosity bump. If HK-II 17435-00532 is ascending the RGB, the luminosity of this star is coincident with the luminosity of the bump for a metallicity of [Fe/H] $= -2.2$. It is possible that the present Li abundance has already been reduced from its maximum abundance. If HK-II 17435-00532 is evolving through the RGB luminosity bump, it is the most metal-poor star for which an extra mixing mechanism has been shown to produce Li enrichment in the stellar envelope.

Enrichment of Li is not expected to occur on the RHB or early-AGB; if HK-II 17435-00532 is on the RHB or early AGB, we are left to postulate that a previously-unidentified efficient extra mixing episode may be operating during this stage of evolution in low-mass, low-metallicity stars.

Like HK-II 17435-00532, several other $r+s$ stars have been found with enhanced Na abundances. The C, Na, and s-process enhancements can all be explained by assuming material was transferred to HK-II 17435-00532 from an undetected binary companion that passed through the thermally-pulsing (TP) AGB phase. In Figure 2 we display the measured [X/Fe] abundance ratios for $6 \leq Z \leq 63$ in HK-II 17435-00532, along with the predicted abundance ratios assuming the transfer of material from a binary companion. These calculations were based on the FRANEC stellar evolution models [37, 38] for low-metallicity stars on the AGB.

The best-fit is obtained with the model that adopts an initial mass of $1.5 M_{\odot}$ for the companion star. We derive a dilution factor of 1.8 dex (i.e., a factor of ≈ 63), which means that one part of accreted material is present in the stellar envelope of HK-II 17435-00532 for every 63 parts of material originally present in HK-II 17435-00532. If we assume an initial [r/Fe] $= +0.3$ for this system, then only ~ 0.2 dex of Eu needs to be acquired from the s-process material of the AGB companion. In contrast to the Eu, which mostly reflects the initial composition of the ISM from which the system formed, the Ba (and other n-capture species observed to be overabundant in this star) was dominantly produced by the companion star during its TP-AGB phase and transferred to HK-II 17435-00532 by s-process enriched winds.

6. CONCLUSIONS

We have performed the first detailed abundance analysis on HK-II 17435-00532. A surprising result of our study is the high Li abundance, $\log \varepsilon\,(\mathrm{Li}) = +2.1$, which is not predicted by standard stellar models for a low-mass, low-metallicity, evolved star such as HK-II 17435-00532. If this star is evolving up the RGB, it is located near the RGB luminosity bump, where extra mixing has been found to cause a short-lived phase of Li enrichment in more massive, metal-rich stars. In this case, HK-II 17435-00532 would represent the most metal-poor object known for which this extra mixing phenomenon occurs. If the star is on the HB or early AGB, any Li that was present at earlier times should have been destroyed. In this case, we propose that an analog of the extra mixing that occurs at the RGB luminosity bump may be occurring on the HB/early AGB. Future theoretical analysis of this point would be necessary to confirm or refute this hypothesis.

Our abundance analysis also finds that C and Na are slightly enhanced. The abundances of O, the α-elements, and the Fe-peak elements all appear typical for stars with metallicity similar to HK-II 17435-00532. This star is enriched with both r- and s-process elements. These enrichments can be explained by assuming mass transfer from

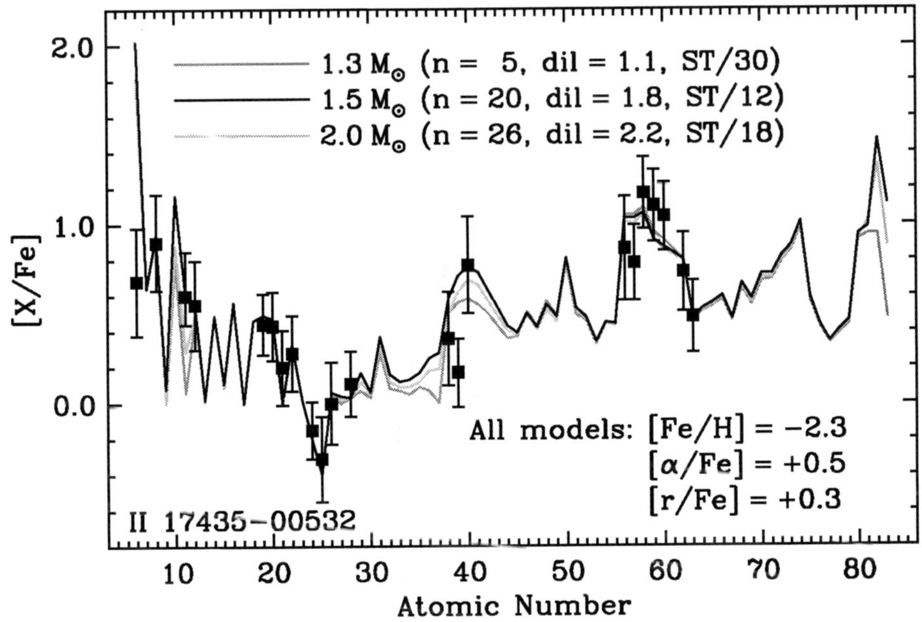

FIGURE 2. Predicted [X/Fe] ratios in HK-II 17435-00532 assuming pollution from a companion star that passed through the TP-AGB phase. The initial mass of the TP-AGB star is the primary variable between the different sets of abundance predictions, although changing the mass also changes the number of thermal pulses ("*n*") and necessitates altering the logarithmic dilution factor ("dil") and ^{13}C pocket efficiency ("ST/"). The black curve reflects our best fit model.

a binary companion with an initial mass of $1.5M_{\odot}$. This scenario would only require that a small initial enrichment of *r*-process material was present in the ISM from which this star formed. Although there is no evidence for radial velocity variations (and hence binarity) at present, radial velocity followup is ongoing.

ACKNOWLEDGMENTS

Funding for this project has been generously provided by the U. S. National Science Foundation (grants AST 06-07708 to C. S.; AST 04-06784, AST 07-07776, and PHY 02-15783 to T. C. B.; and AST 07-07447 to J. J. C.). A. F. acknowledges support through the W. J. McDonald Fellowship of the McDonald Observatory. J. R. gratefully acknowledges partial research support for this work by NASA through the AAS Small Research Grant Program and the GALEX GI grant 05-GALEX05-27. R. G. acknowledges the Italian MIUR-PRIN06 Project "Late phases of Stellar Evolution: Nucleosynthesis in Supernovae, AGB Stars, Planetary Nebulae" for support. T. C. B. acknowledges partial support from the Physics Frontier Center / Joint Institute for Nuclear Astrophysics (JINA).

REFERENCES

1. F. Spite and M. Spite, *A&A* **115**, 357 (1982)
2. C. Charbonnel and F. Primas, *A&A* **442**, 961 (2005)
3. D. N. Spergel, et al. *ApJS* **170**, 377 (2007)
4. A. G. W. Cameron and W. A. Fowler, *ApJ* **164**, 111 (1971)
5. J. Rhee, *PASP* **113**, 1569 (2001)
6. T. C. Beers, G. W. Preston, and S. A. Shectman, *AJ* **90**, 2089 (1985)
7. T. C. Beers, G. W. Preston, and S. A. Shectman, *AJ* **103**, 1987 (1992)
8. R. G. Tull, *Proc. SPIE* **3355**, 387 (1998)
9. N. E. Piskunov and J. A. Valenti, *A&A* **385**, 1095 (2002)
10. R. G. Tull, P. J. MacQueen, C. Sneden, and D. L. Lambert, *PASP* **107**, 251 (1995)
11. A. Alonso, S. Arribas, and C. Martínez-Roger, *A&AS* **140**, 261 (1999)
12. C. A. Sneden, Ph.D. Thesis, Univ. of Texas at Austin (1973)
13. M. Carlsson, R. J. Rutten, J. H. M. J. Bruls, and N. G. Shchukina, *A&A* **288**, 860 (1994)
14. F. Primas, M. Asplund, and D. L. Lambert, *A&A* **390**, 235 (2002)
15. A. E. García Pérez, M. Asplund, F. Primas, P. E. Nissen, and B. Gustafsson, *A&A* **451**, 621 (2006)
16. Y. Takeda, G. Zhao, M. Takada-Hidai, Y.-Q. Chen, Y.-J. Saito, and H.-W. Zhang, *Chinese J. of Astr. Astrophys.* **3**, 316 (2003)
17. R. G. Gratton, E. Carretta, K. Eriksson, and B. Gustafsson, *A&A* **350**, 955 (1999)
18. T. Gehren, Y. C. Liang, J. R. Shi, H. W. Zhang, and G. Zhao, *A&A* **413**, 1045 (2004)
19. Y. Takeda, G. Zhao, Y.-Q. Chen, H.-M. Qiu, and M. Takada-Hidai, *PASJ* **54**, 275 (2002)
20. N. Grevesse and A. J. Sauval, *Adv. Space Res.* **30**, 3 (2002)
21. B. B. Behr, *ApJS* **149**, 101 (2003)
22. R. Cayrel, et al., *A&A* **416**, 1117 (2004)
23. G. W. Preston, C. Sneden, I. B. Thompson, S. A. Shectman, and G. S. Burley, *AJ* **132**, 85 (2006)
24. P. Demarque, J.-H. Woo, Y.-C. Kim, and S. K. Yi, *ApJS* **155**, 667 (2004)
25. S. Cassisi, M. Castellani, F. Caputo, and V. Castellani, *A&A* **426**, 641 (2004)
26. C. A. Pilachowski, C. Sneden, and J. Booth, *ApJ* **407**, 699 (1993)
27. A. McWilliam, G. W. Preston, C. Sneden, and L. Searle, *AJ* **109**, 2757 (1995)
28. K. Jonsell, P. S. Barklem, B. Gustafsson, N. Christlieb, V. Hill, T. C. Beers, and J. Holmberg, *A&A* **451**, 651 (2006)
29. C. Charbonnel, *ApJ* **453**, L41 (1995)
30. C. Charbonnel and S. C. Balachandran, *A&A* **359**, 563 (2000)
31. D. S. P. Dearborn, J. C. Lattanzio, and P. P. Eggleton, *ApJ* **639**, 405 (2006)
32. P. A. Denissenkov and F. Herwig, *ApJ* **612**, 1081 (2004)
33. P. P. Eggleton, D. S. P. Dearborn, and J. C. Lattanzio, *Science* **314**, 1580 (2006)
34. P. P. Eggleton, D. S. P. Dearborn, and J. C. Lattanzio, astro-ph/0706.2710 (2007)
35. C. Charbonnel and J.-P. Zahn, *A&A* **467**, L15 (2007)
36. A. Palacios, C. Charbonnel, and M. Forestini, *A&A* **375**, L9 (2001)
37. O. Straniero, A. Chieffi, M. Limongi, M. Busso, R. Gallino, and C. Arlandini, *ApJ* **478**, 332 (1997)
38. O. Straniero, I. Domínguez, S. Cristallo, and R. Gallino, *PASA* **20**, 389 (2003)

Chemical Diversity of Post-AGB Stars in the LMC

M. Reyniers*,† and H. Van Winckel*

*Instituut voor Sterrenkunde, Departement Natuurkunde en Sterrenkunde, K.U.Leuven, Celestijnenlaan 200D, 3001 Leuven, Belgium
†present address: Koninklijk Meteorologisch Instituut van België, Departement Waarnemingen, Ringlaan 3, 1180 Brussel, Belgium

Abstract. We report on the chemical analysis based on UVES-VLT spectra of an exploratory sample of five post-AGB stars in the Large Magellanic Cloud. The sample reveals a wide chemical diversity, similar to the diversity seen in *Galactic* post-AGB stars. One object is a post-3rd dredge-up object, heavily enhanced in carbon and s-process elements; one object is carbon enhanced, but does not show s-process enhancements, and its chemistry is therefore not understood. The other three objects show some degree of *depletion* in their photosphere: the refractory elements are underabundant with respect to the non-refractories. This depletion pattern is understood as the result of a chemical process in which a gas-dust separation in the circumstellar environment is followed by a photospheric accretion of only the gas. In the Galaxy, the efficiency of this process is dependent on the initial metallicity. Here we show that at LMC metallicities, the depletion process can be very efficient. The most favourable circumstance for this process to occur is, if the circumstellar dust is trapped in a disk. The presence of a disk around those three stars is indeed revealed by IR measurements of Spitzer. The disk is likely related to binarity (see contributions of Gielen and Van Winckel in these proceedings).

Keywords: Stars: AGB and Post-AGB – Stars: Abundances – Stars: Carbon – Magellanic Clouds
PACS: 97.10.Cv, 97.10.Ex, 97.10.Gz, 97.10.Tk, 97.20.Tr, 97.30.Hk, 97.30.Kn, 97.80.Fk

1. CHEMICAL DIVERSITY OF GALACTIC POST-AGB STARS

In this introductory section, we give a very concise overview of the results on the chemical abundance studies of *Galactic* post-AGB stars. During the past decennia, it has been realised that these post-AGB stars are chemically much more diverse than previously thought. A first very clear distinction has to be made between the *single* objects and the *binary* objects: Binary objects appear to have a totally different chemical signature than single objects. Both subgroups are, in their turn, also not chemically homogeneous. In this section, we will first summarize the results on the binary objects, and then give an overview of the results on the single objects.

1.1. Galactic Binary Post-AGB and RV Tauri Stars

Most Galactic binary post-AGB stars show some degree of *depletion* of refractory elements in their photosphere: elements with a high dust condensation temperature are underabundant, e.g. [1, 2]. A self-consistent evolutionary scenario for the depletion phenomenon is not yet developed, but a rough sketch is as follows [3]. When the

CP1001, 9th Torino Workshop on Evolution and Nucleosynthesis in AGB Stars and the 2nd Perugia Workshop on Nuclear Astrophysics, edited by R. Guandalini, S. Palmerini, and M. Busso

AGB star evolves into an orbit which is too small to accommodate its increase in radius, a phase of strong binary interaction will occur which results in the creation of a stable circumstellar or circumbinary disk [4, 5]. In this disk, dust formation will lead to a chemical fractionation in which elements with a high condensation temperature will preferentially condense on dust grains, while elements with a low condensation temperature will stay in the gas phase. The cleaned gas separates from the dust and re-accretes onto the star, making the photosphere devoid of refractory elements. The observational evidence for such a scenario is that around objects in which depletion patterns are observed, a stable and probably circumbinary disk is present indeed [6].

RV Tauri stars are a subclass of the Pop. II Cepheids, probably in the post-AGB stage of evolution, as first argued by [7]. Many of the RV Tauri stars show an infrared excess, which is caused by the circumstellar dust that was likely formed by heavy mass loss during the preceding AGB phase. All members of the RVB spectroscopic subgroup of the RV Tauri stars show also some degree of depletion in their photosphere. Our radial velocity monitoring of these RV Tauri that is still ongoing, proves that the binary fraction is indeed very high [2]. In this picture, the IR excess as observed around RV Tauri stars is coming from a disk and not an outflow. Note that depletion does not seem to occur in globular cluster RV Tauri stars, or in field RV Tauri stars of low metallicity ($[Fe/H] < -1$), even when the disk is present.

1.2. Galactic Single Post-AGB Stars

Other post-AGB objects with a spectral energy distribution (SED) indicative of a cooling and expanding detached shell, but without radial velocity information and hence presumably *single*, are also far from chemically homogeneous. Some objects are the most s-process enriched objects known to date (e.g. [8], and references therein) while others are not enriched at all [9]. This dichotomy is very strict, in the sense that mildly enhanced objects do not exist, except for a few rather atypical objects (e.g. [10, 11]). For the moment it is not clear whether this dichotomy is caused by a selection effect, or if it is the result of two different evolutionary channels. Furthermore, the s-process enriched sub-class of post-AGB objects imposes another unsolved problem. The s-process enriched sub-group exhibits a large spread in s-process efficiency, as these stars do not obey the expected anti-correlation between metallicity and s-process efficiency. In other words, examples exist of post-AGB stars with a very similar metallicity, but with a totally different s-process neutron irradiation.

2. THE LMC SAMPLE

In this contribution, we report on the abundance results of an exploratory sample consisting of the five brightest RV Tauri stars in the LMC found by the MACHO experiment [12]. This LMC sample of RV Tauri stars with their known luminosities is a unique sample to study the nature of these stars and of the post-AGB evolution in general. Moreover, abundance analyses of post-3rd dredge-up stars in the LMC should make it possible to study the yields of the AGB nucleosynthesis in a more metal deficient environment than

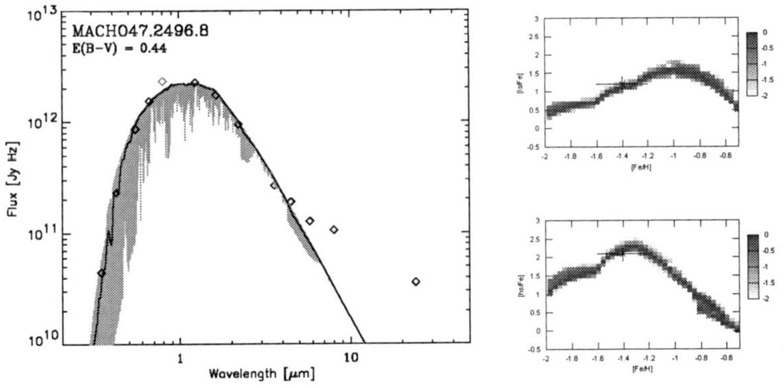

FIGURE 1. The Spectral Energy Distribution (SED) of MACHO 47.2496.8 on the left; the recent fits to its abundances [14] on the right. The SED shows a clear infrared excess, corroborating the post-AGB status of this object in which the excess is a relic of a recent phase of dusty mass loss during the AGB. The colours of this excess are different from the disk sources in Fig. 4. On the right, the [ls/Fe] and [hs/Fe] values are fitted with the recent models of [14]. These models combine stellar population synthesis with s-process nucleosynthesis using a synthetic stellar evolution code, and are described in detail in [15].

the Galaxy. High resolution, high signal-to-noise optical spectra of our exploratory LMC sample were taken with the UVES spectrograph mounted on the VLT-UT2 (Kueyen) telescope. More details on the observations and the reduction can be found in [13].

2.1. MACHO 47.2496.8: a Post-3rd Dredge-up Object

One object of our sample, MACHO 47.2496.8, is a true post-3rd dredge-up object displaying strong carbon and s-process enrichments. Its SED (Fig. 1, *left panel*) shows a clear IR-excess, caused by the mass loss in the preceding AGB-phase. The abundance results of this object are extensively discussed in a dedicated paper [13]. Here, we only note its remarkably low intrinsic metallicity of [Fe/H] = −1.4, and its overall enrichment of [ls/Fe] = +1.2 and [hs/Fe] = +2.1 dex for the light (Zr-peak) and heavy (Ba-peak) s-process elements respectively.

The abundance pattern of MACHO 47.2496.8 has been recently modelled in a letter [14]. The key figure of this letter is shown in the *right panel* of Fig. 1. The grey scale is a logarithmic measure of the number distribution of stars over the plotted s-process index (the number density weighed by the initial mass function and by the time each star spends in an abundance bin, and then normalized for each metallicity). The crosses are the MACHO 47.2496.8 data with their errorbar. The authors obtained a perfect fit for the [ls/Fe] and [hs/Fe] s-process ratios with the same values of $\lambda_{\min}$ (the minimum asymptotic TDU efficiency, see [15]), the mass of the ^{13}C pocket and the minimum core-mass for the 3rd dredge-up that was found for the Galactic post-AGB stars. However, using this set of parameters, no solution was found for the C/O ratio and the ^{12}C/^{13}C ratio, which might be an indication for extra mixing.

TABLE 1. Abundance results for MACHO 79.5501.13, MACHO 81.9728.14 and MACHO 81.8520.15 (the abundance tables of MACHO 47.2496.8 and MACHO 82.8405.15 can be found in resp. [13] and [16]). The explanation of the columns is as follows: N gives the number of lines used; $\log \varepsilon$ is the absolute abundance derived $\log \varepsilon = \log(N(X)/N(H)) + 12$; σ_{ltl} is the line-to-line scatter; and [el/H] is the abundance relative to the sun. For the references of the solar abundances needed to calculate the [el/H] values: see [13]. The dust condensation temperatures are taken from [17], and references therein. They are computed using a solar abundance mix at a pressure of 10^{-4} atm.

	MACHO 79.5501.13 $T_{eff} = 5750$ K $\log g = 0.5$ (cgs) $\xi_t = 3.0$ km s^{-1}				MACHO 81.9728.14 $T_{eff} = 5500$ K $\log g = 1.0$ (cgs) $\xi_t = 3.0$ km s^{-1}				MACHO 81.8520.15 $T_{eff} = 6250$ K $\log g = 1.0$ (cgs) $\xi_t = 3.0$ km s^{-1}					
ion	N	$\log \varepsilon$	σ_{ltl}	[el/H]	N	$\log \varepsilon$	σ_{ltl}	[el/H]	N	$\log \varepsilon$	σ_{ltl}	[el/H]	sun	T_{cond}
C I	6	7.65	0.12	-0.92	2	7.52	0.04	-1.05	8	8.08	0.08	-0.49	8.57	40.
Na I					2	5.07	0.21	-1.26					6.33	958.
Mg I	2	6.09	0.33	-1.45	1	6.78		-0.76	2	6.34	0.05	-1.20	7.54	1336.
Si I	2	6.48	0.20	-1.06	4	6.69	0.12	-0.85					7.54	1310.
Si II									2	6.18	0.06	-1.36	7.54	1310.
S I	5	6.71	0.06	-0.62									7.33	664.
Ca I	2	4.67	0.13	-1.69	9	5.25	0.14	-1.11	12	4.86	0.19	-1.50	6.36	1517.
Sc II	1	0.19		-2.98	2	1.15	0.17	-2.02	5	1.23	0.20	-1.94	3.17	1659.
Ti I									1	3.13		-1.89	5.02	1582.
Ti II	6	2.20	0.25	-2.82	4	3.26	0.29	-1.76	16	3.03	0.18	-1.99	5.02	1582.
Cr I	3	3.85	0.02	-1.82	3	4.39	0.30	-1.28	4	3.91	0.16	-1.76	5.67	1296.
Cr II	10	3.76	0.16	-1.91	11	4.70	0.18	-0.97	12	4.05	0.19	-1.62	5.67	1296.
Mn I	6	4.34	0.21	-1.05	2	4.32	0.31	-1.07	1	3.58		-1.81	5.39	1158.
Fe I	41	5.85	0.15	-1.66	73	6.43	0.16	-1.08	122	5.92	0.17	-1.59	7.51	1334.
Fe II	14	5.56	0.22	-1.95	14	6.47	0.15	-1.04	23	5.95	0.15	-1.56	7.51	1334.
Ni I	2	4.27	0.22	-1.98	10	4.99	0.11	-1.26	6	4.47	0.12	-1.78	6.25	1353.
Zn I	3	3.98	0.04	-0.62	1	3.38		-1.22	1	3.21		-1.39	4.60	726.
Y II	1	-0.26		-2.50	1^{ss}	0.49		-1.75	2	0.11	0.01	-2.13	2.24	1659.
Zr II					1	0.74		-1.86					2.60	1741.
Ba II									4	0.23	0.06	-1.90	2.13	1455.
Ce II	1	-0.28		-1.86									1.58	1478.
Eu II					2	0.12	0.08	-0.40					0.52	1356.

ss: spectrum synthesis

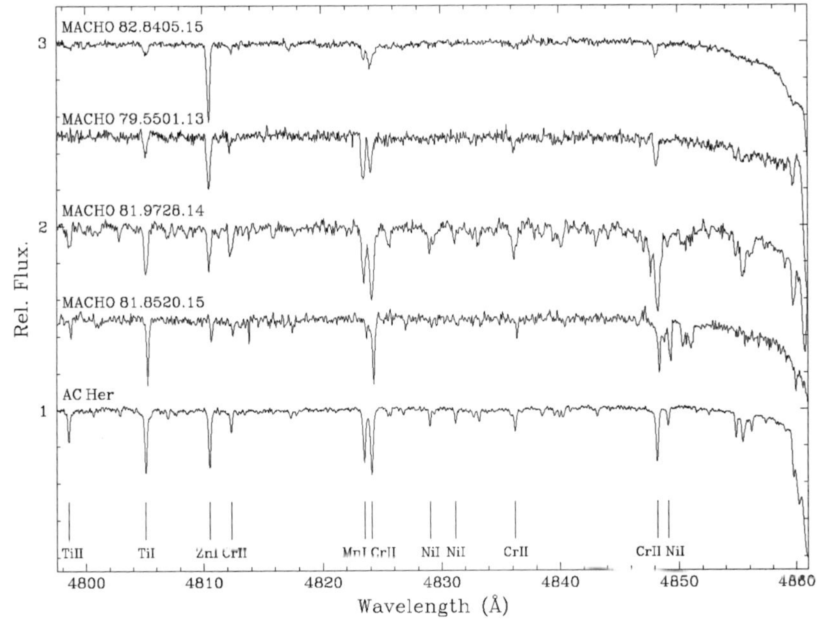

FIGURE 2. Sample spectra of the objects of our exploratory LMC sample (except MACHO 47.2496.8), together with the spectrum of AC Her, a Galactic depleted post-AGB star [18, 19]. [Zn/Fe-peak] and [S/Ti] ratios are diagnostics for depletion, since they are elements with the same chemical history, but with different condensation temperatures. Note the strong Zn line compared to other lines of e.g. Ti or Cr, in the spectra of MACHO 82.8405.15 and MACHO 79.5501.13.

2.2. Three Depleted Objects

In three out of five objects, we found a clear anti-correlation in the abundance results between the abundances of the elements and their condensation temperatures (Fig. 3), indicating that the depletion phenomenon has taken place. In fact, this is already clear by a simple inspection of the spectra shown in Fig. 2. [Zn/Fe-peak] and [S/Ti] ratios are diagnostics for depletion, since they are elements with the same chemical history, but with different condensation temperatures. MACHO 82.8405.15 is the most heavily depleted source of the three, with [Fe/H]=-2.6, in combination with [Zn/Fe] = $+2.3$ and [S/Ti] = $+2.5$. This degree of the depletion is comparable to the strongest depletions seen in field Galactic RV Tauri stars. The results of this star are therefore published in a dedicated letter [16]. MACHO 79.5501.13 is also quite strongly depleted, with [Fe/H] = -1.9 and [Zn/Fe] = $+1.3$. The third object, MACHO 81.9728.14, is only marginally depleted, but still the elements with the highest condensation temperature ($>$ 1500 K) are underabundant by $\sim$0.8 dex. If the Zn abundance is interpreted as the initial metallicity of the object, then it seems that the depletion process is less efficient for more metal deficient objects. This effect is also seen in Galactic depleted objects, although the initial

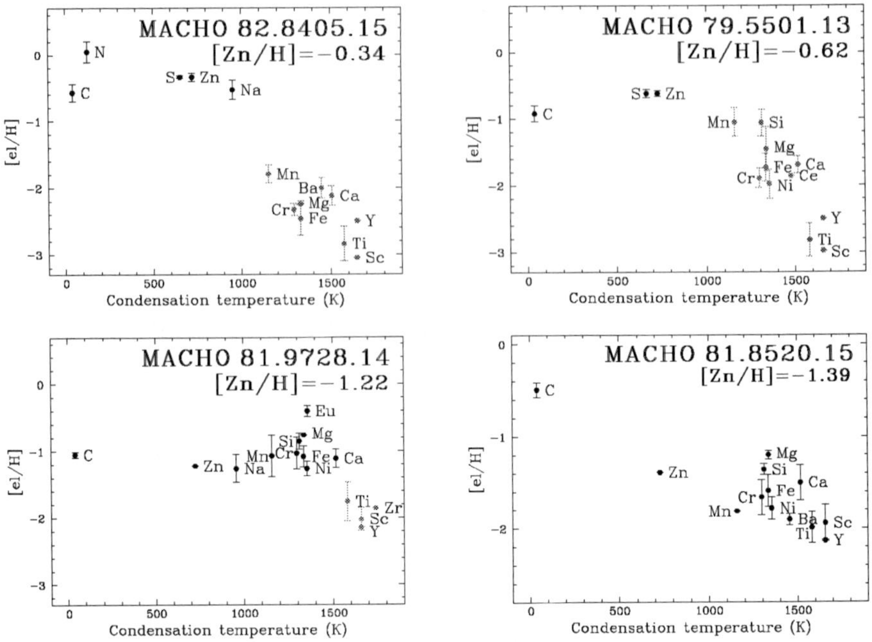

FIGURE 3. Abundances of the three depleted objects (MACHO 82.8405.15, MACHO 79.5501.13 and MACHO 81.9728.14) and MACHO 81.8520.15. The clear anti-correlation between the abundance of an element and its condensation temperature for the first three objects proves that the depletion process took place in these objects. Moreover, the initial metallicity [Zn/H] seems to determine the depletion efficiency, with higher metallicities leading to stronger depletions. The abundance pattern of MACHO 81.8520.15 is not easily understood, as this star shows an enhanced carbon content without s-process enrichment, and probably no depletion.

metallicity is clearly not the only parameter that determines the depletion efficiency. The SEDs of the three depleted objects (Fig. 4) have the typical shape of a *disk* object. They consist of two components: an optical one from the photosphere, and a hot dust excess in the infrared, indicative for the presence of dust near the star likely in a stable disk.

2.3. MACHO 81.8520.15: Not Understood

The abundance pattern of MACHO 81.8520.15 (Fig. 3) is not easily understood. At least two possible interpretations can be put forward: either all elements (including Zn) are depleted, except carbon, or the star is of low initial metallicity and has enhanced its carbon, but not its s-process elements. The scenario in which all elements, except C, are depleted, is less plausible, since in the SED (Fig. 4) there is no strong indication for the presence of a disk. Nevertheless, a very small but significant IR-excess is present at 8 μm. Moreover, carbon stars without s-process enrichment do exist in the Galaxy: the J-type carbon stars [20]. These J-type carbon stars may be related to this object. Unfortunately, we could not derive a $^{12}C/^{13}C$ ratio on the basis of our optical spectrum.

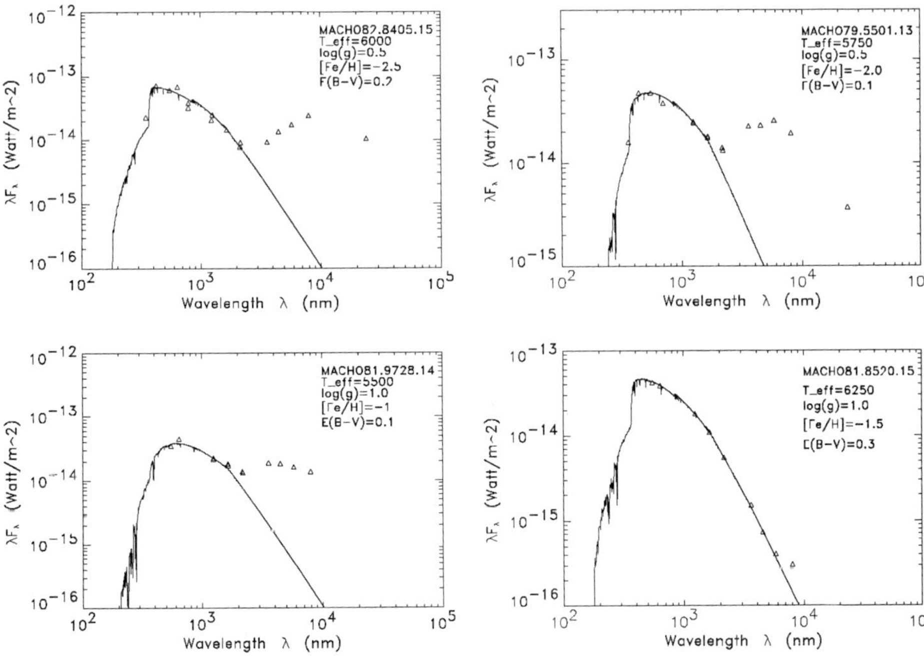

FIGURE 4. Spectral Energy Distributions (SEDs) of the three depleted objects (MACHO 82.8405.15, MACHO 79.5501.13 and MACHO 81.9728.14) and the object MACHO 81.8520.15. The three depleted objects show a double peaked SED, with the infrared excess starting in the near-IR. This hot excess points to the presence of a stable disk containing hot dust around a binary system. The object MACHO 81.8520.15 does not show such an excess, but has only a marginal (but real) excess at 8 μm.

There is not yet a general agreement on possible evolutionary channels leading to these abundance patterns.

3. CONCLUSIONS

This exploratory sample of five post-AGB stars in the Large Magellanic Cloud reveals the same chemical diversity already seen in their Galactic counterparts. One object, MACHO 47.2496.8, is a true post-3rd dredge-up object, heavily enriched in carbon and s-process elements. MACHO 81.8520.15 is C-rich, but not s-process enriched, and is therefore not understood with the current nucleosynthetic AGB models. The other three objects are depleted, showing a stronger depletion for higher initial metallicities. The three depleted objects possess a near IR-excess indicating a stable dust reservoir, probably a disk.

ACKNOWLEDGMENTS

Thank you Clio Gielen for the construction of the SEDs. We would like to thank all people involved in previous collaborations on these data, especially Carlos Abia, Maria Lugaro, Axel Bonačić Marinović and Tom Lloyd Evans. MR acknowledges financial support from the Fund for Scientific Research - Flanders (Belgium).

REFERENCES

1. S. Giridhar, D.L. Lambert, B.E. Reddy, G. Gonzalez, and D. Yong, *ApJ* **627**, 432 (2005)
2. T. Maas, H. Van Winckel, and T. Lloyd Evans, *A&A* **429**, 297 (2005)
3. L.B.F.M. Waters, N.R. Trams, and C. Waelkens, *A&A* **262**, L37 (1992)
4. S. De Ruyter, H. Van Winckel, T. Maas, T. Lloyd Evans, L.B.F.M. Waters, and H. Dejonghe, *A&A* **448**, 641 (2006)
5. P. Deroo, H. Van Winckel, M. Min, L.B.F.M. Waters, T. Verhoelst, W. Jaffe, S. Morel, F. Paresce, A. Richichi, P. Stee, and M. Wittkowski, *A&A* **450**, 181 (2006)
6. H. Van Winckel, *Baltic Astronomy* **16**, 112 (2007)
7. M. Jura, *ApJ* **309**, 732 (1986)
8. M. Reyniers, G.C. Van de Steene, P.A.M. van Hoof, and H. Van Winckel, *A&A* **471**, 247 (2007)
9. H. Van Winckel, *A&A* **319**, 561 (1997)
10. M. Reyniers, and H. Van Winckel, *A&A* **365**, 465 (2001)
11. P. Deroo, M. Reyniers, H. Van Winckel, S. Goriely, and L. Siess, *A&A* **438**, 987 (2005)
12. C. Alcock, R.A. Allsman, D.R. Alves, T.S. Axelrod, A. Becker, D.P. Bennett, K.H. Cook, K.C. Freeman, K. Griest, W.A. Lawson, M.J. Lehner, S.L. Marshall, D. Minniti, B.A. Peterson, K.R. Pollard, M.R. Pratt, P.J. Quinn, A.W. Rodgers, W.Sutherland, A. Tomaney, and D.L. Welch, *AJ* **115**, 1921 (1998)
13. M. Reyniers, C. Abia, H. Van Winckel, T. Lloyd Evans, L. Decin, K. Eriksson, and K.R. Pollard, *A&A* **461**, 641 (2007)
14. A. Bonačić Marinović, M. Lugaro, M. Reyniers, and H. Van Winckel, *A&A* **472**, L1 (2007)
15. A. Bonačić Marinović, R.G. Izzard, M. Lugaro, and O.R. Pols, *A&A* **469**, 1013 (2007)
16. M. Reyniers, and H. Van Winckel, *A&A* **463**, L1 (2007)
17. K. Lodders, *ApJ* **591**, 1220 (2003)
18. H. Van Winckel, C. Waelkens, L.B.F.M. Waters, F.J. Molster, S. Udry, and E.J. Bakker, *A&A* **336**, L17 (1998)
19. C. Gielen, H. Van Winckel, L.B.F.M. Waters, M. Min, and C. Dominik, *A&A* **475**, 629 (2007)
20. C. Abia and J. Isern, *ApJ* **536**, 438 (2000)

Optical Interferometry and C-rich AGB-Stars

Josef Hron[†] , Bernhard Aringer[†,††] , Walter Nowotny[†] and Claudia Paladini[†]

[†]Institute of Astronomy, University of Vienna, Türkenschanzstraße 17, A-1180 Wien, Austria
[††]Department of Astronomy, University of Padua, vic. Osservatorio, 2 I-35122 Padova, Italy

Abstract. We briefly discuss the current capabilities of optical interferometry and its potential to increase our understanding of AGB-stars for such important questions as convection, pulsation, dust formation and mass loss. We then present dynamic model atmospheres for carbon stars and analyze the predicted intensity profiles and visibilities in terms of fits with uniform disks and the current accuracy of visibility measurements. Finally we mention the prospects for real imaging by optical interferometry.

Keywords: Carbon Stars, Optical Interferometry, Stellar Atmospheres, Dynamic Atmospheres.
PACS: 95.75.Kk , 97.10.Ex , 97.30.Hk , 97.30.Jm , 97.10.Sj , 95.30.Ft, 95.30.Jx, 95.30.Lz

1. INTRODUCTION

In the last few years interferometry has become (almost) a mainstream technique in observational astronomy. A major player in this process is the ESO VLT interferometer (VLTI), which is now in routine operation and continuously yields spectacular results (e.g. [1]). The increasing importance of interferometry also comes from a much improved efficiency and flexibility of the instrumentation, the good spectral resolution achievable and the emerging capabilities for true imaging.

AGB stars have always been prime targets of interferometry due to their brightness in the infrared and their large sizes. With respect to the science discussed at this conference, optical interferometry can bring important contributions to the following topics: (i) the measurement of stellar radii which are essential for studying stellar pulsation and evolution, (ii) the derivation of atmospheric stratifications which are needed for proper element abundance determinations and for understanding dust formation and mass loss and (iii) the characterization of surface inhomogeneities which is of direct relevance for convection, dust formation and mass loss.

In this contribution we will concentrate on carbon stars due to several reasons. These objects are of special importance for stellar and galactic evolution and the dust formation is – compared to M type stars – rather well understood, which is an advantage for the construction of model atmospheres. At the same time there are still very few interferometric measurements available (e.g. [2]), thus justifying a thorough investigation of their interferometric properties.

CP1001, 9th Torino Workshop on Evolution and Nucleosynthesis in AGB Stars and the 2nd Perugia Workshop on Nuclear Astrophysics,
edited by R. Guandalini, S. Palmerini, and M. Busso

Finally it should be emphasized that the possibility to simultaneously combine spectroscopy and interferometry with the current instruments is a major step forward as this provides highly complementary information and avoids the usual problems due to the generally pronounced variability of AGB stars.

2. DYNAMIC ATMOSPHERES

While stars evolve along the AGB towards higher luminosities, they become unstable against large amplitude radial pulsations. These may cause shock fronts propagating through the atmosphere, which in turn can lead to a levitation of the whole atmosphere. In cooler regions (polyatomic) molecules and even dust can form. Hydrostatic models are rather limited in reproducing observable properties of such complex and temporally variable atmospheric structures. Satisfactory results (i.e. spectra from the visual to mid IR, narrow band colors) are obtained for red giants with T_{eff} above roughly 3000K, bolometric amplitudes below a few 0.1^{mag} and rather weak mass loss [3]. First results on synthetic intensity profiles and visibilities for such static models have been presented elsewhere [4]. For cooler stars with significant mass loss and more pronounced variability (Miras) dynamic models are needed. Currently, the two most frequently used dynamic models are those developed by Wood et al. (e.g. [5]) and Höfner et al. (e.g. [6]), both being spherically symmetric models. The Wood et al. models are a combination of pulsation models for the stellar interior and model atmospheres but are available only for M-type stars and do not include dust. On the other hand, the Höfner et al. models are based on a combined and self-consistent solution of hydrodynamics, frequency-dependent radiative transfer and a detailed time-dependent treatment of dust formation for C-rich dust but only use a piston to describe the pulsation.

Concentrating on C-stars, we will use the models of Höfner et al. [6]. These models have so far been extensively compared with spectroscopic and photometric data and are approaching quantitative agreement with these observations [7,8]. Common characteristics for this type of models are the following: (i) a (much) larger radial extension of the atmosphere compared to hydrostatic atmospheres, (ii) local density and velocity variations linked with the formation of shock fronts and circumstellar dust, (iii) the need of additional parameters (compared to hydrostatic models) to describe the atmospheres, in particular parameters related to the pulsation and the dust formation and (iv) dust driven mass loss (which works only for strongly pulsating models). The first two points are illustrated in Fig. 1a, which shows the radial density structure of a dynamic model at several instants of the pulsation period.

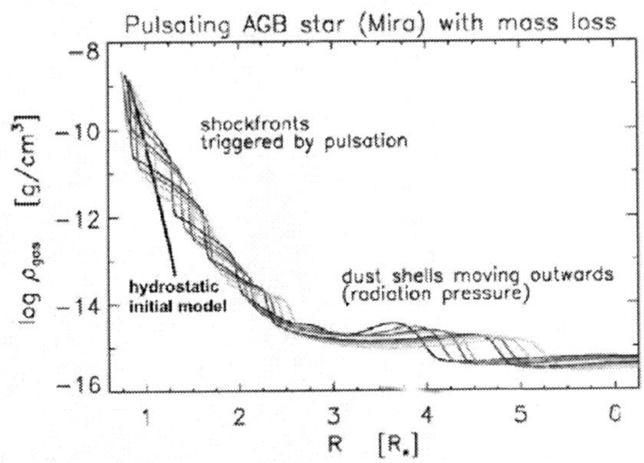

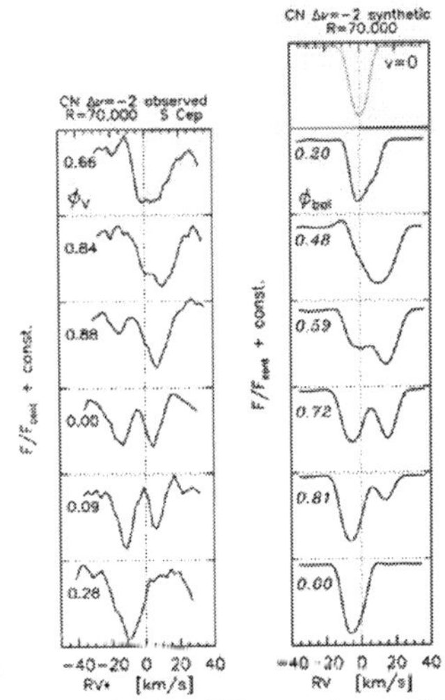

FIGURE 1A (LEFT). Radial density structure of a typical dynamic model for different phases of the pulsation cycle. The structure of the hydrostatic initial model is also shown for comparison.
FIGURE 1B (RIGHT): Comparison of observed (left panel) and synthetic (right panel) line profiles for CN lines in the near IR for different phases of the pulsation cycle. The observations are for the C-rich Mira S Cep, the synthetic spectra are for the model used in this work. The uppermost panel in the right column shows the line profile if the velocity field of the model is not taken into account. Note the good overall agreement between model and observations. Both figures are taken from [9].

For the models the stellar pulsation is simulated by applying a variable inner boundary (piston) moving sinusoidally with a velocity amplitude Δu_P and a period P. A frequency-dependent treatment of the radiation field is used, including both gas and dust opacities. Dust formation for amorphous carbon is treated by the so-called moment method while molecular abundances are computed on the basis of equilibrium chemistry. Opacities for the model construction are treated in the opacity sampling approximation, they include basically all the molecular species relevant for the range of pressures and temperatures, i.e. CO, CH, C_2, CN, C_3, HCN, C_2H_2 and CS. For the dust opacities, only amorphous carbon is considered. The radiative transfer for computing spectra and intensity profiles is done in spherical geometry, the frequency-dependent transfer equation is integrated along characteristic rays. Further details can be found in [6, 8, 9].

For this work we selected one certain dynamic model which can be considered to represent a typical C-rich Mira variable (like e.g. S Cep) and which quite successfully can reproduce KAO and ISO spectra as well as observed infrared CN and CO line profiles of the Mira S Cep (Fig. 1b). The model parameters are the following: 490^d pulsation period, 1.5 solar masses, 7000 solar luminosities, 2600K effective temperature (of the hydrostatic initial model), C/O ratio of 1.4, 1.1^{mag} bolometric luminosity amplitude and 6 km/s piston amplitude Δu_P. The resulting mass loss rate of $2.5 \cdot 10^{-6}$ solar masses per year agrees quite well with the values observed for S Cep while the terminal wind velocity of 7.5km/s is notably lower than observed [9].

3. FROM SPECTROSCOPY TO INTERFEROMETRY

The spectral synthesis tools are also the basis for extending the comparison between models and observations to interferometric data. The solution of the equation of radiative transfer in spherical geometry results in radial intensity profiles for every frequency point of our calculations. The monochromatic visibility profiles are calculated by applying the Fourier transform (which reduces for spherical symmetric distributions to the Hankel transform). Convolving these profiles then with different filter curves yields the needed heterochromatic intensity and visibility profiles. As filters we used the standard Johnson VIJHK filters plus several narrow (20 to 50nm wide) wavelength regions between 1μm and 13μm located inside various molecular bands and in pseudo-continuum regions.

From a comparison of these profiles for different wavelengths and for different phases of the pulsational cycle the following conclusions can be made: (i) the intensity profiles are much more extended than for hydrostatic models (i.e. visibilities are lower at the lower spatial frequencies), (ii) the profiles are markedly different from uniform disks (UD), (iii) the wavelength dependence of visibility within the first lobe of the profile is up to 20% (due to the overall size and the global profile), (iv) the effects due to local density structures are only a few percent in visibility and occur mostly at high spatial frequencies (i.e. beyond the first lobe of the visibility profile) and (v) the phase dependence of the visibilities is less than 10%. In Fig. 2 and 3 we show examples of intensity and visibility profiles illustrating these conclusions. The filter at 1.53μm is representative for all filters mostly affected by CN, CO and C_2 molecular bands, the one at 3.175μm for those dominated by C_2H_2 bands and the one at 9.975μm for regions with strong dust emission (but rather weak molecular absorption).

4. RADII AND THE UNIFORM DISK APPROXIMATION

For stellar evolution and pulsation models, the stellar radius is an important quantity. For such models the most appropriate representation of the stellar radius is the Rosseland radius or a radius measured in the photospheric continuum, corresponding to the deepest layers of the photosphere. However, a comparison of these radii for hydrostatic and dynamic C-star models with radii of (monochromatic) optical depth equals 1 shows, that even for spectral resolutions of several 1000 these deep layers are barely accessible to radius measurements [10]. Furthermore, as discussed above, the intensity profiles of real C-star photospheres are strongly different from uniform disks, complicating the determination of "stellar radii" even further. Thus the radius determination for evolution or pulsation models of C-Miras is a very difficult and perhaps not really solvable problem (see also the extensive discussions for M-Miras e.g. in [11]).

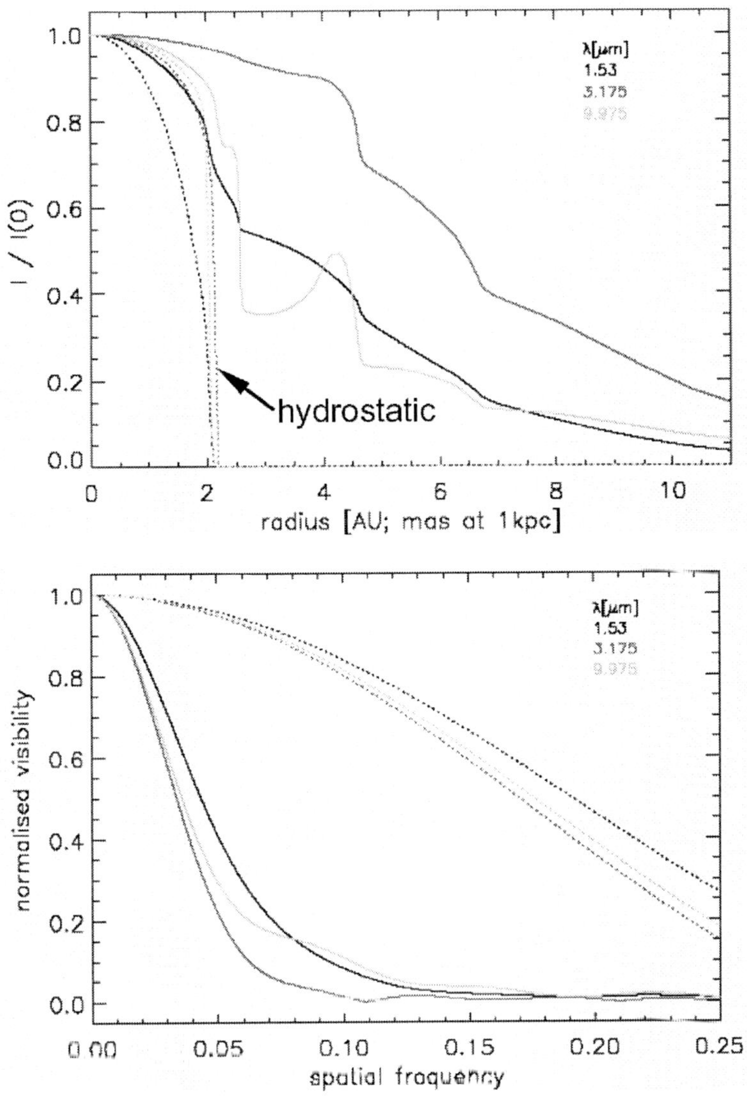

FIGURE 2. Intensity (upper) and visibility (lower) for the maximum phase of our model for narrow-band filters at 1.53μm (black), 3.175μm (dark grey) and 9.975μm (light grey). A hydrostatic model is also shown for comparison (dashed lines). Note the much larger extention of the dynamic intensity profiles (corresponding to lower visibilities) and the effects of the small scale density variations showing up as shells in the intensity profiles and as differences at high spatial frequencies.

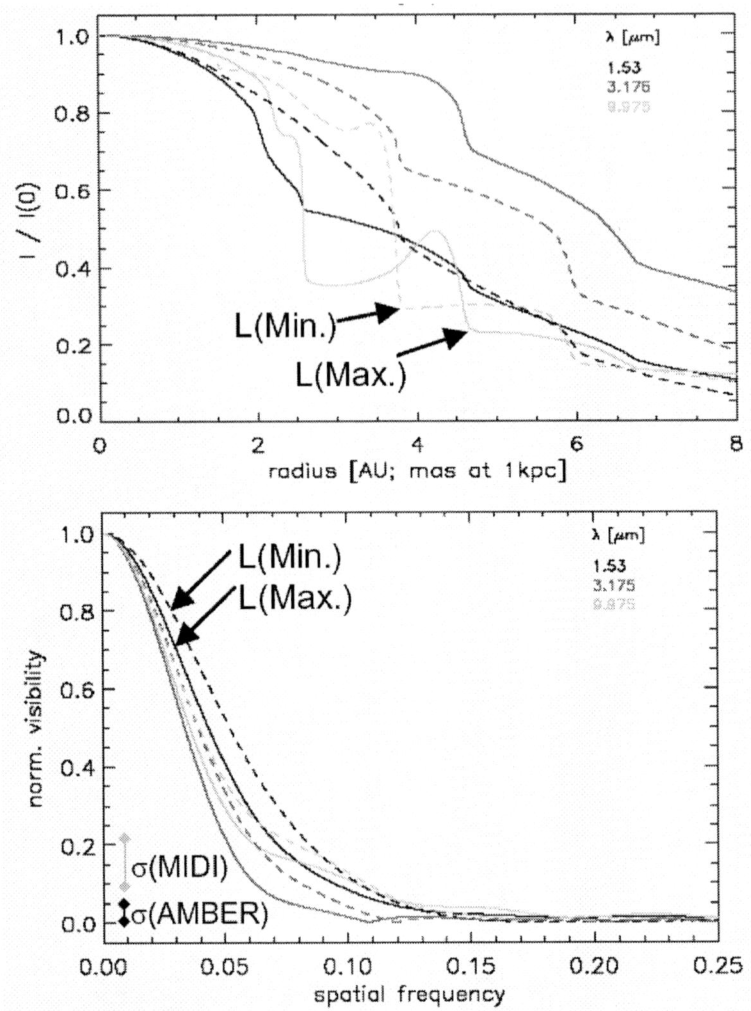

FIGURE 3. Comparison of intensity (upper) and visibility (lower) profiles for the maximum (full lines) and minimum (dashed lines) phase of our model. The same filters as in Fig. 2 are plotted. The error bars in the visibility profile indicate the accuracy currently achievable with the MIDI and AMBER instruments at the VLTI.

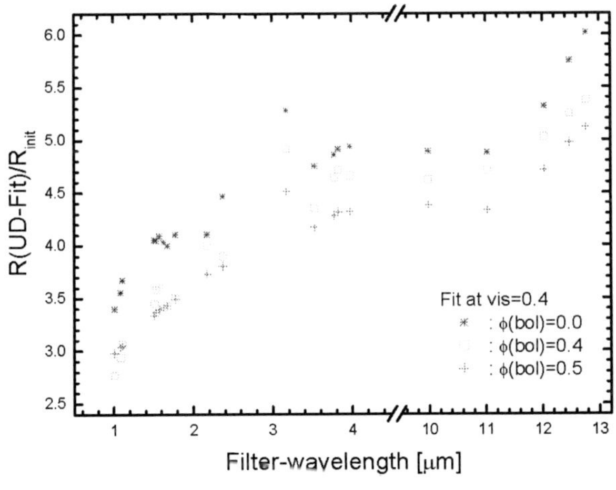

FIGURE 4. Ratio of uniform disk radius (obtained by fitting a UD at visibility 0.4 to the dynamic model) to the radius of the hydrostatic initial model as a function of wavelength. Results for three phases of the pulsational cycle are shown (asterisks, open squares, plus-signs).

Nevertheless, in the past (and partly also now) most interferometric measurements are published in terms of UD radii. For older data, these are also often derived from just one available visibility point. The differences to be expected between such UD diameters and the (hydrostatic) stellar radii for M-Mira models are discussed in [12]. As we want to compare existing interferometric data with our C-Mira models we have started a similar investigation. For this purpose we fitted our visibility profiles for the dynamic model with UDs at single selected visibilities in the first lobe of the visibility profile. Given the difficulty of defining a radius for the dynamic models. the resulting UD radii were then compared to the radius of the hydrostatic initial model of the dynamic model, a procedure similar to the one used by [12].

Figure 4 shows the resulting radius-ratios for the different narrow band filters. Similar trends are found, quite independently from the chosen visibility (0.2, 0.4, 0.8): (i) the UD radius increases with wavelength and decreases with phase, with particularly large ratios found at wavelengths with strong C_2H_2 bands, (ii) the ratios are much larger than found by [12] for M-Mira models which is most likely due to the effects of the dust opacity included in our models. These results demonstrate that UD diameters can serve at best as a first approximation for the stellar radius and that a conversion to the proper radius needs careful modeling.

5. CONCLUSIONS AND OUTLOOK

The dynamic models presented here are ready for a detailed comparison with existing and new interferometric measurements. These models predict a pronounced wavelength and phase dependence of the intensity and visibility profiles. Uniform disks generally are a very bad approximation to the profiles. They also largely overestimate the (hydrostatic) stellar radius and also show a complex wavelength dependence. The currently available observations have (partly) the necessary accuracy to derive atmospheric stratifications and to check the predictions of the models. Observations with the highest possible spectral resolutions will provide the best constraints for the models and more such observations are thus of highest priority.

Currently most of the interferometric data and models are still 1D (i.e. being only sensitive to or assuming spherical symmetry). However, the already existing possibility to combine more than two telescopes (e.g. AMBER at the VLTI) and the ongoing development of new instruments for the VLTI (MATISSE, VSI) will permit real imaging of the surfaces of pulsating red giants in the coming years [13,14]. Furthermore, the efficiency and accuracy of the instruments is continuously improving. In combination with the progress in 2D and 3D models for these stars (e.g. [15,16]), interferometry will thus definitely bring fundamental new insights into the physical processes operating in these stars.

ACKNOWLEDGMENTS

This work has been supported by Projects P19503 and P18939 of the Austrian Science Fund as well as the 6th Framework Program of the EU (JRA4 of the OPTICON I^3).

REFERENCES

1. A. Richichi, F. Delplancke, F. Paresce, and A. Chelli (eds.), *The Power of Optical/IR Interferometry: Recent Scientific Results and 2nd Generation Instrumentation,* ESO Astrophysics Symposia, Springer, (2008)
2. A. Richichi, I. Percheron, and M. Khristoforova, *A&A* **431**, 773 (2005)
3. B. Aringer, F. Kerschbaum, and U.G. Jørgensen, *A&A* **395**, 915 (2002)
4. J. Hron, W. Nowotny, R. Gautschy-Loidl, and S. Höfner, *Ap&SS* **286**, 213 (2003)
5. K.-H. Hofmann, M. Scholz, and P.R. Wood, *A&A* **339**, 846 (1998)
6. S. Höfner, R. Gautschy-Loidl, B. Aringer, and U.G. Jørgensen, *A&A* **399**, 589 (2003)
7. R. Gautschy-Loidl, S. Höfner, U.G. Jørgensen, and J. Hron, *A&A* **422**, 289 (2004)
8. W. Nowotny, T. Lebzelter, J. Hron, and S. Höfner, *A&A* **437,** 285 (2005)
9. W. Nowotny, *PhD Thesis, Univ. of Vienna,* (2005)
10. R. Gautschy-Loidl, *PhD Thesis, Univ. of Vienna,* (2001)
11. M. Scholz, *SPIE* **4838**, 163 (2003)
12. A.P. Jacob and M. Scholz, *MNRAS* **336**, 1377 (2002)
13. B. Lopez, S. Wolf, S. Lagarde, P. Abraham, et al., *SPIE* **6268**, 62680Z (2006)
14. F. Malbet, P. Kern, J.-P. Berger, et al., *SPIE* **6268**, 62680Y (2006)
15. A. Chiavassa, B. Plez, E. Josselin, and B. Freytag, *EAS Publications Series* **18**, 177 (2006)
16. P. Woitke, *A&A* **452**, 537 (2006)

AGB Star Variability Observed from Antarctica

Runa Briguglio [1]

University of Rome, La Sapienza - Third Concordia Winter 2007

Abstract. We present here a short account of the first observations done at the Italo-French base of Dome C, Antarctica, in the field of Mira variability. The observations aimed at improving our knowledge of the variability period of some important Long Period Variables (LPV), to be used in Period-Luminosity criteria for determining in a reliable way the absolute magnitudes and hence the distances of AGB stars. The observations were taken during the 2007 winter campaign, using the small-IRAIT telescope, a 25 cm Cassegrain instrument capable of multiband photometry. The telescope was equipped with a reliable remote control system, which allowed us to get very high duty-cycle observations (the highest efficient was obtained in July 2007, with more than 98% of on-source time, over 9 "solar" days during the long polar night). The experiments done with small IRAIT will be precious as a pathfinder, for defining the observational strategies of the larger IRAIT (International Robotic Antarctic Infrared Telescope [1]), an 80-cm aperture telescope, working mainly in IR, that will be installed in Dome C next year.

Keywords: AGB Stars: Long Period Variables, Photometry, Observational Techniques
PACS: 97.30.Jm; 95.75.De; 95.55.Cs

1. THE SMALL IRAIT TELESCOPE

The Small IRAIT Project was born in Rome in January 2007, during a meeting between the IRAIT coordinators (Busso, Tosti, University of Perugia) and PNRA (Italian Program for Research in Antarctica). The plan was to install a small telescope at the Concordia Station, Dome C (Fig.1), one year before the installation of the major IRAIT telescope. The goal of the Small IRAIT mission was to test the equipment, the procedures and the photometric calibation during the polar night, in the harsh conditions of Antarctic winter. Other experiments are already installed at Dome C, e.g. the site testing activity performed by the AstroConcordia French group [2, 3, 4] and the Australian AASTINO [5, 6].

Small IRAIT (sIRAIT) is a 25cm-Cassegrain telescope, with a focal length of 300cm (F/12). It was assembled by Marcon Telescopes (ITALY), the same company that also built its major brother IRAIT [1]. sIRAIT is equipped with a fork parallactic mount, with large 30 cm gears on both axis: the advantage of oversized mechanics in a polar environment is that they are less influenced by snow deposits and ice.
Two 60W cryogenic motors equip both axes. We can a posteriori say that the choice of cryogenic equipment was a very good idea, as they work very well at the typical Antarctic winter temperatures (-70°C) without the need of any heating system; moreover, they are not affected by self heating, a serious problem to be addressed in Dome C, where thermal dissipation is very low because of the extremely low atmospheric turbulence.

[1] For the IRAIT collaboration

CP1001, *9th Torino Workshop on Evolution and Nucleosynthesis in AGB Stars and the 2nd Perugia Workshop on Nuclear Astrophysics*, edited by R. Guandalini, S. Palmerini, and M. Busso

FIGURE 1. The Italo-French base at Dome C, on the Antarctic plateau.

The Telescope Control Software (TCS) was designed and written by the Perugia University and modified on-site in Dome C. Several functions were added to improve (and facilitate) outdoor operations, as well as the remote control of the instrument.

The custom TCS, the cryogenic motors and an accurate mechanical set-up led after installation to a typical pointing accuracy of 1 arcmin (even in presence of large snow accumulation inside the mechanics) and a good tracking stability so that, without auto-guiding facilities, only few corrections were needed during the 24 hours of a solar day. Automatic acquisition was possible, and remained reliable even at the world seasonal record temperature of -81.9^{o}C, as encountered in Dome C on the eleventh of July, 2007.

Scientific activities in the base of Dome C are carried out in outer laboratories, namely heated containers equipped with wireless LAN connections, computers and maintenance tools. The sIRAIT shelter is placed some 400 m West of the Concordia Base, on a snow dome, 4 m high (see Fig. 2). With the exception of the telescope and of the focal plane instrument, all the equipment is placed inside the shelter: we followed an indoor philosophy to facilitate maintenance and reparation activities.

1.1. The Focal Plane Instrument

The focal plane instrumentation is installed inside an insulated box, named APU (Astronomical Portable Unit), built to protect the electronics and allow it to work outside. Redundancy and easy access were the two keywords for its preparation.
The temperature control is operated by a PID controller, rapidly re-configurable.

Heating is provided by two 100 W heaters, mounted on metallic plates to speed up thermal dissipation; 3 small fans, on the 3 axes, avoid thermal gradients, at any telescope position. Thermal stratification is extremely effective at low humidity and at the pressure conditions of Dome C.
A set of 2 PT100, read by independent controllers, and 2 analogical thermometers, based on AD590 chips, allow continuous temperature monitoring. Analog data acquisition is carried out by a National Instruments AD board, controlled by the supervising software.

FIGURE 2. Observing in the Antarctic night at Dome C (at -71 ^oC) on the sIRAIT snow dome.

Analog boards and software may by-pass the automatic PID temperature controller, thus providing a redundant system in case of fault.
The image acquisition apparatus is made up of a CCD, a filter wheel housing UBVRI standard filters, and a digital focuser. A description is given in Table 1. The imager is connected to the control PC via UBS protocols: this gave us some problems as USB cables for low temperature did not exist. We solved this issue using "recycled" teflon cables (creativity is mandatory in Antarctica) and adding USB repeaters on the cables.
An optical window decouples the heated box from the external environment. In order to prevent ice formation on the window inner surface, due to the high temperature difference, a "cold finger" acts as a trap for humidity. In this way, thanks to the "cold finger" and to strong ventilation inside the heated box, the optical surfaces never suffered from the usually reported problem of icing.

1.2. The Remote Control

As explained before, every scientific activity in Dome C is organized with a laboratory inside the base and a shelter outside, close to the experiment. The remote control is then intended to check every subsystem of the experiment from the base, in a comfortable place, thus reducing the time spent outside, and to increase safety standards. Then, the experimental set up is divided into three parts:

1. the outdoor equipment, such as the telescope, the CCD, sensors, heating and regulating systems; this very last part is completely stand-alone, meaning that it doesn't need any auxiliary system or connection to maintain the instruments at working

conditions;

2. the remote laboratory, inside the shelter, housing the acquisition computers, spare parts, everything needed for working on the experiment; it is also a heated place to allow "on-field" operations in safety conditions and to de-freeze instruments; it is Wi-Fi connected with the base;
3. the base laboratory, housing the supervising computers and all the stuff needed for astronomical activity (star charts, handbooks, datasheets, larger stocks of spare parts and so on).

The small IRAIT remote control is based on two parallel parts: the first is a visualization system, to allow the remote monitoring of the instrument status. At a very low level, the system indicates if the power supply is on, the heaters are powered and active, the instruments are connected. The advantage of this low level set-up is that it doesn't need a supervising PC to work, so it is active even in case of computers faults (a very frequent accident in Dome C). An upper-level monitoring system is based on an acquisition board, reading continuously the thermometers, the supply voltage, the status of fans and heaters.
The second part allows remote switches, useful for reset, to start and stop instruments before/after acquisition, to switch off heating systems in case of a too-high temperature. It also allows to bypass the regulating systems in case of malfunctioning. All the hardware is housed inside the shelter, but it is controlled by the computer in the base through a remote desktop (RDP). The observing schedule is prepared directly in the base.

TABLE 1. KODAK KAF 0402ME and focal plane features.

N. pixel	768x512
Pixel size	9 μm x 9 μm
Sensor size	6.91 mm x 4.6 mm
Image scale	0.62"/pixel
Full well capacity	100000 e
Gain	10 μV/e
Readout noise	15 e
Dark current	< 10pA/cm^2, 25°C
Dynamic range	76 dB
Non linearity (nom)	1.0%
Non uniformity (nom)	0.8%
Power supply	12VDC, 4 A
Filters	UBVRI
Focuser run	2 cm
Focuser accuracy	10μm

2. AGB OBSERVATIONS

2.1. Antarctica: Experimental Problems

The peculiar sky conditions at Dome C notwithstanding (excellent seeing, transparency, stability, long polar night), the harshness of the Antarctic plateau imposes severe prescriptions to the experiments placed outdoor [7]. We want to underline that the researcher's safety must be the first guideline in problem solving. Then, a number of experimental challenges was to be addressed during the telescope preparation in Italy and its installation in Dome C. In both cases technological solutions and procedures were found and improved during the whole duration of the mission. Here we give just a brief description. Only a few problems affected observations during the polar night. With regard to environmental effects, the main difficulty was snow and ice-dust deposit. Especially under the action of wind, even at low speed, snow accumulates on every exposed surface, and fills carters and boxes that are not perfectly sealed. Mechanics, like gears, motors, endless-screws, were completely buried under a deep snow overlay, even few hours after cleaning. Nevertheless, snow on mechanical parts has never been a real challenge: even in presence of thick snow deposits, the overall motion smoothness remained unaltered.
Optical surfaces, like mirrors and lenses, suffered much more. A daily cleaning was mandatory during the whole observing campaign (June-September). Snow and ice dust form a uniform, thin layer on mirrors, made up by small particles (dust) and occasionally by larger ones (1-2 cm, very likely falling from the optical tube) They act as a trap for the incoming snow. When frequently removed, snow doesn't adhere to optical surfaces, hence a simple soft brush is enough for cleaning. On the opposite, after several days of deposition, snow tends to freeze and stick to mirrors: a complex de-icing operation is then needed. We successfully tested a procedure based on infrared light irradiation. using this method mirrors are not significantly heated, thus reducing risks for the optics, and few minutes are enough for complete ice sublimation. The procedure is also very suitable for automatic operations.

With regard to a technological aspects, the main difficulty arose from the motor cables. The original silicon motor cabling was unsuitable at temperatures below -55°C: therefore it was replaced by spare teflon cables, whose features were similar to the original ones. During the use we realized that the new cables caused a decrease in the torque, leading to stops occurring during fast pointing. In late August we replaced them with shielded, larger diameter cables, obtaining much better performances in general, and in particular better results in automatic operations.
In the Antarctic cold atmosphere, the temperature stratification and the low thermal dissipation impose a very effective regulating system, and hamper the operation of large electronic parts placed inside heated boxes situated outdoor, as self-heating becomes extremely hard to prevent. (In the Antarctic conditions heat is a bigger problem than cold!). As a consequence, computers and electronic devices (with the only exceptions of the photometer) were placed inside the laboratory, which fact sometimes created troubles with cables and communications.

TABLE 2. **GSC 05635-00564**

RA (2000)	Dec(2000)	V	B-V	U-B	V-R	R-I
16h 30m 18s	-12°39'8"	10.069	1.568	1.192	1.152	1.498

2.2. Choice of the Sources

We based our data collection on the catalog of AGB light curves published on-line by Maffei and Tosti (hereafter MT) in: http://astro.fisica.unipg.it/atlasmaffei/main.htm). From that sample, we performed VRI photometry on a set of 18 AGB Mira variables. Since our sources are faint at short wavelengths, and the diameter of the telescope is small, only the I magnitudes were quantitatively significant. Data reduction is now in progress and we shall then perform a comparison between observed and expected values, the latter being extrapolated from the mentioned MT catalogue, containing a list of $\sim$200 sources observed for decades, providing a high-precision period estimate. All objects studied are very close to the M16 region: $18h10m < RA < 18h30m; -13^o < Dec < -17^o$. The range of maximum I-magnitudes we obtained is between 10 and 14. The range of minimum I-magnitudes is between 14 and 17.
As a first step we selected stars with a maximum I magnitude brighter than 13, to assure a good field identification and to speed up the pointing procedure. Subsequently, we selected groups of objects (possibly in the same CCD field of view, apart from camera orientation). Only few objects respected this rule: one single FoV with 3 sources (*HV Ser, HU Ser, HT Ser*) and 3 fields with a couple of sources (*GG Ser, FY Ser; IK Ser, HY Ser; V406Sct, V407Sct*). In conclusion, the selected targets are 18, as indicated in Table 4.
A suitable comparison star was chosen from [8]. From this reference we pre-selected a sample of standard stars in the RA range 15h-21h, as the RAs of all our program stars are close to 18h. Then we gave a preference to stars whose declination was as close as possible to the M16 region. This choice is due to a couple of main reasons: first, sky transparency seemed to be very homogenous in azimuth, while we could notice a sharp difference between a strip close to the horizon and the upper levels, due to the presence of a thin fog layer on the northern horizon. Secondly, large declination movements are more risky in terms of pointing accuracy (at least for the very cold August temperatures) than right-ascension movements.
A good compromise, in terms of proximity, luminosity, color index and elevation, is GSC 05635-00564 (please refer to Table 2).

2.3. Data Acquisition and Comments

Data sampling for AGB stars took place during 4 nights: August 11th, 13th, 17th and 21st. Dates were chosen according to other acquisition activities: as the telescope

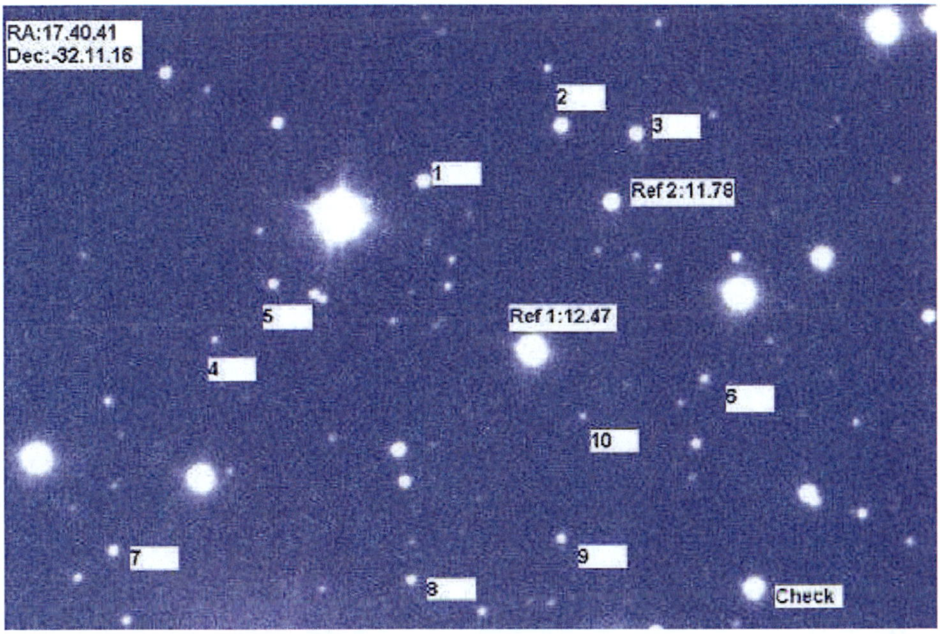

FIGURE 3. A frame in the cluster M6 as observed by small-IRAIT at Dome C.

TABLE 3. ADU and S/N ratio for the I band for a set of sampled AGB stars.

	ADU	SNR
V424 Sct	2970	14.1
HY Ser	3200	15.1
KO Ser	3454	16.3
GR Ser	73741	216.9
V3908 Sgr	24377	94.5

time was devoted mainly to high duty cycle, multi-day runs on clusters (Fig. 3) and variable stars, we selected the half of August because uninterrupted acquisition is by that time impossible (the Sun rises for the first time after the polar night on August, 8th) and because the M16 region culminates near 21:00LT, thus allowing for outdoor activities during the early afternoon, including telescope pointing, and still offering more than 10 hours of profitable time for sampling. Unfortunately, August was often windy and foggy, especially on the North horizon where all objects lie. The effect of sky transparency is still under analysis; a foggy layer was visible with the naked eye, lying just very few degrees above the horizon, but it is difficult to estimate its importance at an elevation of 30° (the typical target elevation near M16). The Moon was below the horizon during the 5 nights.

TABLE 4. List of observed AGB stars, with I-band maximum and minimum expected magnitude.

Name	RA (FK5)	Dec(FK5)	I min	I max
V3940 Sgr	18h 25m 10s	-16° 36.4'	11.6	15.2
GG Ser	18h 08m 12s	-14° 34.3'	10.9	15.9
KO Ser	18h 17m 01s	-15° 18.3'	11.2	15.7
HV Ser	18h 11m 57s	-15° 39.4'	13.0	16.2
V424 Sct	18h 31m 36s	-15° 26.9'	10.0	15.5
HU Ser	18h 11m 55s	-15° 40.2'	13.1	16.0
HT Ser	18h 11m 49s	-15° 37.3'	13.4	16.4
KW Ser	18h 19m 36s	-14° 39.5'	11.6	12.7
GR Ser	18h 09m 24s	-15° 19.5'	10.1	13.4
IK Ser	18h 12m 26s	-15° 45.4'	12.9	15.9
HY Ser	18h 12m 18s	-15° 47.4'	12.3	15.3
V3942 Sgr	18h 25m 40s	-16° 33.8'	11.2	15.3
FY Ser	18h 07m 54s	-14° 31.4'	14.0	16.4
V3908 Sgr	18h 13m 34.7s	-16° 1.9'	8.4	11.0
V407 Sct	18h 29m 33s	-15° 48.7'	11.7	15.2
LQ Ser	18h 20m 46s	-12° 52.8'	13.0	15.8
V417 Sct	18h 30m 16s	-14° 31.4'	11.0	14.9
V406 Sct	18h 29m 24s	-15° 47.5'	11.5	15.4

Observations typically started at 17.00 LT, when the sky was dark enough to allow a fast visual telescope alignment (when needed). Elevation of observed stars was in the range 27°-31°, while the comparison star elevation was in the range 17°-27°; during a single night, we planned to sample the comparison star at the beginning and at the end of the observing session, and with a cadence of 40 to 60 minutes (i.e. after the sampling of two AGBs). During the sampling the elevation range of the comparison star was very small: just on 13th, when sky conditions remained stable for a longer time, we had a comparison sampling with 4 points, at an elevation of 27°, 26°, 22° and 17°. During the 3 remaining nights we have 2 comparison points on 11th (26°, 28°) and just one on 17th and 21st (26° and 21°). This was due mainly to sky conditions worsening during the observing sessions.

An accurate data analysis is still underway: so far we obtained a first check of the images to have an insight of the overall data quality. In Table 3 we present ADU values extracted from a mean image (averaged on 10 single images, with 1-minute exposure time) and the SNR in I band. SNR is calculated as $SNR = \frac{N_{star}}{\sqrt{N_{star}+B\cdot n}}$, where N_{star} is the ADU value extracted from inside a circular aperture of 10 pixel (corresponding to 6.2"), B is the background level, averaged on the whole image and n is the aperture area (pixel). As a first consideration we notice that limit magnitude, after multiple image merging, is not fainter than 14: this is reasonably due to some problems with the optics (they are visibly scratched, after day by day snow removal), so that the typical PSF is larger (roughly 4") than expected [2]. That's why only few stars of the original Maffei&Tosti's catalog were chosen, accordingly to the maximum expected I magnitudes. This problem induced also some difficulties in field identification, as field stars are generally faint. Secondly, in order to sample sufficiently well the PSF, a large aperture is needed, giving rise to a higher noise level. These effects, together with a sky transparency below the mean, reduced the general quality of the data, that was expected to be higher, accordingly to data from the literature [5, 6] and our previous experience during the polar night. Sky transparency during fall nights was dramatically better than in late winter: our future programs should take this into account, especially when sampling low absolute declination objects is necessary (like for Maffei & Tosti's AGBs).

As a final remark, all technological solutions for the telescope were a great success: automation and remote control worked very well all the night long, allowing us to perform multiband photometry from this very peculiar site, that is extremely promising for optical and infrared astronomy.

ACKNOWLEDGMENTS

IRAIT and Small IRAIT projects are funded by PNRA (the Italian Program for Antarctic Research). We wish to thank PNRA and IPEV (the French Polar Institute), that provided logistics for the Concordia Station and a lot of technical help during our observations.

REFERENCES

1. G. Tosti, M. Busso, G. Nucciarelli, et al., *SPIE* **6267**, 47 (2006)
2. A. Agabi, E. Aristidi, M. Azouit, E. Fossat, F. Martin, T. Sadibekova, J. Vernin, and A. Ziad, *PASP* **118**, 344 (2006)
3. E.Fossat, *in Giard et al. (eds.), Dome C, Astronomy, and Astrophysics Meeting, EDP Publ. Ser. 14*, p. 1 (2005)
4. E. Fossat, *in Giard et al. (eds.), Dome C, Astronomy, and Astrophysics Meeting, EDP Publ. Ser. 14*, p. 121 (2005)
5. J.S. Lawrence, M.C.B. Ashley, A. Tokovinin, and T. Travouillon, *Nature* **431**, 278 (2004)
6. S.L. Kenyon, J. Lawrence, M.C.B. Ashley, J.W.V. Storey, A. Tokovinin, and E. Fossat, *PASP* **118**, 924 (2006)
7. K.G. Strassmeier, K. Agabi, L. Agnoletto, et al., *AN* **328**, 451 (2007)
8. K. Schaifers, H.H. Voigt, H. Landolt, R. Boernstein, and K.H. Hellwege, *in "Numerical Data and functional Relationships in Science and Technology. New Series. Astronomy and Astrophysics. B: Stars and Star Clusters", Berlin: Springer* (1982)

SESSION VI:

ABUNDANCES OF s-PROCESS NUCLEI

Analysis of Two Carbon Stars in the Carina Dwarf Spheroidal Galaxy

C. Abia*, P. de Laverny†, R. Wahlin**, I. Domínguez*, S. Cristallo‡ and O. Straniero‡

*Dpto. Física Teórica y del Cosmos. Universidad de Granada, E-18071 Granada, Spain
†Observatoire de la Côte d'Azur, Dpt. Cassiopée UMR6 202, 06 304 Nice Cedex 4, France
**Department of Astronomy and Space Physics, Box 515, 75120 Uppsala, Sweden
‡Osservatorio di Collurania - INAF, Teramo, Italy

Abstract. We present a chemical analysis of two metal-poor carbon stars belonging to the Carina Dwarf Spheroidal galaxy using high resolution optical spectra obtained with the VLT/UVES instrument. Their derived properties and chemical composition are compatible with these stars being in the TP-AGB phase undergoing third dredge-up episodes, although their extrinsic nature cannot be definitively discarded. The analysis indicates that these stars have large *s*-element enhancements, which are compatible with theoretical *s*-process nucleosynthesis predictions in low-mass AGB stars (~ 1.5 $M_\odot$), and also large carbon enhancements (C/O ≥ 5), much larger than the values typically found in galactic AGB carbon stars (C/O $\sim 1-2$). This is also in agreement with the theoretical prediction that carbon stars are formed more easily through third dredge-up episodes as the initial stellar metallicity drops. We speculate on the possibility that the abundance pattern found in the low-mass *s*-elements (Y, Zr) may be revealing the existence of some chemical depletion into dust grains.

Keywords: Stars: Carbon - Stars: AGB - Nucleosynthesis - Galaxies: Dwarfs
PACS: 95.75.Fg, 97.10.Cv, 97.10.Ex, 97.30.Hk, 98.52.Wz

1. THE CARINA GALAXY AND CARBON STARS

Due to their peculiar star formation histories, which translate into complexes stellar metallicity distributions, the satellite galaxies of the Milky Way offer an unique possibility to observe intrinsic metal-poor AGB stars. Another advantage is that because the distance of these stellar systems are relatively well known, it is possible to derive more accurate stellar luminosities, a critical parameter for theoretical comparison purposes. In contrast, the galactic metal-poor AGB counterparts are rare (except may be in the galactic bulge) or do not exist yet (galactic halo) because they have masses too low to be now on the AGB phase.

In this sense, the dwarf spheroidal (dSph) satellite galaxy Carina is particularly interesting among the dSphs in the Local Group because of its unusual, episodic star formation history. [1] found three well-defined, distinct main-sequence turn-offs corresponding to ages of about 2, 3-6 and 11-13 Gyr, respectively. Interestingly, they note that the three main-sequences connect with an unique, narrow red giant branch and conclude that regardless of the age, Carina stars are metal-poor with mean metallicity [Fe/H]$= -1.86 \pm 0.20$. Recently [2] reach the same conclusion, however, they found a much broader spread in the metallicity distribution around the above value

CP1001, *9th Torino Workshop on Evolution and Nucleosynthesis in AGB Stars and the 2nd Perugia Workshop on Nuclear Astrophysics*, edited by R. Guandalini, S. Palmerini, and M. Busso

(FWHM$\sim$ 0.92). The metallicity distribution found by [2] cannot be explained by a simple closed-box model of chemical evolution. More sophisticated models, characterised by more than one episode of star formation and intense galactic winds [3], are required to explain the observed metallicity distribution in this satellite galaxy.

In the present study, we report the chemical analysis of two carbon stars belonging to this metal-poor stellar system. The aim of this work is to test observationally the predicted critical dependence on the stellar metallicity (mass) of the *s*-process nucleosynthesis in AGB stars (and of their carbon enrichment). Also, by studying the *intrinsic index* abundance ratio [hs/ls][1] in AGB stars of different metallicity, it is possible to constraint better the total neutron flux in the *s*-process: the ^{13}C efficiency.

2. OBSERVATIONS AND ANALYSIS

Only a few carbon stars are known in Carina. We selected for this work two of its brightest members (ALW-C6 & ALW-C7) from [4]. The selected carbon stars were observed during 2004 in service mode with the UVES spectrograph attached to the second VLT unit. The available spectral domains are from $\sim$4 200 to $\sim$5 000 Å and from $\sim$6 700 to $\sim$9 000 Å. Individual spectra were added, calibrated in wavelength and then normalised to a continuum using the standard tools within the IRAF software package.

The chemical analysis of the stars was performed by using LTE spectral synthesis of specific spectral regions in the optical. We derived the chemical abundances of the CNO elements, the carbon isotopic ratio, Li, *s*-elements and the mean metallicity [M/H], this later being obtained from the mean value derived from a handful lines of Ca, Ti, V, Mn, Fe and Ni. We used the same atomic and molecular line lists and model atmospheres for cool carbon-rich stars (assuming 2 $M_{\odot}$) as in [5]. A first guess of the effective temperatures was estimated from the calibrations of the $(H\text{-}K)_o$ and $(J\text{-}K)_o$ colour indexes by [6]. The adopted final values (Table 1) were obtained through an iterative process by comparing the observed optical spectra with theoretical ones. This iterative process included also the C/O ratio (or the abundance difference $\varepsilon(C)-\varepsilon(O)$, hereafter C-O) and the metallicity adopted in the model atmosphere. These are the main atmosphere parameters determining the global shape of the synthetic spectrum (gravity has only a minor effect). The gravity was set at log g= 0.0 for both stars. The microturbulence parameter initially adopted was ξ = 2.2 km/s [7], although beyond 6700 Å, larger values were used (3-4 km/s) to fit the strong molecular CN and C_2 lines present in this spectral region. Synthetic spectra were broadened with a Gaussian function to match the instrumental profile and with a macroturbulence velocity ranging from 4 to 7 km/s. Table 1 shows the main atmospheric parameters of the carbon stars together with the C/O and ^{12}C/^{13}C ratios derived.

Similarly to what was found in [5] and in other studies of carbon stars (e.g. [8]), the derived C/O ratio slightly decreases with the wavelength of the analyzed spectral

---

[1] Here we use the standard notation for the chemical abundance ratio of any element X, $[X/H] = \log (X/H)_{\star} - \log(X/H)_{\odot}$, where $\log(H) \equiv 12$ is the abundance of hydrogen by number. In this scale the abundance of any element is noted as $\varepsilon(X)$.

TABLE 1. Main atmospheric parameters of the Carina carbon stars. See text for details.

Star	T_{eff}(K)	[M/H]	C/O	$\varepsilon(C)-\varepsilon(O)$	$^{12}C/^{13}C$
ALW-C6	3 400	−1.70	5-8	7.8	≥ 125
ALW-C7	3 200	−1.90	4-6	7.2	≥ 250

region. However, we found that the difference between the corresponding C/O ratios derived, never exceeds a factor ~ 1.5. Another source of uncertainty in the derivation of the C/O ratio is the fact that synthetic spectra computed with O abundances differing by a large factor cannot be distinguished, which allows a range of C/O ratios (or the difference C-O) giving good fits to the observed spectra. Unfortunately, we cannot derive the absolute O abundance independently from the available spectra. Therefore, we take the O abundance as a free parameter scaled with the metallicity. Due to this, in Table 1 we give a range of values for the C/O ratio and the average abundance difference $\varepsilon(C)-\varepsilon(O)$. This ambiguity in the C/O ratio, however, does not affect the $^{12}C/^{13}C$ ratio derived from ^{12}CN and ^{13}CN features in the ~ 8000 Å range. The best fits to the ^{13}CN lines in this region give 125 and 250 for the carbon isotopic ratio in our stars (see Table 1). Nevertheless, since theoretical spectra computed with an isotopic ratio a factor ~ 2 lower (higher) do not differ significantly, we consider these ratios only as lower limits.

Figure 1 shows a theoretical fit in the region $\lambda 4750-4850$ Å for the star ALW-C7. This is the region from where we derived most of the metal abundances. A full discussion about the sources of error in the absolute abundances and element ratios due to uncertainties in the atmosphere parameters, continuum location and random errors when a given element is represented by a few lines, as it is our case, can be found in [9, 10]. These errors range from ± 0.2 dex for Ba to ± 0.5 dex for Ce and Li. We estimate a typical uncertainty in the mean metallicity ([M/H]) of ± 0.30 dex. Errors for the elemental ratios with respect to the metallicity ([X/M]) range between 0.3-0.4 dex, since some of the uncertainties cancel out when deriving this ratio. For the same reason, the error in the abundance ratio between elements ([X/Y]) is somewhat lower, ± 0.3 dex. These numbers do not include possible systematic errors as N-LTE effects or an uncertainty larger than 5% in the continuum location, as it might be the case in the spectrum of ALW-C6.

Table 2 shows the metallic abundances derived in ALW-C6 and ALW-C7.

3. RESULTS AND DISCUSSION

These stars are not Li-rich; their derived Li abundances are similar to the average value derived in galactic AGB carbon stars [11, 12]. The metallicities (Table 2) agree with the average value previously obtained in the stellar populations of Carina dSph (see above). We found some enrichment of Ti in ALW-C6 ([Ti/M]= +0.3) with respect to the other metallicity indicators and, in contrast, some deficiency in ALW-C7 ([Ti/M]= −0.4). This apparent dispersion in the Ti abundance among the stars of Carina is also found by [13] in several RGB stars. Nevertheless, the low Ti abundance in ALW-C7 might have another interesting explanation (see below).

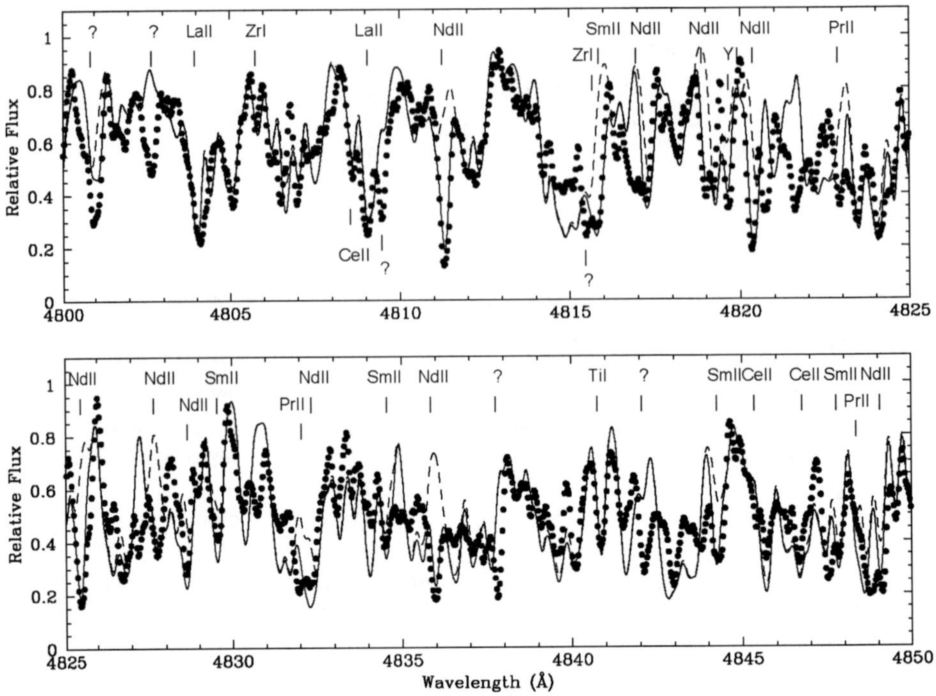

FIGURE 1. Synthetic fit to the spectrum of ALW-C7 in the region around λ4825 Å. Some atomic lines, main contributors to the specific spectral feature, are marked. Black dots represent the observed spectrum. Lines are synthetic spectra: only molecules and metals (dashed line), and including *s*-elements (continuous line) using the mean values for the abundances shown in Table 2. Some spectral (probably molecular) features, not reproduced by the synthetic spectrum, are marked.

A relevant issue is the extrinsic or intrinsic nature of the carbon stars analyzed. The available photometric indexes and the derived luminosities and stellar parameters (see Table 1) suggest that these stars are on the AGB phase. In fact, normal (spectral type N) carbon rich AGB stars are commonly placed in the colour diagram (J-H) vs.(H-K) in the region (H-K)>0.4 and (J-H)>0.8 [14]. The infrared colours of ALW-C6 and ALW-C7 fulfil this requirement. Unfortunately, the detection of Tc lines or/and the study of the [Nb/Zr] ratio in our stars, two powerful nucleosynthesis indicators of an intrinsic nature, were not possible (e.g. [15]). The average *s*-element enhancement in ALW-C6 and ALW-C7 (Table 2) is similar to that found in galactic metal-poor ([Fe/H]≤ -1.5) carbon stars (most of them being extrinsic, e.g. [16]). At the metallicity of our stars, a larger enrichment of the heavy *s*-elements compared to the light-*s* elements is expected. This is because the number of neutrons captured per seed nuclei (^{56}Fe) increases with decreasing metallicity and the *s*-process overcomes the ls peak (Sr, Y, Zr) favoring the formation of *s*-elements of the second peak (Ba, La, Ce etc). This is what we observe

TABLE 2. Summary of the abundances derived in the programme stars. N, indicates the number of lines utilized for a specific element when more than one line is used (we then give the mean abundance together with its dispersion). [el/M] is the abundance ratio with respect to the mean metallicity derived [M/H] (see text) and is only quoted for elements with A$\geq$ 30. The [el/M] ratios are calculated with respect to the solar photospheric abundances according to [17].

Element	N	ALW-C6 log ε	[el/M]	N	ALW-C7 log ε	[el/M]
Li		0.00			−0.50	
Ca		4.60			4.30	
Ti	2	3.50±0.10		2	2.60±0.15	
V	2	2.30±0.20			2.10	
Mn	2	3.70±0.10		2	3.10±0.20	
Fe		-			5.50	
Ni	2	4.40±0.02		2	4.20±0.05	
Zn		3.00	+0.10		-	
Rb		-			<1.00	< +0.3
Sr		-			<1.90	< +0.9
Y	2	1.90±0.00	+1.4	3	0.80±0.05	+0.5
Zr	3	1.90±0.10	+1.0	3	0.90±0.10	+0.2
Ba		2.50	+2.0		2.10	+1.8
La	2	0.90±0.00	+1.5	3	0.90±0.15	+1.7
Ce		<1.80	< +1.9	2	1.80±0.05	+2.1
Nd	4	1.75±0.25	+2.0	5	1.70±0.10	+2.1
Sm	3	1.00±0.03	+1.7	4	1.10±0.00	+2.0
Hf		<1.20	< +2.0		<0.60	< +1.6

in the carbon stars of Carina (see Table 2), [hs/ls]> 0. Moreover, at the metallicity of our stars, it is predicted that also the second *s*-peak will be overcome by the *s*-process fluency, feeding the termination point of the *s*-path at ^{208}Pb. Unfortunately, the available Pb lines at $\sim$ 4000 Å cannot be detected in C-rich AGB stars since they are too faint at this wavelength.

On the other hand, the derived *s*-element abundance pattern in our stars can be reproduced theoretically by *s*-process nucleosynthesis models in AGB stars. In Fig. 2 we compare the *s*-element abundance enhancement relative to Ba found in ALW-C6 and ALW-C7 with the prediction of a 2 M$_{\odot}$, Z= 10^{-4} TP-AGB model by [18] after the 9th thermal pulse. In this computation it is assumed that some protons are mixed-down from the envelope into the He inter-shell throughout a exponentially decreasing profile of the convective velocity at the inner border of the convective envelope at the third dredge-up. Then, these protons are burnt via the CN-cycle and the ^{13}C produced provides the source of neutrons for the *s*-process via the ^{13}C(α,n)^{16}O reaction. In that way, the ^{13}C-pocket formed varies pulse by pulse. Physics and chemist are fully coupled in the model.

As can be seen in Fig. 2, the agreement between observations and model predictions is quite good (considering the errors in the abundances), except for the elements belonging to the first *s*-peak, in particular in ALW-C7 (filled circles). If we exclude a severe error in the abundance analysis, this figure has not an easy explanation. According to the standard theoretical calculations of the main component of the *s*-process in low mass

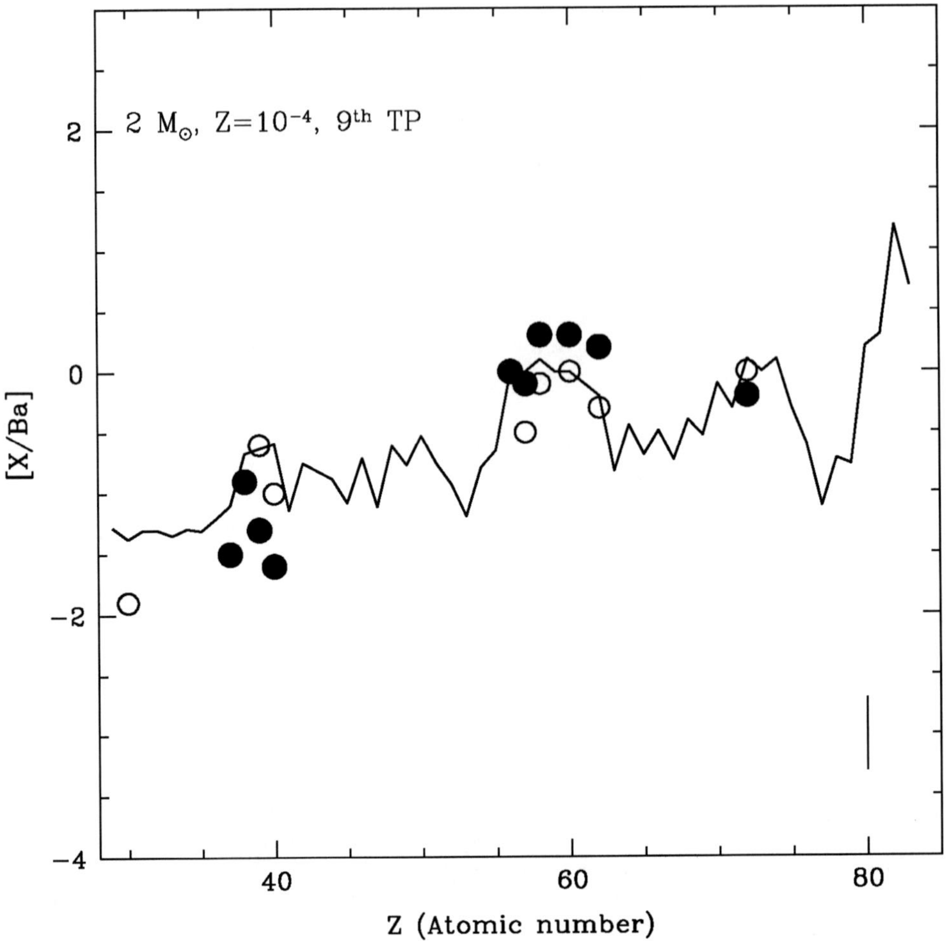

FIGURE 2. Comparison of the abundance patterns found in ALW-C6 (open circles) and ALW-C7 (filled circles) with the *s*-process distribution (continuous line) predicted by a 2 $M_\odot$, Z= 10^{-4} AGB model after the 9th thermal pulse [19, 18]. The abundances of all elements are referred to that of Ba in each star. The line in the right-lower corner indicates a typical error bar in the abundance ratio. Note that some of the data points have to be considered as upper limits (see Table 2).

AGB stars, the enhancement should be similar for Y and Zr, i.e. [Zr/Fe]$\sim$[Y/Fe] (actually, slightly higher for Zr). However, we have found that the [Y/M] ratio significantly exceeds the [Zr/M] one. The differences are larger than the expected error in the [X/M] ratios ($\sim$ 0.3 dex). Thus, except assuming a wrong derivation of the model atmosphere parameters, or a conspiracy with molecular blends, this difference may be real. A possible explanation of this difference would be to admit some Zr depletion into dust grains formed in the circumstellar envelope around these stars (in ALW-C7 in particular). In C-rich circumstellar envelopes (C/O> 1), a refractory element as Zr, condenses into ZrC for any pressure, at a temperature about 500 K higher than Y does in YC_2 (e.g.[20, 21]). According to these authors, the temperature condensation sequence for trace elements (as Zr) is TaC, WC, NbC, ZrC, HfC, TiC and then, YC_2 at much lower temperature than the former molecules. Other refractory *s*-elements as Ba, La, Ce condense even at much lower temperatures. Therefore, an important depletion of these heavy-*s* elements in the gas phase is not expected. Theoretical calculations show that the condensation temperature increases with the C/O ratio and with increasing elemental abundances: i.e. ZrC will form more easily in the circumstellar envelope of a carbon-rich AGB star, the larger the C/O ratio and the larger the Zr enhancement. Interestingly enough, Ti has a condensation temperature only $\sim$ 150 K lower than Zr independently of the actual C/O ratio and partial pressure. Thus, the moderate Ti deficiency found in ALW-C7 (see Table 2) might be another indication of some photospheric depletion in this star. This might explain why apparently some depletion may be present in ALW-C7 (with a large C/O ratio) and not in less carbon enhanced stars as their galactic counterparts [10]. In any case, we have to emphasise that this is only preliminary and that it requires further and careful studies to try to reduce the abundance uncertainties. In particular, the typical depletion pattern showing lower abundances for elements with higher condensation temperature should be found. Also, quantitative calculations on the expected Zr depletion factor and how it may vary with the C/O ratio and average metallicity are required before extracting any conclusion.

4. SUMMARY

The observational properties and derived chemical abundances of ALW-C6 and ALW-C7 suggest that both stars are currently in the AGB phase undergoing third dredge-up and thermal pulses episodes. Despite to the large uncertainties in the analysis, their derived metallicities are compatible with the average metallicity of the main stellar component of Carina derived in previous studies. The analysis of molecular bands reveals that ALW-C6 and ALW-C7 have large carbon enhancements (C/O$\geq$ 5), which makes these stars the more C-rich AGB stars ever observed. Both stars also show large *s*-element enhancements, which are in agreement with theoretical *s*-process nucleosynthesis predictions in low-mass metal-poor AGB models on the basis that the $^{13}C(\alpha,n)^{16}O$ reaction is the main source of neutrons.

ACKNOWLEDGMENTS

Part of this work was supported by the Spanish grant AYA2002-04094-C03-03 from the MEC. P. de Laverny acknowledges the financial support of Programme National de Physique Stellaire (PNPS) of CNRS/INSU, France.

REFERENCES

1. T.A. Smecker-Hane, P.B. Stetson, J.E. Hesser, M.D. Lehvert, and A. VanderBerg, *'From stars ti Galaxies: The impact of stellar physics on galaxy evolution', ed. C. Leitherer, V. Fritze-von Aliensleben, J. Huchra, AIP Conference Series*, 328 (1996)
2. A. Koch, E.K. Grebel, R.F. Wyse, et al., *AJ* **131**, 895 (2006).
3. G.A. Lanfranchi, F. Matteucci, and G. Cescutti, *A&A* **453**, 67L (2006)
4. M. Azzopardi, J. Lequeux, and B.E. Westerlund, *A&A* **161**, 232 (1986)
5. P. de Laverny, C. Abia, I. Domínguez, et al. , *A&A* **446**, 1107 (2006)
6. J. Bergeat, A. Knapik, B. Rutily, *A&A* **369**, 178 (2001)
7. D.L. Lambert, B. Gustafsson, K. Eriksson, and K.H. Hinkle, *ApJS* **62**, 373 (1986)
8. S. Uttenthaler, T. Lebzelter, S. Palmerini, M. Busso, B. Aringer, and M.T. Lederer, *A&A* **471**, L41 (2007)
9. C. Abia, M. Busso, R. Gallino, I. Domínguez, O. Straniero, and J. Isern, *ApJ* **559**, 1117 (2001)
10. C. Abia, I. Domínguez, R. Gallino, S. Masera, M. Busso, O. Straniero, P. de Laverny, B. Plez, and J. Isern, *ApJ* **578**, 817 (2002)
11. G.R. Denn, R.E. Luck, and D.L. Lambert, *ApJ* **377**, 657 (1991)
12. C. Abia, H.M.J. Boffin, J. Isern, and R. Rebolo, *A&A* **272**, 455 (1993)
13. M. Shetrone, K.A. Venn, E. Tolstoy, et al., *AJ* **125**, 684 (2003)
14. G.R. Knapp, D. Pourbaix, and A. Jorissen, *A&A* **371**, 222 (2001)
15. I.I. Ivans, et al., *ApJ* **627**, 145 (2005)
16. T.C. Beers and N. Christlieb, *ARA&A* **43**, 531 (2005)
17. M. Asplund, N. Grevesse, and A.J. Sauval, *'Cosmic Abundance as Records of Stellar Evolution and Nucleosynthesis', Astron. Society of the Pacific, Eds. T.G. Barnes, F.N. Bash Conference Series* **336**, (2005)
18. S. Cristallo, *PhD. Thesis, Università di Teramo*, (2006)
19. S. Cristallo, I. Domínguez, et al., *in preparation*, (2007)
20. K. Lodders and B. Fegley, *Meteoritics* **30**, 661 (1995)
21. K. Lodders and S. Amari, *Chemie der Erde* **65**, 93 (2005)

s-Process Enhancement During the AGB Phase

E. C. Wylie de Boer[*,†] and P. L. Cottrell[†]

[*]*Research School of Astronomy and Astrophysics, Australian National University, Cotter Rd, Canberra, ACT, Australia*
[†]*Dept of Physics and Astronomy, University of Canterbury, Private Bag 4800, Christchurch, New Zealand*

Abstract. The results of a study of the AGB phase are presented. Abundances have been determined for Fe, C, O, the light s-process elements, Y and Zr, the heavy s-process elements, La and Nd, and the r-process element, Eu. The expected relationship between enhanced C, increasing C/O ratio and enhanced s-process elements has been quantified. Results are presented to provide observational data with which to compare theoretical predictions. This paper gives a brief overview of results, full details of the analysis and thorough discussion can be found in Wylie de Boer and Cottrell, 2008 [1].

Keywords: AGB Stars, C Stars, s-Process Elements, Abundance Analysis
PACS: 97.20.Jg, 97.30.Hk

1. INTRODUCTION

The surface of AGB stars can be enriched in both carbon and s-process elements as a direct result of the third dredge-up (TDU) phenomenon. Stars can either enrich themselves via their own internal TDU, in which case they are known as intrinsic carbon stars, and display technetium in their spectra, or are enriched from transfer of carbon-rich material from a binary companion, in which case they are known as extrinsic carbon stars. In order to compare observed s-process enhancements with those predicted by theory, it is desirable to have a set of results along the AGB phase that are internally self-consistent. Relatively few studies have been devoted to large samples of AGB stars, usually studying only one or two spectral classes at a time. Light and heavy s-process elements are hereafter referred to as 'ls' and 'hs' respectively.

A sample of 22 M and MS stars have been analysed by [2, 3]. These studies used high-resolution, infra-red spectra to obtain abundances for C, N, O, ls and hs elements. They found s-process enhancements in seven of the eleven M stars but no measurable enhancements in four of the eight MS stars. The most detailed study of S and SC stars has been undertaken by [4] using high-resolution spectra over the optical wavelength range. This study of S and SC stars derived abundances for C, N, O and a large number of heavy elements, from Rb to Hf. They found considerable s-process elements enhancements in all S and SC stars, of values as much as [s/Fe]$\sim$+1.4 dex. Both of these studies measured equivalent widths for atomic lines, including s-process lines, which are often weak and are affected heavily by molecular blending.

This present study of eleven stars has eight stars in common with these previous studies, and includes two, HD 49368 and HD 35155, not previously analysed. It provides a self-consistent set of results over a range of spectral classes, M, MS, S and SC using

CP1001, 9th *Torino Workshop on Evolution and Nucleosynthesis in AGB Stars and the 2nd Perugia Workshop on Nuclear Astrophysics*, edited by R. Guandalini, S. Palmerini, and M. Busso

TABLE 1. Basic Stellar Quantities.

Star	Spectral Type	B	V	K	Mbol	Variable Type [5]	Tc [6]
HD7351	M3 / S	8.07	6.38	1.64	-3.02	SR	no
HD30959	M3 / S	6.53	4.75	-0.66	-3.87	SRB (30 days)	yes
HD35155	S	8.64	6.87	2.14	-4.45	EA/GS/WD	no
HD44478	M3	4.53	2.91	-1.86	-3.31	LB	...
HD49331	M2 / S	6.91	5.11	0.56	-4.12	susp	no
HD49368	S3	9.53	7.78	2.46	-3.57	SRB	no
HD64332	S4	9.34	7.64	2.31	-2.42	LB	yes
HD102212	M1	5.58	4.05	0.16	-2.10	SRB	...
HD112300	M3	4.96	3.38	-1.19	-2.37	susp	...
HD216672	S5	8.21	6.47	1.04	-3.42	SRB (50 days)	yes
HD286340	SC7					SRB (370 days)	...

high-resolution (R$\sim$40,000), high signal to noise (S:N>100) spectra over an optical wavelength range of 4500 to 7500Å. Instead of equivalent width measurement for abundance determination it uses spectrum synthesis, which enable abundances to be determined for heavily blended lines. The self-consistency of these results minimises all discrepancies due to differing methods or systematic offsets between studies and allows for an internally accurate abundance analysis of stars relative to each other along the AGB phase.

2. ANALYSIS

The eleven field AGB stars analysed are listed in Table 1, with previously published spectral types [3], recent magnitudes (SIMBAD database[1] and 2MASS survey[2]), and calculated bolometric magnitudes, variability type [5] and previous detection of the radioactive element, technetium [6]. Bolometric magnitudes were calculated using the bolometric correction approximation from Alonso et al. [7] and absolute K magnitudes using SIMBAD's apparent magnitudes and distances from the HIPPARCOS catalogue [8]. Magnitudes are not given for HD286340 due to its extreme variability preventing reliable measures of its brightness.

Due to severe molecular blending and difficult continuum placement in these stars, it becomes increasingly difficult to find unblended Fe I and Fe II lines for equivalent width measurement with which to determine the atmospheric parameters. However, the standard techniques were adopted for all stars to obtain a spectroscopic effective temperature. A photometric temperature was also calculated using the colour-index calibration published by [7], assuming an [Fe/H] of 0.0 for this calibration. With three exceptions, HD7351, HD49331 and HD64332, the spectroscopic and photometric temperatures agreed to within 50K. All gravities fell between the expected values of log g=0.5 to 1.5. For one star, HD286340, no reliable B-V values could be found due to

[1] http://simbad.u-strasbg.fr

[2] http://irsa.ipac.caltech.edu

the star's extreme variability. In this case, typical values of T_{eff}=3500K, log g=1.0 and ν=1.50 km s^{-1}were chosen. All atmospheric parameters derived can be found in [1].

The abundance analysis for these field stars included using atomic line lists from Kurucz [3] with refined oscillator strength values for the s-process lines via a reverse solar analysis. Due to the fact that at these cooler temperatures and in these C enhanced stars, molecular features become very strong, all molecular features of CO, CN, CH, ZrO, VO, and TiO were included. With the exception of TiO line data, which was taken from Plez [9], all molecular line data was taken from Kurucz's linelists. This study does not employ equivalent width measurement for any lines due to this strong molecular blending and difficult continuum placement. All abundances are obtained using the spectrum synthesis program MOOG [10]. Model atmospheres are taken from the Bell et al. grid [11] for the M stars. For the C stars, as C/O approaches unity, C-rich model atmospheres were used, provided by the MARCS model grid (Eriksson, private communication).

3. RESULTS

All detailed abundance results for this sample of field AGB stars can be found in the detailed paper, [1]. Specific results of interest are discussed further below.

Iron abundances were obtained for all stars for which spectroscopic atmospheric parameters could be determined. The solar iron abundance MOOG adopts was used for all relative abundance analyses. This value is low, at log ε (Fe) = 7.52, close to the meteoritic abundance published by [12]. For HD286340, for which standard parameters were adopted, an iron abundance of [Fe/H]=0.0 was assumed. [Fe/H] values ranged from -0.26 dex to +0.17 dex. Thus all stars analysed can be considered of solar or near-solar metallicity. All stars are members of the disk population, and the spread in [Fe/H] results from both uncertainties and genuine differences.

3.1. C and O Abundances

The C abundance was determined from the C_2 bands in the 5060 to 5100Å region. The nature of C_2 results in extreme sensitivity to any change in C abundance, making it a reliable molecule to use in deriving any C enhancements. The C_2 molecular line parameters were taken from Kurucz's linelists. The O abundance was obtained from spectrum synthesis of the forbidden O lines at 6300 and 6363Å. As mentioned earlier, the abundances of C and O in AGB stars are of interest when comparing with theoretical predictions. It is expected that the C abundance will increase with recurring thermal pulses as fresh carbon is dredged to the surface. Consequently, the C/O ratio changes as the star moves up the AGB, from C/O$\sim$0.5 in M stars to C/O>1.0 in S stars.

The results of the C and O analysis are shown in Fig. 1, with C/O shown as a function of the frequency distribution among the sample. The C/O ratio is an indication of the quantity of carbon brought to the surface by the third dredge-up process. The M stars

[3] http://kurucz.harvard.edu

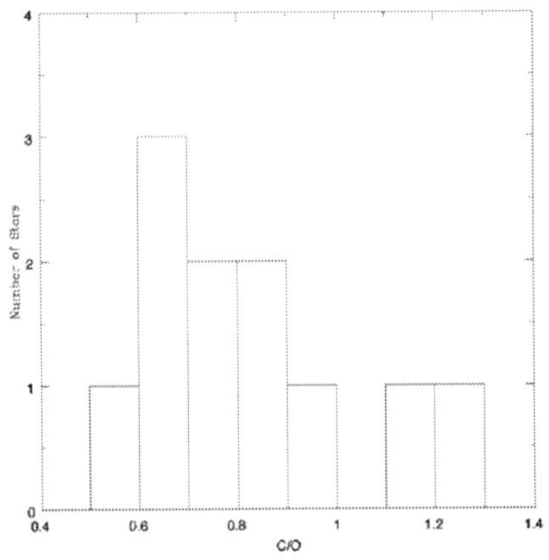

FIGURE 1. Distribution of C/O ratio for the field stars studied.

have C/O ratios ranging from C/O=0.52 to 0.76 with a mean of C/O=0.66, while the S and SC stars show values of C/O=0.60 to 1.32 with a mean of C/O=0.96. There is quite clearly a distinct spread in the C/O values in this sample.

3.2. s-Process Element Abundances

The M stars show small or no enhancements of s-process elements, Y, Zr, La and Nd. This is consistent with theory, in that there is not expected to be any process in M stars that would cause enhanced s-process element abundances as there have not been any thermal pulses this early in the AGB phase. The star HD7351 (M3) gives more enhanced values of [ls/Fe]=+0.62 and [hs/Fe]=+0.58 and also displays a higher C/O ratio of 0.72, which suggests it is a more evolved spectral class. Figure 2 shows a schematic of a Zr region and can be used to compare the relative strengths of Zr lines in a K giant (top) compared to M, S and SC (bottom) stars, indicated by the vertical lines.

There is a definite separation, marked by an increase in the s-process elements in the S, SC and C stars. These stars show values of [ls/Fe]=+0.55 to +1.12 and [hs/Fe]=+0.42 to +0.96. A single star, HD216672 (S5) has considerably lower values of [ls/Fe]=+0.39 and [hs/Fe]=+0.39 dex. With its lower s-process element enhancements and lower C/O ratio, it and HD7351 lie between the two extreme groups or M and SC stars.

4. DISCUSSION OF TRENDS IN AGB STARS

In an attempt to quantify the relationship between enhanced [C/Fe] and C/O values and s-process element enhancement, this research examines the [ls/Fe] and [hs/Fe] values as a function of increasing AGB spectral type (Fig. 3), [C/Fe] value and C/O ratio (Fig. 4).

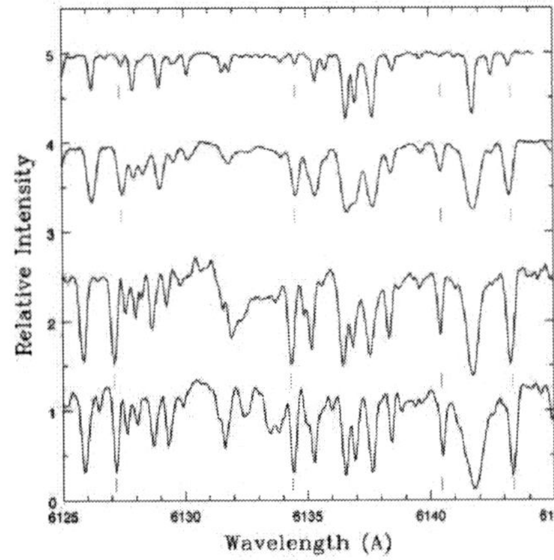

FIGURE 2. Spectra of a K giant (top), and an M, S and SC star. The increase in the strength of the Zr lines, from top to bottom, is marked with a vertical line.

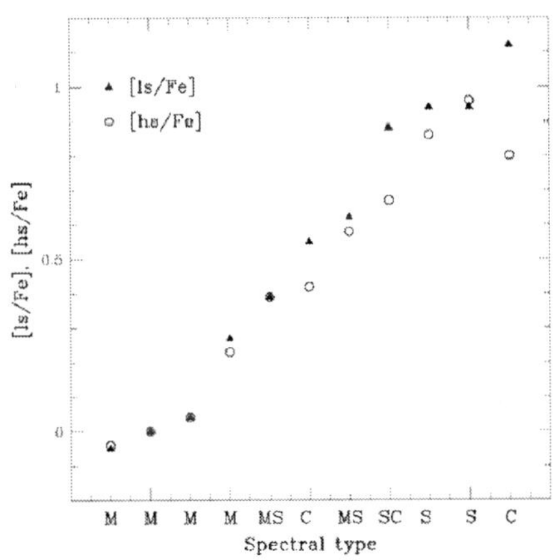

FIGURE 3. Trends in [ls/Fe], [hs/Fe] and [hs/ls] as a star progresses up the AGB - M, MS, S, SC, C.

Figure 3 shows the enhancements of the light s- (as triangles) and heavy s- process (as circles) for each individual star as the AGB phase progresses.

The M stars are clearly minimally enhanced, and as the spectral type advances the enhancements of the light and heavy s-process elements clearly increases. However, this figure relies heavily upon spectral classification, which can often be difficult or ambiguous. A more accurate way of quantifying s-process enhancements is through the [C/Fe] and C/O values.

Figure 4 shows the [ls/Fe] (as triangles) and [hs/Fe] (as circles) values as a function

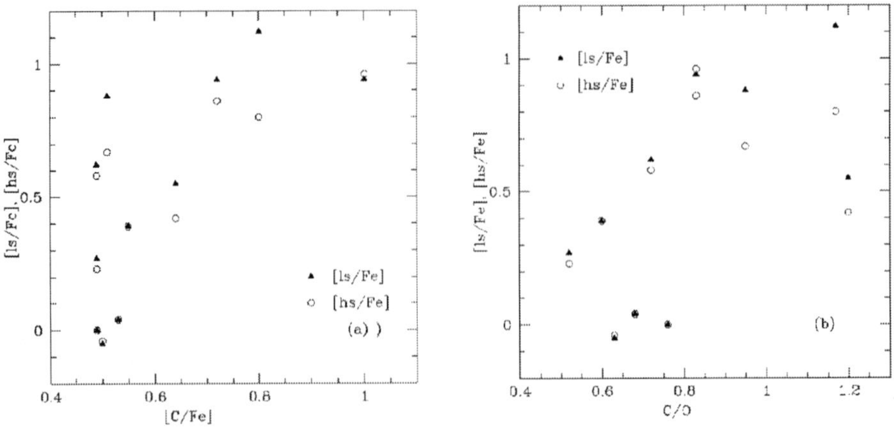

FIGURE 4. Trends in [ls/Fe], [hs/Fe] as a function of (a) [C/Fe] and (b) C/O.

of [C/Fe] value and a function of C/O ratio. In general, it is clear that as the [C/Fe] value increases, the related s-process enhancements are seen.

In addition, as the C/O ratio increases the difference between the [ls/Fe] and [hs/Fe] gets larger. This can be confirmed by looking at the individual [hs/ls] values as a function of C/O, or AGB type. The [hs/ls] values is a way of quantifying the relative strengths of the light s- and heavy s- process peaks, and indicates how far down the valley of stability the s-process has progressed. The M and MS stars have a mean [hs/ls] value of -0.01 while the S and SC stars show a smaller value of [hs/ls]=-0.06. This implies that while the s-process enhancements are minimal the two peaks are enhanced by the same amounts. However, as the enhancements increase in value, the two peaks separate and begin to be enhanced at different rates, with the light s-process peak being favoured.

Two stars, HD49368 and HD64332 show large Eu enhancements. The AGB evolutionary phase is not expected to produce r-process elements in any way, so it is not expected that AGB stars would show Eu enhancements due to their stellar evolution. These observed enhancements are more likely to come from the primordial enrichment of the material from which the stars formed or external pollution from other sources. It is possible that these two stars have undergone pollution from material rich in r-process elements. This may have involved extrinsic pollution from supernova remnants or other ejecta in which r-process element formation has previously occurred. This would explain the observed Eu enhancement in these stars. It may also help to explain the observed Nd enhancement. Nd has five stable isotopes, four of which are formed by both the s- and r-process and one of which is r-process only. If these two stars had undergone pollution from r-process element rich material, then it is possible that the source of this pollution is responsible for the observed enhancement of Nd. La, which has one stable isotope, can also be produced by both the s- and r-process, thus providing more support for this theory. The large enhancements of La, Nd and Eu observed in these stars may not be due solely to AGB nucleosynthesis, but partially due to pollution from previously r-processed matter.

5. SUMMARY

The field AGB stars studied in this paper give a self-consistent overview of nucleosynthesis during the AGB phase of evolution. The predicted s-process enhancements have been observed and the expected relationships between enhancements of both C and the s-process elements together appear to have been quantified.

ACKNOWLEDGMENTS

ECW acknowledges the support of a University of Canterbury Research Award, a Departmental Scholarship and the Dennis William Moore Scholarship, all held for the duration of this research. We would like to thank the MARCS group for their generosity in providing a few C-rich models for the analysis. This research has made use of the SIMBAD database, operated at the Centre de Donées Astronomiques de Strasbourg (CDS), France. This publication makes use of data products from the Two Micron All Sky Survey, which is a joint project of the University of Massachusetts and the Infrared Processing and Analysis Center, California Institute of Technology, funded by the National Aeronautics and Space Administration and the National Science Foundation.

REFERENCES

1. E.C. Wylie de Boer and P.L. Cottrell, *MNRAS*, in prep. (2008)
2. V. Smith and D. Lambert, *ApJ* **294**, 326 (1985)
3. V. Smith and D. Lambert, *ApJ* **311**, 843 (1986)
4. C. Abia and G. Wallerstein, *MNRAS* **293**, 89 (1998)
5. N. Samus and O. Durlevich, *Combined General Catalogue of Variable Stars, 2004yCat.2250....0S*
6. V. Smith and D. Lambert, *ApJ* **333**, 219 (1988)
7. A. Alonso, S. Arribas, and C. Martínez-Roger, *AAS* **140**, 261 (1999)
8. The HIPPARCOS and TYCHO Catalogues, *European Space Agency Publication Division, The Netherlands*, (1997)
9. B. Plez, *A&A* **337**, 495 (1998)
10. C. Sneden, *Carbon and Nitrogen Abundances in Metal-Poor Stars, PhD Thesis, University of Texas at Austin*, (1973)
11. R. Bell, K. Eriksson, B. Gustafsson, and A. Nordlund, *A&A* **23**, 37 (1976)
12. E. Anders and N. Grevesse, *Geochim.Cosmichem.Acta.* **53**, 197 (1989)

s-Process Element Abundance Results for 47 Tuc Stars Using SALT

C.C.Worley*, P.L.Cottrell* and E.C.Wylie de Boer†

*Department of Physics & Astronomy, University of Canterbury, Private Bag 4800, Christchurch 8020, New Zealand
†Research School of Astronomy & Astrophysics, Mount Stromlo Observatory, Cotter Road, Weston ACT 2611, Australia

Abstract. Eleven giant branch stars in 47 Tucanæ were observed using the Robert Stobie Spectrograph (RSS) on the Southern African Large Telescope during the performance verification phase of this instrument. These stars were analysed as part of a quest to investigate *s*-process element abundances throughout the colour-magnitude diagram of 47 Tucanæ. No abundance variations of Zr, Ba, Nd and Eu were found in these eleven stars, such that [X/Fe] = 0.0 ± 0.5 dex. Further, theoretical analysis indicates that the maximum resolution on RSS and AAOmega on the Anglo-Australian Telescope is sufficient to detect *s*-process element abundance variations. More detailed discussion on the abundance analysis of these stars and the theoretical analysis of spectrograph resolution can be found in [1].

Keywords: Globular Clusters, 47 Tuc, *s*-process, Robert Stobie Spectrograph, SALT
PACS: 95.55.Qf, 95.85.Kr, 97.10.Tk, 97.20.Li, 98.20.Gm

1. INTRODUCTION

The stars in a globular cluster (GC) are assumed to have all formed from the same gas cloud. Therefore, all the stars have the same initial chemical composition and the main sequence (MS) stars directly reflect this in their spectra. Observed abundance variations in evolved stars in globular clusters have been attributed to products of core nuclear burning being mixed into the stellar atmosphere. However, as GCs have been observed in progressively greater detail abundance variations have been found that cannot not be explained solely by nuclear processing and mixing. The review by [2] gives a comprehensive overview of observed GC abundance variations.

The four scenarios which could explain the abundance variations are:

- Mixing, the mixing of processed material to the surface of the star;
- Primordial, two gas clouds of differing chemical composition merged prior to star formation;
- Self-enrichment, previous generations of cluster stars enriched the gas cloud either through supernova or AGB stellar winds; and
- Pollution, material ejected from AGB stars in the cluster is accreted on to the surface of the observed stars polluting their atmospheres [3].

Extensive research has been carried out investigating the light element (C,N,O, etc) abundances anomalies and the results are favouring the pollution and self-enrichment

CP1001, *9th Torino Workshop on Evolution and Nucleosynthesis in AGB Stars and the 2nd Perugia Workshop on Nuclear Astrophysics*, edited by R. Guandalini, S. Palmerini, and M. Busso

scenarios. What can be added by looking at the heavier *s*-process elements? The *s*-process elements are formed during the Third Dredge Up (TDU) in the thermally pulsing stage of AGB stars. This process is theorized to only occur for AGB stars of greater than 1 $M_\odot$ [4]. A comparison of the *s*-process element abundances of AGB and RGB stars within a globular cluster should give an indication of whether TDU is taking place. If AGB and RGB *s*-process element abundances both show variations this is evidence for contamination by primordial or pollution theories.

2. GLOBULAR CLUSTERS & *S*-PROCESS ELEMENTS

Recent high resolution studies have concluded that *s*- and *r*-process element abundances are homogeneous within a cluster. In particular [5] found uniform enrichment of *s*- and *r*-process elements in NGC 6397 ([Fe/H] = -2.0), NGC 6752 ([Fe/H] = -1.5) and 47 Tuc ([Fe/H] = -0.7). Also [6] found that *n*-capture elements are homogeneously distributed in NGC 6388 ([Fe/H] = -0.44).

There have been several high resolution studies of stars in 47 Tuc where *s*- and *r*-process element abundances have been derived. While within each study there is no convincing evidence of *s*-process element abundance anomalies, there is disagreement in the abundance values between the studies. Table 1 shows mean values and uncertainties from four key studies, Brown and Wallerstein [7] (BW92), Alves-Brito et al. [8] (AB05), James et al. [5] (J04) and Wylie et al. [9] (W06). The elements that are compared are: Y & Zr (light *s*-process); Ba & La (heavy *s*-process); and Eu (*r*-process).

TABLE 1. Mean values and uncertainties for *n*-capture element abundances for 47 Tuc stars from four high resolution studies.

	BW92	AB05	J04	W06
No. Stars	4	5	12	7
[Fe/H]	-0.81±0.02	-0.67±0.12	-0.69±0.06	-0.62±0.07
[Y/Fe]	+0.45±0.11		-0.11±0.10	+0.64±0.19
[Zr/Fe]	-0.39±0.05	-0.17±0.12		+0.67±0.15
[Ba/Fe]		+0.31±0.22	+0.35±0.12	
[La/Fe]	+0.18±0.06	+0.05±0.10		+0.31±0.19
[Nd/Fe]				+0.42±0.13
[Eu/Fe]	+0.25±0.06	+0.33±0.10	+0.17± ...	+0.14±0.12
hs/ls	+0.06	+0.35	+0.23	-0.28±0.11

[8, 5, 9] are in line with the currently accepted metallicity value for 47 Tuc of [Fe/H] = -0.7 [2]. However, Y and Zr show a large variation between studies and there is some discrepancy for all the elements except Ba which agrees to the limits of the uncertainties. hs / ls is a ratio of the average of the heavy to light *s*- abundances and gives an indication of the relative strengths of the two *s*-process peaks.

The most recent study of RGB and AGB stars in 47 Tuc indicated that there may be a genuine star-to-star scatter in *s*-process element abundances [9]. Y and particularly Zr and La were found to have an average abundance that is much enhanced compared with previous values. Overall both the light *s*- and heavy *s*- elements are enhanced giving a hs / ls ratio that is very different to past studies. Y and Zr both being enhanced to a similar degree is a more consistent result than the previous studies.

It is unlikely that these enhancements in 47 Tuc are due to TDU. The AGB stars in 47 Tuc have mass of at most 0.85 $M_\odot$ [10], and according to TDU theory this mass is too low for *s*-process element formation to occur. [9] shows a spread in *s*-process abundance values for both AGB and RGB stars, supporting the argument that TDU is not the source of the enhancements. But the scatter in abundance values still indicates an inhomogeneous distribution of *s*-process element abundances in stars in 47 Tuc.

3. SALT & 47 TUC

A large scale survey of giant branch stars in 47 Tuc would establish the nature and origin of *s*-process element abundance variations. As *s*-process element lines are typically quite weak features in a stellar spectrum the observed spectra need to be sufficiently resolved that variations in abundance can be detected. A high resolution survey would provide the necessary spectral detail but would require extensive amounts of telescope time. However, multiple object observations are possible on several medium resolution instruments: AAOmega on the Anglo-Australian Telescope (AAT); Flames on the VLT; and the Robert Stobie Spectrograph (RSS) on the Southern African Large Telescope (SALT).

The University of Canterbury is a partner in the SALT consortium and current research programmes at Canterbury are being developed to maximize the use of this impressive facility. The Robert Stobie Spectrograph, which has been commissioned on SALT, has a maximum operating resolving power of $\sim$ 10 000, and multi-object capabilities via a slit mask system. HRS is the high resolution spectrograph, which is currently under construction and due for commissioning in late 2009. The 47 Tuc colour magnitude diagram in Fig. 1 was generated from [3, 11, 12] and includes the magnitude limits of current (solid lines) and future (dashed lines) observations on both SALT and the AAT. Both RSS and HRS cover a useful range of apparent magnitudes for 47 Tuc and the multi-object mode of RSS makes it ideally suited to large scale surveys of globular clusters.

However, is a resolving power of $\sim$ 10 000 sufficient to detect variations in *s*-process element abundances? A theoretical exploration of *s*-process element line strength in 47 Tuc stars at different evolutionary stages showed that a resolving power of 10 000 was sufficient to observe line strength variations of at least 0.3 dex [1]. This resolving power can be obtained on RSS and a comparable resolving power of $\sim$ 8 000 on AAOmega.

During 2006 RSS was installed on SALT for performance verification and eleven giant branch stars in 47 Tuc were observed for this study [1]. The location of the eleven stars on the 47 Tuc CMD is shown in Fig. 1.

The observed wavelength regions included key features for *s*- and *r*-process elements. The resolving power was calculated to be $\sim$ 5 000 and an *s*-process abundance analysis was carried out. Figure 1 shows both the observed spectra of the eleven stars and the synthesized spectra that were generated to determine element abundances. The effective temperatures ($T_{\rm eff}$) and surface gravity ($log\ g$) were calculated for each star from their apparent magnitude (V) and colour index ($B-V$) using the calibration equations in [13]. These were used to generate the synthesized spectra for the abundance analysis [14]. The spectra in Fig. 1 are presented with the temperature increasing down the panel

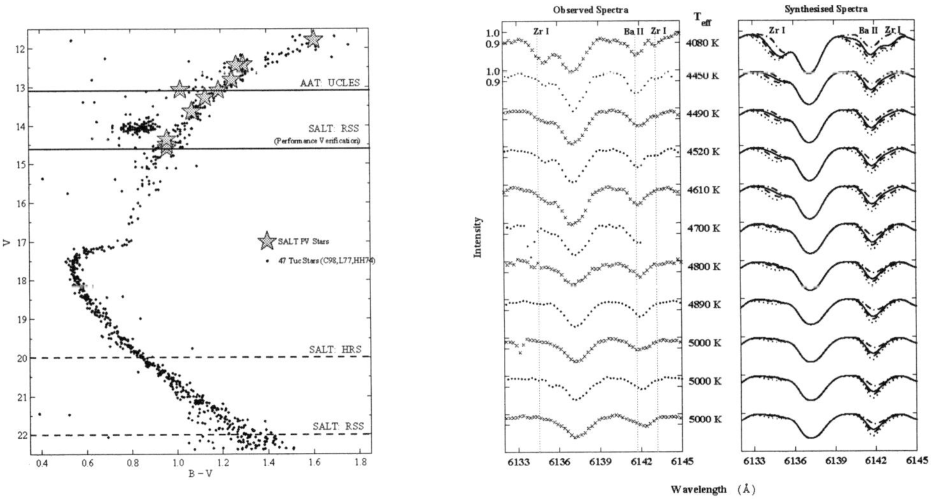

FIGURE 1. LEFT: CMD of 47 Tuc generated from [3, 11, 12] with V magnitude limits of current (solid line) and future (dashed line) observations on SALT and the AAT. Also shown are the stars observed using SALT RSS during performance verification ($\star$). RIGHT: Temperature sensitivity of Zr I & Ba II features. Observed spectra for the eleven 47 Tuc stars obtained on SALT RSS (left panel) & corresponding synthesized spectra (right panel). Vertical lines trace the key features in the left panel. For the synthesized spectra the dotted line indicates none of the element is present, the solid line indicates the model abundance, the dashed line indicates +0.5 dex enhancement and the dot-dash line indicates +1.0 dex enhancement.

replicating the traverse down the giant branch.

This region contains two Zr I features and one Ba II feature which are traced through the spectra by vertical lines for identification. In the coolest stars the Zr I features are quite distinct but as the temperature increases they diminish considerably in strength. The Ba II feature weakens only slightly as the spectra become hotter. As expected from theory the neutral lines are more sensitive to the change in temperature than the ionized lines. This temperature sensitivity shown by the Zr I lines confirms the temperature scale derived from the Alonso et al. [13] equations. This is replicated in the synthetic spectra.

Figure 2 shows four spectra in two key regions that were used in the abundance analysis. Error bars are included to give an indication of the uncertainty due to noise in the four spectra.

The left panel of Fig. 2 shows the region from Fig. 1 which includes the Zr I and Ba II features. The coolest star provides the best details as the Zr I and Ba II features are quite prominent. In terms of abundance this spectrum best fits the model synthesis indicating no Zr or Ba enhancement.

For the next two spectra, both of $T_{\rm eff} \sim 4500$ K, there maybe an indication of Zr enhancement although the error bar brings it within model abundance. The Ba II feature indicates model abundance.

In the spectra of the hottest stars, the Zr I features have diminished too much to

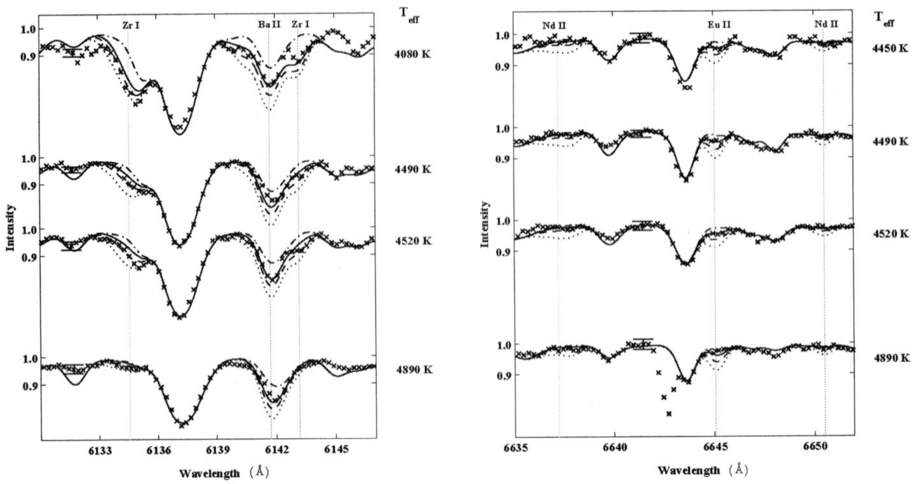

FIGURE 2. Abundance Analysis. LEFT: Zr I & Ba II Features. RIGHT: Nd II & Eu II Features. Key as for Fig. 1 except the observed spectrum is overlaid on the synthesized spectra for four of the eleven stars. An absorption artifact from a cosmic ray is present in the hottest star of the right panel.

observe any abundance variations. The Ba II feature is still prominent but again best fits the model abundance.

The right panel of Fig. 2 shows a region which has Nd II and Eu II features. However, the Nd II features are very weak and difficult to distinguish above the noise. The Eu II feature is prominent initially but diminishes with the hotter temperatures. A model abundance is the best fit here.

The abundance analysis on these 47 Tuc stars indicates an [Zr/Fe] = 0.0 with a possible uncertainty to +0.5 dex, while [Ba/Fe],[Nd/Fe],[Eu/Fe] = 0.0.

4. CONCLUSION & FUTURE WORK

The analysis of abundance anomalies in globular clusters must be extended to *s*- and *r*-process elements in order to develop GC formation and structure theories. Current theory anticipates a homogeneous distribution of these elements in GCs, but recent discrepancies between studies of 47 Tuc stars must be further investigated.

Abundance analysis of the eleven stars observed using RSS on SALT found no variations in Zr, Ba, Nd or Eu, such that $[X/Fe] = 0.0 \pm 0.5$ dex. However, the spectra were not obtained at the maximum resolution of RSS so only upper limits could be placed on these abundances. The temperature sensitivity of the neutral lines did provide a temperature scale that confirms the one calculated from the [13] calibrations.

A large scale survey of 47 Tuc is planned to take place in 2008 and will provide definitive results on the nature of *s*-process element abundance variations. This work will be extended to other globular clusters with a range of metallicities in order to determine whether the distribution of *s*-process element abundances is a characteristic of all globular clusters.

Observations of main sequence stars in globular clusters will further extend this analysis as the atmospheres of main sequence stars can be considered to reflect solely the chemical composition of the initial gas cloud. Comparison between main sequence and giant branch stars will identify which abundance variations are primordial in origin. HRS on SALT will be capable of observing these faint stars and a follow up to the large scale survey using RSS is planned for that instrument.

ACKNOWLEDGMENTS

This research has been supported by the University of Canterbury Scholarship programme, the Dennis William Moore Scholarship Fund and a 2007 Fellowship from the New Zealand Federation of Graduate Women. Attendance at the IXth Torino Workshop 2007 was made possible through funding from the Royal Society of New Zealand, the Canterbury Branch of the Royal Society of New Zealand, and the Department of Physics & Astronomy at the University of Canterbury. Financial support was also gratefully received from the organizers of the IXth Torino Workshop 2007.

REFERENCES

1. C.C. Worley, P.L. Cottrell, and E.C. Wylie de Boer, *s*- and *r*-process element abundances in the CMD of 47 Tucanæ using the Robert Stobie Spectrograph on SALT, in preparation (2007)
2. R. Gratton, C. Sneden, and E. Carretta, *ARA&A* **42**, 385 (2004)
3. R.D. Cannon, B.F.W. Croke, R.A. Bell, J.E. Hesser, and R.A. Stathakis, *MNRAS* **298**, 601 (1998)
4. M. Busso, R. Gallino, D. L. Lambert, C. Travaglio, and V.V. Smith, *ApJ* **557**, 802 (2001)
5. G. James, P. François, P. Bonifacio, E. Carretta, R.G. Gratton, and F. Spite, *A&A* **427**, 825 (2004)
6. E. Carretta, A. Bragaglia, R.G. Gratton, Y. Momany, A. Recio-Blanco, S. Cassisi, P. François, G. James, S. Lucatello, and S. Moehler, *A&A* **464**, 967 (2007)
7. J.A. Brown and G. Wallerstein, *AJ* **104**, 1818 (1992)
8. A. Alves-Brito, B. Barbuy, S. Ortolani, Y. Momany, V. Hill, M. Zoccali, A. Renzini, D. Minniti, L. Pasquini, E. Bica, and R.M. Rich, *A&A* **435**, 657 (2005)
9. E.C. Wylie, P.L. Cottrell, C.A. Sneden, and J.C. Lattanzio, *ApJ* **649**, 248 (2006)
10. R.L. Gilliland, G. Bono, P.D. Edmonds, F.Caputo, S. Cassisi, L. D. Petro, A. Saha, and M.M. Shara, *ApJ* **507**, 818 (1998)
11. S.W. Lee, *A&AS* **27**, 381 (1977)
12. F.D.A. Hartwick and J.E. Hesser, "Photometry To The Main-Sequence of 47 Tucanae." in *BAAS*, vol. 6 of *BAAS*, p. 216 (1974)
13. A. Alonso, S. Arribas, and C. Martínez-Roger, *A&AS* **140**, 261 (1999)
14. C. Sneden, *MOOG: An LTE Stellar Line Analysis Program*, Ph.D. thesis, University of Texas at Austin (1973)

s-Process Abundances in Globular Clusters: M4 vs. M5 and NGC 1851

David Yong*, Frank Grundahl†, Amanda Karakas* and David L. Lambert**

**Research School of Astronomy and Astrophysics, Australian National University, Weston, ACT 2611, Australia*
†*Department of Physics and Astronomy, University of Aarhus, Denmark*
***The W.J. McDonald Observatory, University of Texas, Austin, TX 78712, USA*

Abstract. We present s-process abundances for giant stars in the globular clusters M4, M5, and NGC 1851. We find that all s-process elements are overabundant in M4 relative to M5. Curiously, the abundances of Cu and Zn are also overabundant in M4 which may offer clues to the synthesis of these elements. For NGC 1851, we find that the abundances of the s-process elements Zr and La exhibit a large dispersion. The abundances of Zr and La also appear to have a bimodal distribution reinforcing the idea that there are two stellar populations in this cluster.

Keywords: Abundances, Chemical Composition: Giant and Subgiant Stars: Globular Clusters in the Milky Way
PACS: 97.10.Tk,97.20.Li,98.20.Gm

1. INTRODUCTION

Globular clusters are ideal laboratories for testing many aspects of stellar evolution and stellar nucleosynthesis. Observational and theoretical attention has focused primarily on the star-to-star abundance variations for light elements found in every well studied cluster (see contributions by D'Antona and Ventura). However, measurements of the s-process elements also offer great insight into the chemical evolution of globular clusters and stellar nucleosynthesis (see contributions by Gallino and Worley).

2. M4 VS. M5

The globular clusters M4 and M5 have been well studied by [1, 2]. They have essentially identical metallicities, [Fe/H] $\simeq -1.2$, and the abundance similarities extend to many α and Fe-peak elements. The s-process elements reveal the most striking abundance comparison between M4 and M5. Specifically, Ba and La are overabundant in M4 relative to M5. Of great interest is the identification of additional elements whose abundances differ between these two clusters. Such elements, if they exist, offer further clues to the different chemical evolutionary histories of these clusters.

We measured the abundances for a large number of additional elements in 12 bright giants in M4 and 2 bright giants in M5 using spectra obtained from the Magellan telescope following the analysis by [3]. Our abundances were supplemented by Cu measurements from [4] and Ba measurements from [1, 2]. We show the results in Fig. 1 and 2.

CP1001, *9th Torino Workshop on Evolution and Nucleosynthesis in AGB Stars and the 2nd Perugia Workshop on Nuclear Astrophysics,*
edited by R. Guandalini, S. Palmerini, and M. Busso

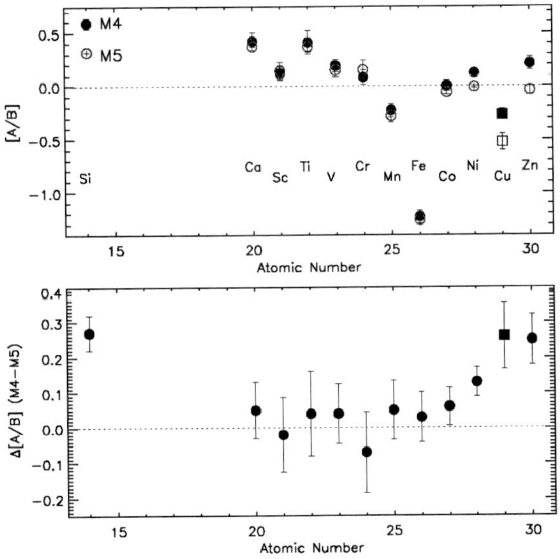

FIGURE 1. The upper panel shows [X/Fe] and [Fe/H] ratios for M4 and M5. The lower panel shows the abundance ratios for M4 − M5.

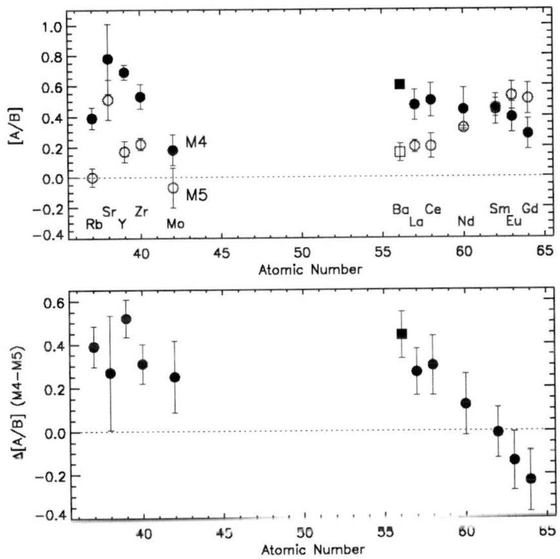

FIGURE 2. Same as Fig. 1 but for different elements.

For α and Fe-peak elements, we extend the measurements to additional elements not studied by [1, 2]. For Ca to Co, M4 and M5 have identical ratios [X/Fe]. For all *s*-process elements (Rb, Sr, Y, Zr, Mo, Ba, La, Ce), M4 has overabundances compared to M5. For all *r*-process elements (Sm, Eu, Gd), M4 and M5 have essentially identical abundances. For Nd, which is 50% *s*- and 50% *r*-process, the abundance difference between M4 and M5 lies midway between the *s*- and the *r*-process elements. For Cu and Zn, M4 is overabundant compared to M5. The abundance difference for Cu and Zn is comparable to the *s*-process abundance difference in these clusters. The similarities in abundance differences may indicate that Cu and Zn are produced in the same stars that synthesize the *s*-process elements, presumably AGB and/or massive stars. (See contributions by Chieffi and Primas who discussed Cu).

However, we confirm that Si is overabundant in M4 relative to M5. The amplitude of the abundance difference is comparable to that seen for Cu, Zn and the *s*-process elements. Since Si is an α element, the chemical differences between M4 and M5 are presumably caused by different nucleosynthetic processes occurring in different types of stars.

3. NGC 1851

The globular cluster NGC 1851 has a bimodal color distribution of horizontal branch stars. The sole spectroscopic analysis of NGC 1851 by [5] found extremely CN-strong stars and the hint of Sr and Ba abundance variations. [6] have recently found that this cluster has a double subgiant branch. However, no detailed abundance analyses have been performed upon this cluster.

We measured the abundances for Fe, O, Na, Mg, Al, Zr, La and Eu in 8 bright giants using spectra obtained from the VLT. (Details can be found in [7].) We show the abundance results in Fig. 3.

NGC 1851 shows star-to-star abundance variations of O, Na, and Al. The abundances of O and Na are negatively correlated and the abundances of Na and Al are positively correlated. Such abundance patterns are found in every well studied Galactic globular cluster.

There are also large star-to-star abundance variations of the light *s*-process element Zr and the heavy *s*-process element La. Further, the Zr and La abundances are positively correlated with Al and negatively correlated with O. The Zr and La abundances appear to have a bimodal distribution adding further weight to the notion that there are two stellar populations in this cluster. The abundance variation of light (O, Na, and Al) and heavy (Zr and La) elements and the double subgiant branch are characteristics also shared by ω Cen, the most massive globular cluster now regarded as the nucleus of an accreted dwarf galaxy. NGC 1851 has experienced a complicated star formation history and may be regarded as a bridge between ω Cen (large variation of all elements) and NGC 6752-like clusters (constant Fe and heavy elements but large variation of the light elements C-Al). Therefore, NGC 1851 will play a prominent role in our understanding of the formation of the Galactic globular cluster system.

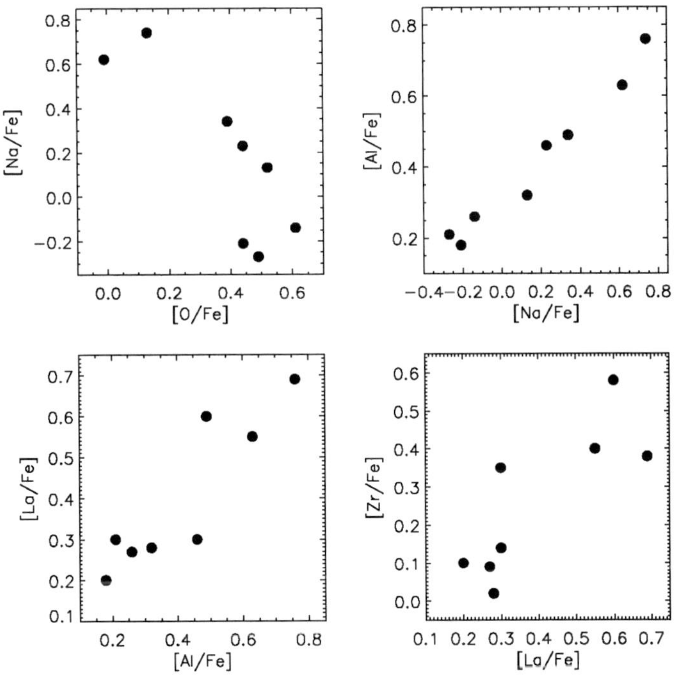

FIGURE 3. Clockwise from upper left, [Na/Fe] vs. [O/Fe], [Al/Fe] vs. [Na/Fe], [Zr/Fe] vs. [La/Fe], and [La/Fe] vs. [Al/Fe].

ACKNOWLEDGMENTS

DY thanks the local organizing committee for their hard work and the workshop participants for many interesting talks and stimulating conversations.

REFERENCES

1. I.I. Ivans, C. Sneden, R.P. Kraft, N.B. Suntzeff, V.V. Smith, G.E. Langer, and J.P. Fulbright, *AJ* **118**, 1273 (1999)
2. I.I. Ivans, R.P. Kraft, C. Sneden, G.H. Smith, R.M. Rich, and M. Shetrone, *AJ* **122**, 1438 (2001)
3. D. Yong, D.L. Lambert, D.B. Paulson, and B.W. Carney, *ApJ in press (ArXiv 0710.2367)* **710** (2007)
4. J. Simmerer, C. Sneden, I.I. Ivans, R.P. Kraft, M.D. Shetrone, and V.V. Smith, *AJ* **125**, 2018 (2003), arXiv:astro-ph/0303176
5. J.E. Hesser, R.A. Bell, G.L.H. Harris, and R. D. Cannon, *AJ* **87**, 1470 (1982)
6. A.P. Milone, L.R. Bedin, G. Piotto, J. Anderson, I.R. King, A. Sarajedini, A. Dotter, B. Chaboyer, A. Marin-Franch, S. Majewski, A. Aparicio, M. Hempel, N.E.Q. Paust, I.N. Reid, A. Rosenberg, and M. Siegel, *ApJ in press (arXiv:0709.3762)* (2007)
7. D. Yong and F. Grundahl, *ApJL in press (ArXiv 0711.1394)* **711** (2007)

Copper Abundances in Metal-Poor Stars

F. Primas and J. Sobeck

European Southern Observatory, Karl-Schwarzschildstr. 2, 85748 Garching (Germany)

Abstract. By studying the trends and possible intrinsic dispersion of element abundance ratios in extremely metal-poor (EMP) stars it is possible to uncover information about the degree of homogenization of the early Galaxy, how the Galactic mixing processes behaved with time, and to test theoretical models predictions (e.g. Type II supernovae yields). Here, we report on the preliminary results of a new study of copper abundances in a large sample of metal-poor stars.

Keywords: Stellar Abundances, Low Metallicity Stars, Copper, Nucleosynthesis
PACS: 95.30.Ky, 95.55.Cs, 95.85.Kr, 95.85.Ls, 97.10.Cv

1. INTRODUCTION

It had been thought for a long time that the iron-group elements (usually Sc through Zn) exist at roughly their solar ratios with respect to iron over a wide range in metallicity (from solar down to one/thousandth solar). In the last decade, however, previously "unnoticed" departures of iron-peak elements Cr, Mn, Co, and Cu from their solar ratios have been detected and confirmed at [Fe/H]≤ -2.5 (e.g. [1] for the first findings). These findings suggested that a distinct phase of non-solar nucleosynthesis occurred before the Galaxy reached that metallicity and raised the question of what truly constitutes the high-Z end of the iron-group elements.

Among all iron-peak elements, copper remains the least studied. Observational efforts date back to the early 90s (e.g. [2]) and due to the weakening of the optical Cu transitions as more metal-poor stars are investigated, they did not succeed in exploring the Cu abundance trend below [Fe/H]=−3.0. The yet unknown trend of Cu with metallicity at the very low metallicities, routinely probed nowadays for other elemental ratios, was the main trigger of our new observational campaign.

2. THE NEW PROJECT

Our aim was to gather a large sample of field stars (dwarfs and giants), pushing the detection limit below [Fe/H]=−3.0. The main challenge remains the same as the one Sneden and collaborators were confronted with back in the early 90s, i.e. the fact that the (relatively few) observable atomic transitions of copper in the optical spectra of stars become vanishingly weak at low metallicities. Copper is usually observed via two Cu I lines at λ510.5 nm and λ578.2 nm. In order to make some significant progress, we decided to target the resonant Cu I lines in the near-UV (λ 324.7 nm and λ 327.4 nm), now easily accessible from high quality spectra obtained at 8m class telescopes equipped with efficient UV spectrographs.

CP1001, *9th Torino Workshop on Evolution and Nucleosynthesis in AGB Stars and the 2nd Perugia Workshop on Nuclear Astrophysics*, edited by R. Guandalini, S. Palmerini, and M. Busso

2.1. The Observed Data-Set

Our main data-set was observed at the ESO VLT, making use of the high-resolution Echelle spectrograph UVES, mounted on the Nasmith platform of the Kueyen 8m telescope, and it includes 50 metal-poor dwarf and subgiant stars, spanning the metallicity range between [Fe/H]=-3.3 and -0.5. All spectra cover both the ultraviolet and optical Cu transitions, and are of high quality: R=40-50,000 and S/N ratios ranging from 80-150 (in the UV) to 250-500 (in the red). The data were reduced following standard echelle reduction prescriptions.

This sample was then combined with a data-set secured earlier at the McDonald Observatory, using the 2d-coudé spectrograph mounted on the 2.7m telescope, which includes another 50-60 stars (mostly giants) spanning the same metallicity range. This data set is also characterized by high quality standards: R=60,000 and S/N$\geq$100. Standard reduction procedures were applied.

2.2. Stellar Parameters

Since the VLT sample has already been employed in other studies (e.g. [3]), no new effort was devoted to the determination of the stellar parameters, which were taken directly from these investigations. In short, final effective temperatures were derived from Strömgren uvby-β and K photometry (respectively from [4,5]) and the InfraRed Flux Method calibrations derived by [6]. Hipparcos parallaxes, available for all target stars, were used to calculate the surface gravities. Microturbulent velocities were determined by requiring that the derived iron abundances do not exhibit a dependence on the reduced equivalent width. Metallicities were derived from measuring the equivalent width of a set of Fe II lines, the oscillator strength of which came from the same laboratory work [7]. Some Fe I lines were also checked and found to provide consistent values.

For the McDonald sample, stellar parameters were taken for the time being from the literature. However, we note that the sample consists of very well known and extensively studied metal-poor halo giants, for which very consistent stellar parameter sets have been reported.

The metallicity ranges covered by the two samples are very similar and go from -3.3 (-3.1) to -0.5 (-0.6) for the VLT (MDO) samples. Temperature and gravity ranges instead differ, because of the different evolutionary stages probed by the two datasets: main-sequence and turn-off stars from the VLT (hence, $T_{\rm eff}$=5500-6500 K and log g=3.7-4.6), subgiants and giants, with only few turn-off stars from MDO (hence, $T_{\rm eff}$=4200-6500 K and log g=0.5-4.5).

2.3. Preliminary Results

Equivalent width measurements do not properly account for the presence of line contaminants and hyperfine structure, hence the most current version of the spectral

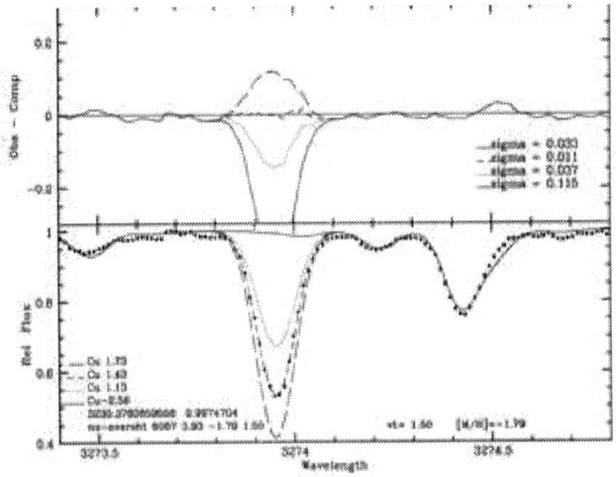

FIGURE 1. The spectral region around one of the Cu I resonant doublet lines as observed in a [Fe/H]=−1.8 star.

synthesis code MOOG [8] was employed for the derivation of the abundances. One-dimensional plane parallel stellar atmospheric models were used under the assumption of local thermodynamic equilibrium conditions. OSMARCS model atmospheres [9,10] were computed for the VLT sample, whereas ATLAS models [11] have been preliminarly generated for the MDO sample.

Figure 1 shows a portion of the near-UV spectrum (filled circles) of a metal-poor star around the Cu I resonant transition at 327.4 nm, together with our best-fit synthesis (dashed line). Three more syntheses are overlapped, that were computed respectively with no copper (continuous line), and with a copper abundance equal to ±0.3 dex with respect to the best-fit value (long-dashed and dotted lines respectively).

3. RESULTS AND DISCUSSION

The synthesis of copper is accounted for by two main production mechanisms: one which is Fe-independent, i.e. a primary process such as explosive nucleosynthesis which allows for a simultaneous enrichment of Cu and Fe, and one Fe-dependent, i.e. a secondary process such as neutron captures which require an iron seed nucleus. Our current understanding of the nucleosynthesis of copper considers slow neutron capture processes the main production channel. Further, the vast majority of copper is generated by the weak s-process that transpires in massive giants during the He-core burning and C-shell burning phases [12,13], with a limited amount (5.2%) coming from the main component of the s-process, i.e. from syntheses taking place in low mass asymptotic giant branch stars [14]. Very little copper, if any, can be accounted for by Type Ia supernovae (SNe, cf [15,16]), whereas 5-10% at most may be abscribed to core collapse SNe (via the radiogenic decay of ^{63,65}Zn, cf. [12]).

Figure 2 displays our derived copper abundances as a function of metallicity for a subset (VLT data only) of our total data sample. The top panel shows the [Cu/Fe] values

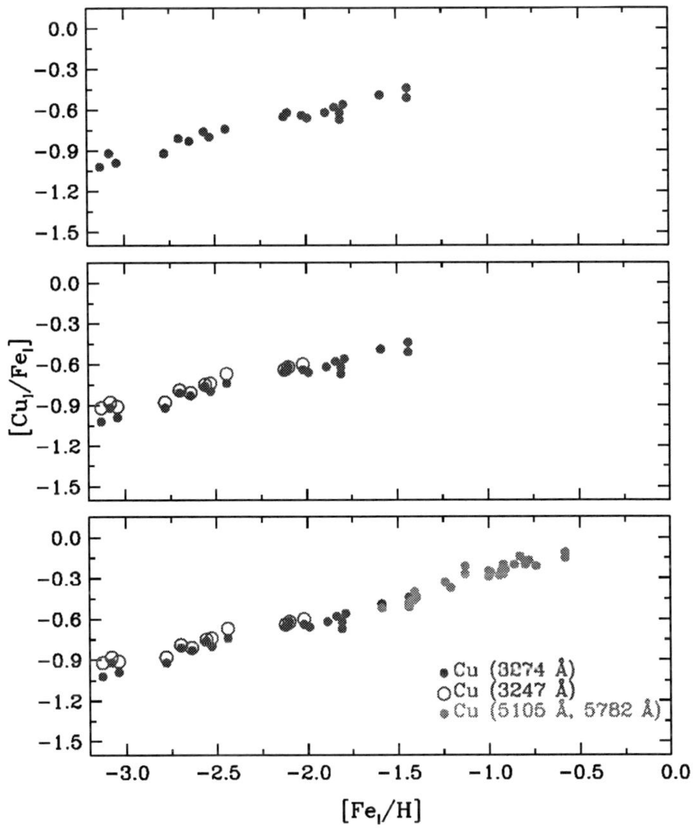

FIGURE 2. [Cu/Fe] vs [Fe/H], as derived from different Cu transitions. See text for more details.

as derived from the 327.4 nm transition, whereas the middle and bottom panels show the inclusion of the abundances derived from the 324.7 nm (open circles) and 510.5, 578.2 nm lines (filled data points above [Fe/H]=−1.5). The agreement obtained from the different abundance indicators is very satisfactory, though the copper abundances derived from the 327.4 nm line are found to be on average slightly higher than those derived from the 324.7 nm transition. The difference is marginal and well within the observational errors (estimated to range between 0.15 and 0.25 dex, respectively for the more metal-rich and the more metal-poor stars).

Even just a quick look at Fig. 2 shows that the copper abundance continuously decreases over the entire range of metallicity. Moreover, this data-set does not support the presence of any scatter because the very little differences are well within the observational errors.

Despite the preliminary nature of these results and the fact that only a sub-set of our final sample is presented here, we can already draw some tentative conclusions. Our derived trend of Cu versus metallicity reinforces the results found by [12], i.e. a strong signature of secondary-like production in massive stars by n-captures with a very small primary contribution by explosive nucleosynthesis , if any. Our derived trend does not require any contribution from Type Ia SNe, which had been suggested in the past in order to account for the possible flattening of [Cu/Fe] at the lowest metallicity end.

Next steps include the completion of the analysis of the entire sample, and the increase of the number of data points at $[Fe/H] \leq -3.0$.

ACKNOWLEDGMENTS

We are very grateful to J. Lawler for extensive discussion regarding atomic parameters for the neutral species of copper.

REFERENCES

1. A. McWilliam, et al, *AJ* **109**, 2757 (1995)
2. C. Sneden, R. G. Gratton, and D. A. Crocker, *A&A* **246**, 354 (1991)
3. P. E. Nissen, F. Primas, M. Asplund, and D. L. Lambert, *A&A* **390**, 235 (2002)
4. W. J. Schuster and P. E. Nissen, *A&AS* **73**, 225 (1988)
5. M. F. Skrutskie, et al., *AJ* **131**, 1163 (2006)
6. A. Alsonso, S. Arribas, and C. Martinez-Roger, *A&AS* **117**, 227 (1996)
7. E. Biémont, M. Baudoux, R. L. Kurucz, W. Ansbacher, and E. H. Pinnington, *A&A* **249**, 539 (1991)
8. C. Sneden, *ApJ* **184**, 407 (1973)
9. B. Gustafsson, R. A. Bell, K. Eriksson, and A. Nordlund, *A&A* **42**, 407 (1975)
10. M. Asplund, B. Gustafsson, D. Kiselman, and K. Eriksson, *A&A* **318**, 521 (1997)
11. R. L. Kurucz, *CD-ROM No. 13* (1993)
12. S. Bisterzo, L. Pompeia, R. Gallino, M. Pignatari, K. Cunha, A. Heger, and V. V.Smith, *Nuc. Phys. A* **758**, 284c (2005)
13. D. Romano and F. Matteucci, *MNRAS* **378**, L59 (2007)
14. C. Travaglio, R. Gallino, E. Arnone, J. Cowan, F. Jordan, and C. Sneden, *ApJ* **601**, 864 (2004)
15. K. Iwamoto, F. Brachwitz, K. Nomoto, N. Kishimoto, H. Umeda, W. R. Hix, and F. K. Thielemann, *ApJS* **125**, 439 (1999)
16. C. Travaglio, W. Hillebrandt, and M. Reinecke, *A&A* **443**, 1007 (2005)

SESSION VII:

SOLAR SYSTEM MATERIALS, PRESOLAR GRAINS, AND NUCLEOSYNTHESIS

Presolar SiC Grains and Rare Earth Element Production in AGB Stars

Andrew M. Davis[a] and Roberto Gallino[b,c]

[a]Department of the Geophysical Sciences, Enrico Fermi Institute and Chicago Center for Cosmochemistry, University of Chicago, Chicago, IL 60637
[b]Dipartimento di Fisica Generale, Universitá di Torino, Via P. Giuria 1, 10125 Torino, Italy
[c]Center for Stellar and Planetary Astrophysics, School of Mathematical Sciences, Monash University, PO Box 28M, Victoria 3800, Australia

Abstract. The relative abundances of many isotopes of the rare earth elements are strongly affected by *s*-process nucleosynthesis in low-mass asymptotic giant branch (AGB) stars. We briefly explore the nucleosynthesis of the rare earth elements in AGB stars and their implications for measurements of presolar SiC grains from those stars.

Keywords: Rare Earth Elements, *s*-process Nucleosynthesis, AGB stars, Presolar Grains
PACS: 97.10.Cv

1. INTRODUCTION

Circumstellar presolar grains from meteorites formed in stellar outflows and survived a variety of potentially destructive processes in the interstellar medium, in the solar nebula, in the asteroidal parent bodies of their primitive meteorite hosts, and in the laboratory. Presolar grains provide an isotopic record of nucleosynthesis in individual stars of a variety of types, including asymptotic giant branch (AGB) stars, core collapse supernovae, and novae [1,2]. Silicon carbide is the most widely studied type of presolar grain and most (~90%) SiC grains have isotopic properties consistent with formation in low mass AGB stars. Such stars are thought to be the primary stellar site for *s*-process nucleosynthesis.

A number of heavy elements have been measured in presolar SiC grains, first in aggregates of hundreds to thousands of grains in the early 1990s, and then with the development of new analytical techniques, principally resonant ionization mass spectrometry (RIMS) but also high sensitivity secondary ion mass spectrometry (SIMS), in single μm-sized SiC grains beginning around ten years ago. Heavy elements measured in individual grains include strontium [3], zirconium [4,5,6], molybdenum [5,6,7], ruthenium [8], and barium [5,6,9,10]. The nucleosynthesis record of strontium, zirconium, molybdenum and barium in presolar grains confirm the theoretical picture of nucleosynthesis of *s*-process isotopes in the helium intershell of thermally pulsing low mass AGB stars with episodic mixing of this material into the stellar envelope by third dredge up events [11]. Recent measurement of several heavy elements in each grain suggest that the amount of ^{13}C in the ^{13}C pocket [11,12,13] is restricted to a fairly narrow range within the range of values that are allowable in models [5,6].

CP1001, *9th Torino Workshop on Evolution and Nucleosynthesis in AGB Stars and the 2nd Perugia Workshop on Nuclear Astrophysics*, edited by R. Guandalini, S. Palmerini, and M. Busso

Many of the rare earth elements (REE) have elevated concentrations in presolar SiC [14,15]. We summarize here predictions of *s*-process nucleosynthesis calculations for low mass AGB stars for the rare earth elements in anticipation of future measurements in individual presolar grains. We focus on RIMS, which has the highest sensitivity and greatest reduction of isobaric interferences, although in a few cases, lack of interferences allows analysis by SIMS.

2. *S*-PROCESS NUCLEOSYNTHESIS OF THE REE

The REE comprise 15 elements with atomic numbers from 57 to 71. We begin with a brief review of *s*-process nucleosynthesis in AGB stars. A thermally pulsing AGB star consists of a carbon-oxygen core consisting of the ashes of helium burning, a helium intershell, and a convective envelope that at the beginning of the thermally pulsing stage has the elemental and isotopic abundances that the star was born with (in this case we have assumed that it was the same as the Sun). There are two sources of energy in the star. Most of the time, hydrogen burning occurs at the base of the convective envelope, gradually increasing the mass of the radiative helium intershell. When the temperature at the base of the helium intershell becomes high enough to drive the triple alpha reaction, helium burning occurs in a thermal pulse of short duration (a few years), causing the star to expand and shutting down hydrogen burning. After the thermal pulse ceases, the star contracts, the convective envelope reaches down into the helium intershell and dredges ^{12}C and *s*-process products up into the envelope, and hydrogen burning begins once again. This cycle is repeated many times. The *s*-process is driven by two neutron sources: $^{13}C(\alpha,n)^{16}O$ during the interpulse period, and $^{22}Ne(\alpha,n)^{25}Mg$ during thermal pulses. Immediately after each thermal pulse, some hydrogen is ingested into the helium intershell and makes some ^{13}C via $^{12}C(p,\gamma)^{13}N(\beta^{+}\nu)$ ^{13}C, forming a layer within the helium intershell referred to as the ^{13}C pocket. The $^{13}C(\alpha,n)^{16}O$ reaction is then activated over a period of several thousand years, giving a low flux, high fluence neutron irradiation and producing *s*-process isotopes, primarily from an iron seed. During thermal pulses, the *s*-process-product-rich ^{13}C pocket is mixed with the rest of the helium intershell and subjected to a high flux, low fluence neutron irradiation from the marginally activated $^{22}Ne(\alpha,n)^{25}Mg$ reaction. This second irradiation favors the neutron capture channel over the β decay channel along the *s*-process neutron capture path for some isotopes with half-lives of a few tens of days or more. Examples of such branch points include ^{141}Ce and ^{147}Pm. The amount of ^{13}C in the ^{13}C pocket is a free parameter in models. The so-called "standard case", denoted as "ST" in subsequent discussion, corresponds to the ^{13}C amount that best matches the solar system *s*-process isotopic composition for an AGB star of one-half solar metallicity [12]. The possible range of ^{13}C amount in the ^{13}C pocket ranges from one-twelfth the ST case ("D12"), which corresponds to no hydrogen ingestion after thermal pulses, to twice the ST case ("U2"), which is a practical upper limit (further hydrogen ingestion leads to the formation of ^{14}N, a neutron poison). Recent measurements of the isotopic compositions of zirconium, molybdenum, and barium suggest that most SiC grains come from AGB stars with ^{13}C pocket amounts within 50% of the ST case [5,6]. See references [11–13] for far more detailed descriptions of the workings of the *s*-process in low mass AGB stars. All calculations described here make use of the neutron capture cross section compilation of Bao et al. [16].

Shown in Fig. 1 is the *s*-process path in the REE region of the chart of the nuclides. Although the *s*-process path is simple in parts of the chart, there are a number of branch points that are sensitive indicators of neutron density during thermal pulses. We will discuss these in the treatment of individual elements below.

We now examine the production of the REE as a group. Here, it is easiest to express this as an overproduction factor. In Fig. 2, we show the overproduction of all of the REE isotopes in the envelope of a 2 $M_{\odot}$ AGB star of solar metallicity. A number of interesting features become readily apparent. The *s*-only isotopes in Fig. 1, designated by bold lines around them, have higher overproductions than the other isotopes in Fig. 2, and all have approximately the same degree of overproduction, 20–25 times the solar value for the U2 case. Those isotopes that are made only by the *r*- and *p*-processes show only a very slight depletion compared to the solar value, as the convective envelope of the AGB star reflects the original composition of the star (in this case, solar metallicity and isotopic composition). Although these isotopes are completely destroyed by neutron capture in the helium intershell, the convective envelope is little affected: it simply has *s*-process isotopes and ^{12}C added by dredge up after each thermal pulse.

Of the 15 REE, promethium has no long-lived or stable isotopes, praseodymium, terbium, holmium, and thulium have only one stable isotope, and, although lanthanum has two stable isotopes, one is of very low abundance. Since isotopic ratios of these elements cannot be measured, we will only consider the remaining REE individually.

2.1 Cerium

Cerium is an *s*-process element whose elemental abundance is dominated by neutron magic ^{140}Ce, which comprises more than 88% of solar system cerium. ^{136}Ce and ^{138}Ce are rare *p*-process isotopes whose abundance is too low to practically measure; ^{142}Ce is mostly produced by the *r*-process, but there is a small production when the neutron capture channel at the ^{141}Ce branch point is activated during thermal pulses.

2.2 Neodymium

Neodymium has no *p*-only isotopes. The lightest isotope, neutron magic ^{142}Nd, is an *s*-only isotope and comprises 27% of solar system neodymium. ^{143}Nd, ^{144}Nd, ^{145}Nd and ^{146}Nd are of mixed *r*- and *s*-process origin (of these, the odd ones are most *r*-process and the even ones mostly *s*-process) and ^{148}Nd and ^{150}Nd are *r*-only isotopes. The most reliable isotope for normalization is ^{144}Nd, because ^{142}Nd can be partially bypassed by the neutron capture channels at the ^{141}Ce branch point. ^{143}Nd, ^{144}Nd, ^{145}Nd and ^{146}Nd are little affected by branching, but ^{148}Nd can be slightly enhanced by the neutron capture channel at ^{147}Nd ($T_{1/2}$ = 11 days). Measurements of neodymium in aggregates of presolar grains have been made by SIMS [17] and thermal ionization mass spectrometry (TIMS) [18]. RIMS offers the opportunity to measure neodymium isotopes in individual presolar grains without isobaric interferences from ^{142}Ce, ^{148}Sm and ^{150}Sm.

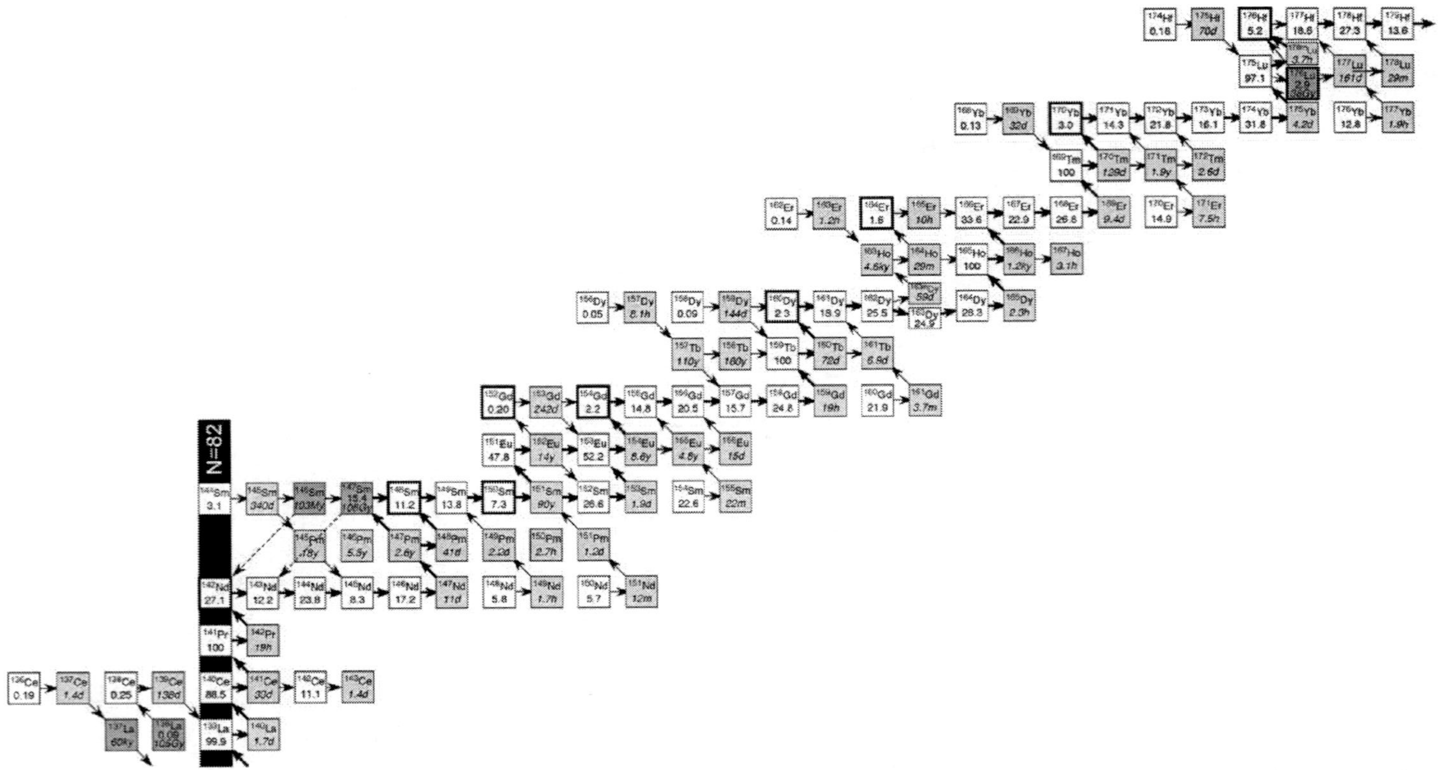

FIGURE 1. The lanthanide region of the chart of the nuclides is shown with the *s*-process path. Stable isotopes are shown in white, and their percent abundance in the solar system is given for each. Short-lived and long-lived unstable isotopes are shown in light and dark gray, respectively, and their half-lives at room temperature are given (for a few, half-lives at stellar temperatures can be dramatically different). *s*-Only isotopes are shielded from the *r*-process and are shown with a bold outline. The main *s*-process path is shown as a bold line, with fainter lines indicating branches. ^{146}Sm and ^{147}Sm are long-lived α emitters and their decay mode is shown as dashed arrows. Isotopes made by the *r*- or *p*-processes alone do not have any lines feeding them, but they have lines leading away from them as they are destroyed in the helium shell by neutron capture reactions.

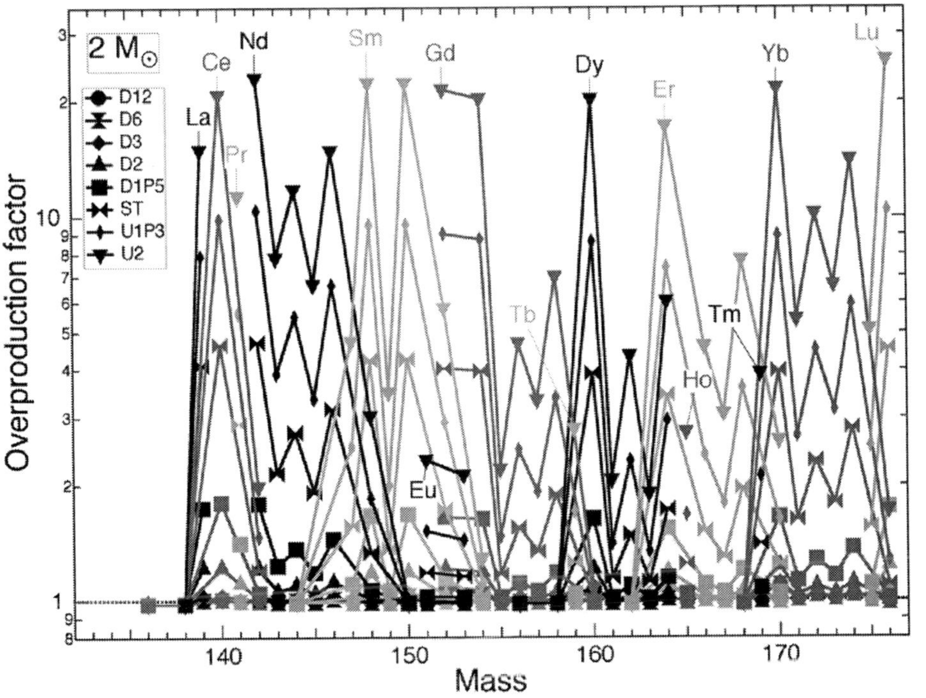

FIGURE 2. Overproductions of the REE isotopes in the envelope of a solar metallicity AGB star with an initial mass of 2 $M_\odot$. The overproduction is calculated from the element fraction in the envelope normalized to the element fraction in bulk solar composition. Note that the *s*-only isotopes all have about the same overproduction and that the *r*- and *p*-process isotopes have no overproduction (i.e., their values are ~1).

2.3 Samarium

The lightest isotope of samarium is neutron magic ^{144}Sm, which is *p*-only. ^{146}Sm ($T_{1/2}$ = 103 My) is predominantly produced by the *p*-process also, but small amounts can be made in AGB stars by neutron capture on ^{144}Sm followed by the neutron capture channel at the ^{145}Sm ($T_{1/2}$ = 340 days) branch point. ^{148}Sm and ^{150}Sm are *s*-only, but ^{148}Sm can be bypassed if the neutron capture channels at the ^{147}Nd and ^{148}Pm branch points are activated during thermal pulses. Thus, the ^{148}Sm/^{150}Sm ratio can be a sensitive indicator of the strength of the ^{22}Ne neutron source in AGB stars. The remaining samarium isotopes are little affected by branchings. There is a branch point at ^{151}Sm ($T_{1/2}$ = 93 y) that has a small effect on ^{152}Sm and also controls the ^{151}Eu/^{153}Eu ratio and *s*-process production of ^{152}Gd; the neutron capture cross section was recently measured [19]. Measurements of neodymium in aggregates of presolar grains have been made by SIMS [17] and TIMS [18]. RIMS allows samarium isotopic measurement without isobaric interferences from ^{148}Nd, ^{150}Nd, ^{152}Gd and ^{154}Gd.

2.4 Europium

Approximately 90% of solar system europium is made by the *r*-process, but the remainder is made by the *s*-process. Europium isotopes in SiC grains are of considerable interest, as europium is one of the few elements for which isotopic ratios can be measured directly in stars by spectroscopy [20]. Using the standard neutron capture cross sections [16], calculations for AGB stars predict a 2–10% enhancement in the ^{151}Eu/^{153}Eu ratio compared to the solar system value. However, use of more recent cross sections from the KADoNiS database [21], the effect largely goes away, with a prediction of <1% difference between AGB stars and the solar system ^{151}Eu/^{153}Eu ratio. Europium isotopes have been measured by SIMS in individual presolar SiC grains [22], but the precision of a few tens of percent is not sufficient to detect any deviation from solar system ratios.

2.5 Gadolinium

Gadolinium has two *s*-only isotopes, rare ^{152}Gd and minor ^{154}Gd (0.2% and 2.2% of solar gadolinium, respectively). ^{160}Gd is almost entirely *r*-process, but is slightly fed by neutron capture at the ^{159}Gd branch point during thermal pulses. The remaining isotopes, ^{155}Gd, ^{156}Gd, ^{157}Gd, and ^{158}Gd, are predominantly of *r*-process origin, receiving *s*-process contributions of 5, 20, 14 and 30%, respectively [23], and are little affected by branchings. Although it is rare, ^{152}Gd would be of interest to measure, because of its complex origin. It is produced by β^- decay of ^{151}Sm ($T_{1/2}$ = 93 y) to ^{151}Eu, followed by neutron capture to ^{152}Eu, which has a branched decay, with the β^- decay branch to ^{152}Gd. The half-lives of ^{151}Sm and ^{152}Eu are strongly reduced relative to terrestrial values at stellar temperatures; these isotopes also have very high neutron capture cross sections. The final distribution of these isotopes is determined by freeze-out at the end of the short ^{22}Ne exposure, and the abundance of ^{152}Gd is governed by the half-lives and neutron capture cross sections of the unstable isotopes [19]. RIMS allows gadolinium isotopic measurement without isobaric interferences from ^{152}Sm, ^{154}Sm, ^{156}Dy, ^{158}Dy and ^{160}Dy.

2.6 Dysprosium

Dysprosium consists of two rare *p*-process isotopes, ^{156}Dy and ^{158}Dy, minor *s*-only ^{160}Dy and ^{161}Dy, ^{162}Dy, ^{163}Dy and ^{164}Dy, which are of mixed *s*- and *r*-process origin. ^{163}Dy has a complex origin: it is normally stable, but at astrophysical temperatures, it decays by bound-state β decay [24]. It becomes highly depleted during thermal pulses, but is reestablished by electron capture on ^{163}Ho at the end of the high temperature phase. ^{160}Dy is somewhat affected by the ^{160}Tb branch point. Dysprosium isotopic measurements on presolar SiC aggregates by TIMS [25] are highly discrepant with AGB predictions but more recent measurements on a SiC-bearing presolar diamond fraction by inductively coupled plasma mass spectrometry [15] are consistent with model predictions. It would be of great interest to measure dysprosium isotopes on single SiC grains. RIMS allows dysprosium isotopic measurement without isobaric interferences from ^{156}Gd, ^{158}Gd, ^{160}Gd, ^{162}Er and ^{164}Er.

2.7 Erbium

Erbium has a rare *p*-process isotope, ^{162}Er, a minor *s*-only isotope, ^{164}Er, an *r*-only isotope, ^{170}Er, and three isotopes, ^{166}Er, ^{167}Er, and ^{168}Er that are of mixed *r*- and *s*-process origin. There is a small amount of production of ^{170}Er due to the branch point at ^{169}Er as well as the very low neutron capture cross section of ^{170}Er. ^{164}Er would appear at first sight to be a *p*-process isotope, but it is actually most produced by the *s*-process. As discussed previously, at high temperatures ^{163}Dy becomes unstable against β^- decay, producing ^{163}Ho. The latter isotope can either β^+ decay back to ^{163}Dy or neutron capture to ^{164}Ho, which can either β^+ decay to ^{164}Dy or β^- decay to ^{164}Er. The other erbium isotopes are not significantly affected by branchings. RIMS allows erbium isotopic measurement without isobaric interferences from ^{162}Dy, ^{164}Dy, ^{168}Yb, and ^{170}Yb.

2.8 Ytterbium

Ytterbium has a rare *p*-only isotope, ^{168}Yb, a minor *s*-only isotope, ^{170}Yb, an *r*-only isotope, ^{176}Yb; the remaining isotopes, ^{171}Yb, ^{172}Yb, ^{173}Yb and ^{174}Yb are of mixed *r*- and *s*-process origin. The *s*-only isotope ^{170}Yb is affected by the branch point at ^{170}Tm, which can cause the *s*-process flow to bypass ^{170}Tm. The branch at ^{171}Tm can also affect the *s*-process production of ^{171}Yb. RIMS allows ytterbium isotopic measurement without isobaric interferences from ^{168}Er, ^{170}Er, ^{176}Lu, and ^{176}Hf.

2.9 Lutetium

Lutetium has only one stable isotope, ^{175}Lu, but *s*-only isotope ^{176}Lu has half-life of 38 Gy. As is clear from Fig. 1, the *s*-process nucleosynthesis of ^{176}Lu is rather complicated. The half-life of ^{176}Lu decreases dramatically at stellar temperature because transitions between the long-lived ground state and the short-lived isomeric state can occur through higher lying states. Lutetium nucleosynthesis in AGB stars has recently been modeled in great detail in light of new cross section measurements [26]. RIMS allows lutetium isotopic measurement without isobaric interferences from ^{176}Yb and ^{176}Hf.

3. CONCLUSIONS

REE are known to be present in presolar SiC grains [14,15,17] and a number of isotope ratios are affected by branchings along the *s*-process path. Thus, isotopic analysis of REE will deepen our understanding of *s*-process nucleosynthesis in AGB stars. The main difficulty will be that the concentrations of heavy REE in SiC are likely to be low, as the most abundant isotopes are of mostly *r*-process origin. Furthermore, the chemical behavior of the REE appears to inhibit condensation of samarium and ytterbium into SiC in AGB outflows [15], so these elements may be particularly difficult to measure.

ACKNOWLEDGMENTS

This work was supported by the National Aeronautics and Space Administration Cosmochemistry Program and by the Italian MIUR-PRIN06 Project "Late phases of stellar evolution: nucleosynthesis in supernovae, AGB stars, planetary nebulae".

REFERENCES

1. M. Lugaro, *Stardust from Meteorites*, Singapore, World Scientific Publishing, p. 209 (2005)
2. E. Zinner, "Presolar Grains" in *Meteorites, Planets and Comets*, edited by A. M. Davis, Volume 1 of Treatise on Geochemistry, edited by H. D. Holland and K. K. Turekian, 2nd Edition, Oxford, Elsevier, online at http://www.sciencedirect.com/science/referenceworks/0080437516
3. G.K. Nicolussi, M.J. Pellin, R.S. Lewis, A.M. Davis, R.N. Clayton, and S. Amari, *Phys. Rev. Lett.* **81**, 3583 (1998)
4. G.K. Nicolussi, A.M. Davis, M.J. Pellin, R.S. Lewis, R.N. Clayton, and S. Amari, *Science* **277**, 1281 (1997)
5. J.G. Barzyk, M.R. Savina, A.M. Davis, R. Gallino, M.J. Pellin, R.S. Lewis, S. Amari, and R.N. Clayton, *New Astron. Rev.* **50**, 587 (2006)
6. J.G. Barzyk, M.R. Savina, A.M. Davis, R. Gallino, F. Gyngard, S. Amari, E. Zinner, M.J. Pellin, R.S. Lewis, and R.N. Clayton, *Meteorit. Planet. Sci.* **42**, 1103 (2007)
7. G.K. Nicolussi, M.J. Pellin, R.S. Lewis, A.M. Davis, S. Amari, and R.N. Clayton, *Geochim. Cosmochim. Acta* **62**, 1093 (1998)
8. M.R. Savina, A.M. Davis, C.E. Tripa, M.J. Pellin, R. Gallino, R.S. Lewis, and S. Amari, *Science* **303**, 649 (2004)
9. M.R. Savina, A.M. Davis, C.E. Tripa, M.J. Pellin, R.N. Clayton, R.S. Lewis, S. Amari, R. Gallino, and M. Lugaro, *Geochim. Cosmochim. Acta* **67**, 3201 (2003)
10. K.K. Marhas, P. Hoppe, and U. Ott, *Meteorit. Planet. Sci.* **42**, 1077 (2007)
11. M. Lugaro, A.M. Davis, R. Gallino, M.J. Pellin, O. Straniero, and F. Käppeler, *ApJ.* **593**, 486 (2003)
12. R. Gallino, C. Arlandini, M. Busso, M. Lugaro, C. Travaglio, O. Straniero, A. Chieffi, and M. Limongi, *ApJ.* **497**, 388 (1998)
13. O. Straniero, R. Gallino, and S. Cristallo, *Nucl. Phys. A* **777**, 311 (2006)
14. S. Amari, P. Hoppe, E. Zinner, and R.S. Lewis, *Meteoritics* **30**, 679 (1995)
15. Q.-Z. Yin, C.-T. A. Lee, and U. Ott, *ApJ.* **647**, 676 (2006)
16. Z.Y. Bao, H. Beer, F. Käppeler, F. Voss, K. Wisshak, and T. Rauscher, *Atom. Data Nucl. Data Tables* **76**, 70 (2000)
17. E. Zinner, S. Amari, and R.S. Lewis, *ApJ* **382**, L47 (1991)
18. S. Richter, U. Ott, and F. Begemann, In *Nuclei in the Cosmos II*, edited by F. Käppeler and K. Wisshak, Bristol, Institute of Physics, p. 127 (1993)
19. K. Wisshak, F. Voss, F. Käppeler, M. Krtička, S. Raman, A. Mengoni, and R. Gallino, *Phys. Rev. C* **73**, 015802, 1 (2006)
20. W. Aoki, S.G. Ryan, N. Iwamoto, T.C. Beers, J.E. Norris, H. Ando, T. Kajino, G.J. Mathews, and M.Y. Fujimoto, *ApJ.* **592**, L67 (2003)
21. http://nuclear-astrophysics.fzk.de/kadonis/.
22. K. Terada, K. Itoh, H. Hidaka, T. Yoshida, N. Iwamoto, W. Aoki, and I.S. Williams, *New Astron. Rev.* **50**, 582 (2006)
23. C. Arlandini, F. Käppeler, K. Wisshak, R. Gallino, M. Lugaro, M. Busso, and O. Straniero, *ApJ.* **525**, 886 (1999)
24. K. Takahashi and K. Yokoi, *Atom. Data Nucl. Data Tables* **36**, 375 (1987)
25. S. Richter, U. Ott, and F. Begemann, *Meteoritics* **29**, 522 (1994)
26. M. Heil, N. Winckler, S. Dababneh, F. Käppeler, K. Wisshak, S. Bisterzo, R. Gallino, A.M. Davis, and T. Rauscher, *ApJ* **673**, 434 (2008)

News from r-Process Nucleosynthesis: Consequences from the ESS to Early Stars

K. Farouqi[*,†], K.-L. Kratz[**], L. I. Mashonkina[‡], B. Pfeiffer[**] and F.-K. Thielemann[§]

[*]*Institut für Kernchemie, Universität Mainz, D-55128 Mainz, Germany.*
[†]*Department of Astronomy and Astrophysics, University of Chicago, Chicago, IL60637, USA.*
[**]*Max-Planck-Institut für Chemie, Otto-Hahn-Institut, D-55128 Mainz, Germany.*
[‡] *Institute of Astronomy, Russian Academy of Science, RU-119017 Moscow, Russia.*
[§]*Departement für Physik und Astronomie, Universität Basel, CH-4056 Basel, Switzerland.*

Abstract. The exact conditions for the supernova high-entropy wind (HEW) as one of the favored sites for the rapid neutron-capture (r-) process still cannot be reproduced self-consistently in present hydrodynamic astrophysical models. Therefore, we have performed large-scale dynamical network calculations within a parameterized HEW model to constrain the necessary conditions for a full r-process and to compare our results with recent observations. A model-inherently weighted superposition of entropy trajectories results in an excellent reproduction of the overall Solar-System isotopic abundances ($N_{r,\odot}$) of the "main" r-process elements beyond Sn. For the lighter r-elements in the range $26 \leq Z \leq 42$, our HEW model supports earlier qualitative ideas about a multiplicity of nucleosynthetic processes leading to the total $N_{r,\odot}$ distribution. In the HEW scenario, these suggestions are quantified, and the origin of the missing primary process is confirmed to be a rapid charged-particle process, thus excluding a "weak" neutron-capture component. This explains the recent halo-star observations of a Z-dependent non-correlation, respectively partial correlation of the lighter r-elements with metallicity and/or enrichment of "main" r-process elements.

Keywords: r-process Nucleosynthesis, Low Metallicity Stars, Core Collapse Supernovae
PACS: 20.40.Lw, 24.10.-i, 26.30.+k, 97.10.Cv, 99.60.Bw

1. THE HEW MODEL OF SN II

Recent observations in ultra-metal-poor (UMP), neutron-capture-element rich halo giants and astrophysical models suggest a supernova (SN) origin of the classical r-process. A promising stellar environment is the so-called high-entropy wind (HEW), expelled from newly formed hot neutron stars in core-collapse type II supernovae (SN II), which continue to release neutrinos after the SN shock wave is launched. These neutrinos interact with matter of the outer neutron star layers leading to moderately neutron-rich ejecta with high entropies (see, e.g. [1, 2, 3, 4, 5, 6, 7, 8]). Despite the promise of this HEW model, in the past there have been severe theoretical problems (e.g. failures in obtaining successful SN explosions, or lack of sufficient entropy (S) to synthesize the heaviest elements) that have prevented a definitive confirmation of this site with the classical r-process nucleosynthesis. Therefore, we still consider it useful to perform dynamical network calculations within a *parameterized* HEW model in order to derive tight constraints on the necessary astrophysical conditions for a full r-process. Here, we basically follow the description of adiabatically expanding homogeneous mass zones as outlined in [5], and make use of an extended version of the Basel network code. Details about

CP1001, *9th Torino Workshop on Evolution and Nucleosynthesis in AGB Stars and the 2nd Perugia Workshop on Nuclear Astrophysics*, edited by R. Guandalini, S. Palmerini, and M. Busso

the actual astrophysical input is given in [9, 10, 11, 12]. The nuclear-physics input is the same as in our earlier studies within the site-independent "waiting-point" approximation [13, 14, 15, 16, 17, 18].

After the charged-particle freezeout at $T_9 \simeq 3$, the expanding and eventually ejected mass zones have different initial entropies ($S \sim T^3/\rho$ [k_b/Baryon]), so that the overall explosion represents a superposition of S components. In the HEW scenario, in principle S is a "measure" of the ratios of matter locked into α-particles, heavy ($A \simeq 100$) "seed" nuclei and free neutrons (see, e.g. Fig. 1 in [12]). Furthermore, the main parameters of our HEW model – the abundance ratio Y_n/Y_{seed}, the electron abundance Y_e, the entropy S and the expansion velocity V_{exp} – are correlated by the simple expression $Y_n/Y_{seed} \simeq$ const. V_{exp} $(S/Y_e)^{3.33}$, which in principle is a *"r-process strength function"* [9, 10, 19].

2. RESULTS AND DISCUSSION

While for the classical r-process approach a range of neutron densities ($10^{20} \leq n_n \leq 10^{28}$) is necessary to reproduce the full $N_{r,\odot}$ distribution up to the Th, U region [13, 14, 15, 16, 17], within the HEW model at a given Y_e and V_{exp}, the entropy S (or the correlated Y_n/Y_{seed} ratio) will play this role. In the following, as in our earlier papers [10, 11, 12, 18], we will present nucleosynthesis results as a function of entropy S for the choice of Y_e=0.45 and V_{exp}=7500 [km/s], and compare these model predictions to recent astronomical observations.

We first want to examine what abundance distributions can be produced with the HEW model when utilizing the above parameter combination. In Fig. 1 of [12] we have already shown the abundance correlations of α-particles (Y_α), heavy "seed" nuclei (Y_{seed}) and free neutrons (Y_n) as a function of entropy S at the freezeout of charged-particle reactions. We have seen from this approach that over the whole entropy range of $10 \leq S \leq 350$ most of the matter is locked into α-particles. From roughly equal abundances at low entropies, the distribution of Y_{seed} is steadily decreasing with S, whereas the abundance fraction Y_n is increasing. From this behavior it became immediately evident that our HEW model predicts (at least) two different primary nucleosynthesis mechanisms for the elements beyond Fe; i.e. for low S a charged-particle process and for high S a neutron-capture process. For the low-S region ($1 \leq S \leq 100$), the amount of free neutrons relative to the available "seed" nuclei is negligible; hence the nucleosynthesis mechanism in this region cannot be a classical "weak" r-process, but instead is a primary, rapid charged-particle process. This can also be seen from Table 1, where the correlated values of selected entropies S and Y_n/Y_{seed} ratios are listed together with the most abundant isotopes produced *before* (3rd column) and *after* (4th column) freezeout under the respective astrophysical conditions.

It is interesting to note that for very low entropies (for S=10, see the first line of Table 1), the initial and final nuclei are identical, indicating a pure charged-particle process at the β-stability line. When slightly increasing the entropy, besides stable nuclides also long-lived, neutron-rich isotopes are produced before freezeout (for S=50, see line 2 in Table 1). In principle, this picture confirms earlier ideas of [3, 4, 20] where a so-called "α"-process involving charge-particle reactions (CPR) has been suggested for the production of the elements between Sr and Cd. Within the parameter space

TABLE 1. Correlated values of entropy (S) and abundance ratio of free neutrons to seed nuclei (Y_n/Y_{seed}), and the appropriate three most abundant progenitor isotopes *before* freezeout (3rd column) and stable nuclei *after* freezeout (4th column). The other parameters contained in the *"r-process strength function"* are Y_e=0.45 and V_{exp}=7500 [km/s] [9, 10, 12, 18]. For discussion, see text.

		Most abundant isotopes	
Entropy S	Y_n/Y_{seed}	*before* **freezeout**	*after* **freezeout**
10	7 x 10^{-21}	^{62}Ni, ^{66}Zn, ^{88}Sr	^{62}Ni, ^{66}Zn, ^{88}Sr
50	2.5 x 10^{-15}	^{86}Kr, ^{90}Sr, ^{87}Rb	^{86}Kr, ^{90}Zr, ^{87}Sr
80	1.3 x 10^{-3}	^{93}Rb, ^{102}Zr, ^{94}Sr	^{93}Nb, ^{91}Zr, ^{102}Ru
110	1.0	^{102}Sr, ^{92}Kr, ^{106}Zr	^{102}Ru, ^{92}Zr, ^{106}Pd
150	13	^{101}Rb, ^{107}Y, ^{127}Rh	^{101}Ru, ^{100}Mo, ^{106}Pd
190	33	^{129}Ag, ^{130}Cd, ^{131}In	^{129}Xe, ^{130}Te, ^{131}Xe
230	63	^{150}Te, ^{190}Gd, ^{142}Sn	^{154}Sm, ^{150}Nd, ^{192}Os
270	105	^{194}Er, ^{195}Tm, ^{193}Ho	^{195}Pt, ^{196}Pt, ^{194}Pt
300	154	^{248}Ra, ^{260}Th, ^{246}Rn	^{232}Th, ^{206}Pb, ^{209}Bi

chosen for our HEW-model calculations [9, 10, 11, 12, 18], we now can quantify these results by predicting even *isotopic* abundances as function of Y_e for the CPR elements. For Y_e=0.45 and low S, for example, apart from so-called "r-only" and standard *"r+s"*-nuclei also *"s-only"* isotopes are produced. And when increasing the electron abundance to $Y_e \simeq$0.495 even light *"p"*-isotopes (such as ^{92}Mo and ^{94}Mo in the correct SS proportions of 1/0.6) are formed in the HEW scenario [19]. Since the network code used in our studies so far does not include neutrino-nucleon/nucleus interactions, our above results for low S values will, however, have to be carefully compared to the recently postulated new "νp" nucleosynthesis process (see, e.g. [21, 22]).

With increasing S, the "seed" composition at freezeout is gradually shifted from the stability line to neutron-rich isotopes. Already around S$\simeq$100, well-known β-delayed neutron (βdn) precursors in the Se to Y region with half-lives in the second-range act as "seed" nuclei at freezeout. For S=110 (see line 4 of Table 1), most of these isotopes have already been measured as r-process relevant "progenitor" nuclei for the low-n_n components of the classical "waiting-point" model (see, e.g. [14, 15, 16, 23, 24]). However, even in this S-range the total n_n from free neutrons and βdn-emission is too low to operate a "weak" r-process. The neutron-to-seed conditions might rather resemble a low neutron-density, s-like *"Light Element Primary Process"* (LEPP), as suggested by [25]. It is certainly a question how the Y_n/Y_{seed} conditions of a classical r-process should be defined. Within our chosen HEW parameter combination, one might take as a possible criterion the actual neutron-capture / β-decay matter flow from the initial "seed" composition at charged-particle freezeout on the onset of the N=82, A$\simeq$130 r-abundance peak. With this definition, in the HEW scenario for Y_e=0.45 a real r-process would start around S=130 with a corresponding $Y_n/Y_{seed} \simeq$6, where most of the matter is still concentrated in the A$\simeq$100 region, but small amounts are already shifted up to the N=82 "waiting-point" isotopes ^{127}Rh and ^{128}Pd in the rising wing of the 2nd $N_{r,\odot}$ peak. The full synthesis of this peak is reached at S$\simeq$195 with $Y_n/Y_{seed} \simeq$35, where the maximum

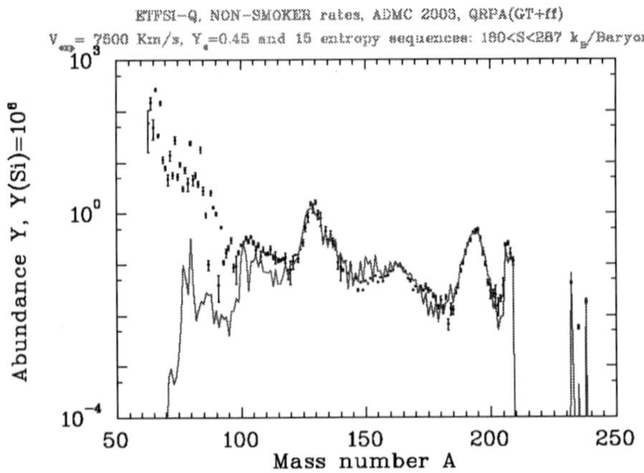

FIGURE 1. Comparison of the $N_{r,\odot}$ distribution (data points [27]) with predicted isotopic abundances (full line) from a weighted superposition of 15 HEW entropy components in the range 160≤S≤287. For further details and discussion, see text.

yields of the "waiting-point" isotopes ^{129}Ag, ^{130}Cd and ^{131}In are produced (compare lines 5 and 6 in Table 1). It is interesting to note in this context that under the above HEW conditions, the formation of the full A≃130 $N_{r,\odot}$ peak only requires about 145 ms. This is roughly an order of magnitude less compared to the process duration needed for the classical "waiting-point" model with an Fe-group seed (see, e.g. [13, 17]).

For entropies of S≃210, where the matter flow has already passed the N=82 bottleneck of the 2nd peak, the r-process speeds up and proceeds to the rare-earth elements (REE). This whole region of the REE "pygmy peak" is produced within a quite small entropy interval of 220≤S≤250. At S≃250, already the rising wing of the 3rd $N_{r,\odot}$ peak with the N=126 "waiting-point" isotopes ^{191}Tb and ^{192}Dy is reached. To produce the maximum of the A≃195 peak with the neutron-magic nuclides ^{193}Ho, ^{194}Er and ^{195}Tm, conditions of S≃270 and Y_n/Y_{seed}≃105 are requested (compare lines 7 and 8 in Table 1). The process duration up to this point is about 330 ms, again about an order of magnitude less than for the classical "waiting-point" approach.

Finally, after having passed also the N=126, A≃195 bottleneck, at entropies of S≥290 the r-process again speeds up and forms very heavy actinide precursor isotopes which either fission or decay back to the Th, U r-chronometers. In the last line of Table 1, for S=300 also the dominant production of Pb and Bi is indicated, originating from the β-plus α-decays of progenitor isotopes in the 230≤A≤250 mass region.

As a function of time, the HEW will eject matter with possibly varying S, Y_e and V_{exp}; for a recent 3D core-collapse SN calculation leading to promising conditions for an r-process, see e.g. [26]. If one assumes that the amount of ejected material per equidistant entropy interval is constant, the sum of the abundance contributions – weighted according to Y_n/Y_{seed} as a function of S – results in an SS-like isotopic abundance distribution ($N_{r,calc}$) as displayed in Fig. 1 together with the standard r-process "residuals" ($N_{r,\odot}=N_\odot-N_{s,\odot}$ [27]). The $N_{r,calc}$ distribution shown in the figure represents a superposition of 15 equidistant S-components in the range 160≤S≤287.

The lower limit of S=160 has been chosen to restrict the calculations to true "main" r-process conditions. Consequently, with this parameter choice, no "best fit" to the LEPP region below the A$\simeq$130 peak is anticipated. The upper limit of S=287 has been chosen to ensure that fission recycling remains negligible. The big advantage of the above S to Y_n/Y_{seed} correlation is the introduction of a "model-inherent" weighting of the S-components, which – if successfully reproducing the $N_{r,\odot}$ pattern – can replace our earlier, algebraic expressions of the weights as a function of neutron density n_n in the "waiting-point" approximation [16, 17]. However, it has to be admitted that the scaling of the different S-components still is based on the above assumption of equal fractions of ejecta from each mass zone, which may not occur in reality. Nevertheless, it is very satisfactory to see that the decreasing slope of Y_{seed} with increasing S from our parameterized HEW model (see, e.g. Fig. 1 in [12]) seems to agree well with the recent hydrodynamic 3D SN results of [26]. In both models, between S=100 and S=300 – roughly representing the S-range for a full r-process – the respective matter fractions drop by a factor 20.
In any case, with our above Y_{seed}=f(S) weighting, an excellent overall agreement of the $N_{r,calc}$ distribution with the observed $N_{r,\odot}$ pattern from the rising wing of the A$\simeq$130 peak up to the Pb, Bi "spike" is obtained. In particular the positions of the maxima as well as the widths and the relative heights of the 2nd and 3rd $N_{r,\odot}$ peaks are perfectly reproduced (see Fig. 1), much better than in any other recent r-process calculation (see, e.g. [28, 29, 30, 31, 32]). Only minor local overabundances at A$\simeq$150 and under-abundances at A$\simeq$175 relative to the SS "residuals" are observed, which are, however, understood in terms of nuclear-physics model deficiencies (see, e.g. [13, 14, 15, 16]). It should be emphasized that only with this quality of agreement between observations and calculations, meaningful quantitative conclusions about elemental or isotopic abundance signatures in UMP halo stars – including r-chronometric age determinations – can be derived.

Given the success in reproducing the isotopic $N_{r,\odot}$ distribution in great detail, we now can confront our HEW model predictions with recent astronomical observations. In the present paper, we will focus on *elemental* abundances in the LEPP region between Cu and Ag (with special emphasis on the three Z-neighbors Sr, Y and Zr for which the largest data sets exist so far [33, 34, 35, 36, 37, 38, 39]), covering the full range of "r-process poor" stars (like HD 122563) up to "r-process rich" stars (like CS 22892-052). These data have shown that the abundances of Sr, Y and Zr viewed as separate elements are not tightly correlated with the (almost pure) "main" r-process element Eu. Analogously, the elemental abundance ratios of log(Sr/Y/Zr) from Galactic disk and halo stars, e.g. as a function of "metallicity" [Fe/H], have been shown to exhibit a rather robust, constant pattern [20, 25, 39]. As already qualitatively discussed earlier (see, e.g. [20, 25, 36, 39]), this behavior seems to indicate a nucleosynthetic origin different from that of the heavier neutron-capture elements.
These conclusions about the robustness of the nucleosynthetic origin of Sr, Y and Zr are clearly confirmed and can be further quantified by our HEW model calculations [12, 18]. In Fig. 2 we show two plots for the element abundance ratios log(Sr/Y/Zr) as functions of "r-process enrichment" in the range -3.5$\leq$[Eu/H]$\leq$-0.5 and as a function of "metallicity" in the range -3.3$\leq$[Fe/H]$\leq$-0.9. The mean values of the observed linear abundance ratios

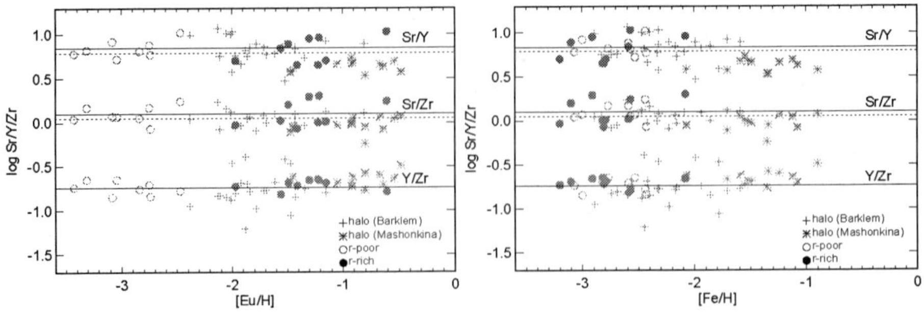

FIGURE 2. Elemental abundance ratios of log(Sr/Y/Zr) as a function of "r-process enrichment" [Eu/H] (left side) and "metallicity" [Fe/H] for halo stars in our Galaxy. The observational data [33, 34, 35, 36, 37, 38, 39] (with their mean values shown as full lines) are compared to the abundance predictions of our HEW model (dotted lines, partly indistinguishable from the full lines) with Y_e=0.45 and V_{exp}=7500 for the low-S range ($1 \leq S \leq 110$) of the CPR component [9, 10, 11, 12, 18, 19]. For discussion, see text.

(Sr/Y=6.17±1.06, Sr/Zr=1.12±0.54 and Y/Zr=0.18±0.04) are very well reproduced by the low-S ($1 \leq S \leq 110$) CPR component of our HEW model (Sr/Y=6.80, Sr/Zr=1.34 and Y/Zr=0.20), and are partly indistinguishable in the log-plots. Moreover, a comparison to the predictions for a neutron-capture r-process in the S-range $110 \leq S \leq 280$ (Sr/Y=9.74, Sr/Zr=0.042 and Y/Zr=0.004) clearly indicates that this classical nucleosynthesis mode in the HEW scenario can be excluded. Further summarizing, we can conclude that the Sr/Y/Zr abundance ratios are very *robust* and over wide ranges *independent* of both "r-process enrichment" and "metallicity".

In a next step, we want to extend the range of the lighter trans-Fe elements from Cu to Ag (see Fig. 3, and Table 1 in [12]), and confront our HEW model results to earlier nucleosynthesis predictions. On the one hand, [25] – by comparing the contributions of classical abundance fractions from "AGB+weak-s" and "r-residuals" with the observational data of the r-process rich / α-process poor star CS 22892-052 – consider an additional neutron-capture LEPP component increasing from about 18% for Nb (Z=41) to tentatively 41% for Cd (Z=48). On the other hand, [20] consider these elements in the $90 \leq A \leq 110$ mass region as dominantly CPR nuclei, however (with reference to [4]), admitting an *"incipient r-process"* with $Y_n/Y_{seed} \leq 10$. As can be seen from Table 1 in [12], our HEW results are somewhere "in between" the two above model predictions. From Sr (Z=38) up to Nb (Z=41), we disagree with both, the classical "r-residual" $N_{r,\odot}$ picture [27] and with the "n-fraction" LEPP prediction [25]. For these elements, our HEW model shows only a negligible neutron-capture component of $\leq$6%; hence, here we rather support the CPR hypothesis [20]. From Mo (Z=42) on upwards, however, our HEW model predicts strongly increasing fractions of ("weak" plus "main") r-process material (see columns 3 and 4 in Table 1 in [12]). For example, for Pd (Z=46) and Ag (Z=47) already between 88% and 96% of the $N_\odot$ values should be of $S \geq 110$ neutron-capture origin, now in qualitative agreement with the LEPP approach, but – from Mo (Z=42) upwards – in disagreement with the CPR picture.

In any case, an answer to the questions about the main nucleosynthesis origin of the

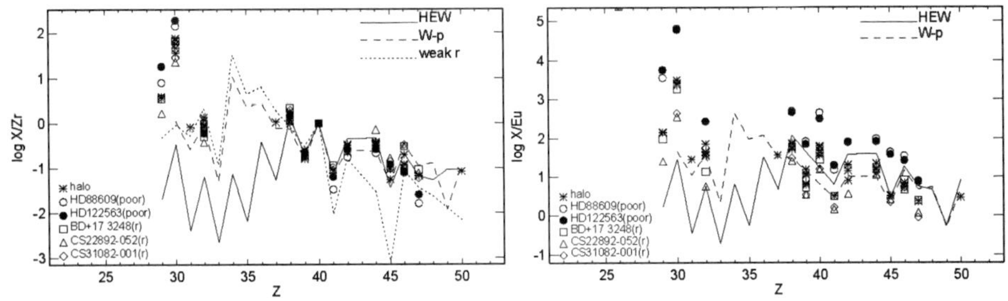

FIGURE 3. Elemental abundance ratios of log(X/Zr) (left figure) and log(X/Eu) (right figure) as a function of atomic number Z. The observational data in the LEPP region between Cu (Z=29) and Ag (Z=47) for selected halo stars are given by different symbols in the lower left corner of both figures. These data are compared with predictions from three different nucleosynthesis approaches; (i) the present HEW model [9, 10, 11, 12] (full line), (ii) the classical "waiting-point" model with the standard Fe-seed and the full n_n-range [13, 14, 15, 16, 17] (dashed line), and (iii) a special "weak" r-process model with low n_n and a solar-like Si – Cr seed composition. For discussion, see text.

elements between Sr and Cd – with the chance of even disentangling the components of the discussed charged-particle and neutron-capture processes – must come from future higher quality observational data. However, very recent, but still scarce abundance data on Ru and Pd for "r-poor" relative to "r-rich" stars seem to provide first tentative evidence for the existence of two separate nucleosynthesis components with different coupling to Eu. The first observation is a slightly lower ratio for both, Ru/Sr and Pd/Sr by about a factor 2 for the [Eu/H]$\simeq$-3.5 "r-poor" stars than for the -1.$\leq$[Eu/H]$\leq$-0.7 "r-rich" stars. This indicates at least that Ru and Pd are not formed under exactly the same nucleosynthetic conditions as Sr. The second observation is a definitely higher ratio again for both Ru/Eu and Pd/Eu by roughly a factor 5 for the "r-poor" stars relative to the "r-rich" stars. These very interesting indications will have to be compared carefully with our HEW model predictions of multiple nucleosynthesis components.

As already noticed above, in Fig. 3 we show two plots of LEPP abundance ratios, log(X/Zr) in the left part and log(X/Eu) in the right part, as functions of atomic number in the range 29$\leq$Z$\leq$50. For log(X/Zr), we compare the observational data with the predictions from three different nucleosynthesis approaches: (i) the present HEW model, (ii) the classical "waiting-point" model with the standard Fe-seed and the full n_n-range, and (iii) a special "weak" r-process model with low $n_n \simeq 10^{19}$ and a solar-like Si – Cr seed composition. For log(X/Eu), the latter approach cannot be applied because Eu is not produced in a low-n_n "weak" r-process.

The observed log(X/Zr) pattern shows a decreasing pattern with Z. From Sr (Z=38) upwards, this trend is best reproduced by the HEW model, but also reasonably well by the two neutron-capture approaches. The latter two models also seem to reproduce the lower-Z abundances of Rb (Z=37) down to Ga (Z=31), where the CPR component of our HEW model predicts considerably lower abundances. Finally, the high abundances with the large scatter observed for Cu (Z=29) and Zn (Z=30) obviously cannot be reproduced by any of the three models. At the moment, we only can conclude from our HEW predictions that the major abundance fractions of the elements with Z<38 are not produced together with the elements with Z$\geq$38. For the log(X/Eu)=f(Z) plot of the observational data we see a similar trend with Z, however, with a much larger scatter than found for the log(X/Zr) ratios. Again, also in this depiction for the elements beyond Sr there is no clear preference for the CPR or the classical ("weak" plus "main") r-process model. And again, both approaches strongly underestimate the observed abundances in the low-Z region between Cu and Ge. From this picture, the minimal conclusion to be drawn right now is that these light trans-Fe elements are not produced under the same nucleosynthetic conditions as Eu, and should therefore be un-correlated with the "main" r-process.

3. SUMMARY AND CONCLUSION

We have performed large-scale dynamical network calculations within a parameterized model of an adiabatically expanding high-entropy wind (HEW) of core-collapse type II supernovae (SN II). We find that over a wide range of the three correlated main parameters Y_e, S and V_{exp}, which determine the ratio of Y_n/Y_{seed} in terms of a simple *"r-process strength function"*, a surprisingly robust picture is obtained for the production of heavy elements beyond Fe. Our results request astrophysical conditions for the HEW scenario which are more realistic than most of the recent parameter studies for more exotic SN scenarios which are based on partly extreme nuclear and / or astrophysical conditions – such as rather unlikely low Y_e values [30, 31], unusually low freezeout temperatures [29], very high expansion velocities with extremely short process time scales [32], or strong effects of neutrino or active-sterile neutrino reactions combined with dominant fission recycling [31].

For the light-Z region between Cu and Rb, the HEW scenario seems to be of minor importance for their total production observed in halo stars. For Sr to about Mo the dominating nucleosynthesis mode is a low-S (1$\leq$S$\leq$110) CPR ("α"-) process with S-dependent "seed" nuclei smoothly shifting from β-stability to the neutron-rich side, starting to include well-known, short-lived βdn-precursor isotopes. Rapid neutron-capture nucleosynthesis only sets in at higher entropies, where the range 110$\leq$S$\leq$150 produces elements up to the rising wing of the A$\simeq$130 $N_{r,\odot}$ peak under "weak" r-process conditions, whereas for 150$\leq$S$\leq$300 a robust "main" r-process between A$\simeq$120 and the actinide region occurs. This HEW model is able to reproduce the overall historical $N_{r,\odot}$ pattern, as well as all major recent observations from UMP halo stars. With high electron abundances (Y_e=0.49), in the low-S range in addition to r-nuclei also classical s-only isotopes as well as p-nuclei are produced in the range ^{54}Fe up to ^{106}Cd. In contrast to earlier phenomenological approaches, the HEW model allows to substantially revise the abundance estimates of different primary nucleosynthesis processes for elements in the historical weak-s- / weak-r process region and to quantify their correlation with the "main" r-process.

As far as future model developments are concerned, what still remains to be done is the coupling of the present HEW network calculations with more realistic trajectories of astrophysical parameters (including neutrino interactions), derived from full hydrodynamical SN II models, such as e.g. [21, 26, 28, 40].

ACKNOWLEDGMENTS

We would like to thank Roberto Gallino for helpful discussions. This work has been supported by the Deutsche Forschungsgemeinschaft (DFG) under contract KR 806/13-1, the Helmholtz Gemeinschaft under grant VH-VI-061, and the Physics Frontier Center (JINA) under grant PHY 02-16783.

REFERENCES

1. S.E. Woosley, et al., *ApJ* **433**, 229 (1994)
2. K. Takahashi, J. Witti, and H.-T. Janka, *A&A* **286**, 857 (1994)
3. Y.-Z. Qian and S.E. Woosley, *ApJ* **471**, 331 (1996)
4. R.D. Hoffman, et al., *ApJ* **482**, 951 (1997)
5. C. Freiburghaus, et al., *ApJ* **516**, 381 (1999)
6. T.A. Thompson, A. Burrows, and B.S. Meyer, *ApJ* **562**, 887 (2001)
7. S. Wanajo, et al., *ApJ* **636**, 842 (2006)
8. A. Burrows, S. Reddy, and T.A. Thompson, *Nucl. Phys. A* **777**, 356 (2006)
9. K. Farouqi, PhD Thesis, Univ. Mainz (2005)
10. K. Farouqi, et al., *Nucl. Phys. A* **758**, 631c (2005)
11. K. Farouqi, et al., *AIP Conf. Proc.* **819**, 419 (2006)
12. K. Farouqi, et al., *AIP Conf. Proc. "First Stars III"*, in print (2007)
13. K.-L. Kratz, et al., *ApJ* **403**, 216 (1993)
14. K.-L. Kratz, et al., *AIP Conf. Proc.* **769**, 1365 (2006)
15. K.-L. Kratz, et al., *AIP Conf. Proc.* **819**, 409 (2006)
16. K.-L. Kratz, K. Farouqi, and B. Pfeiffer, *Prog. Particle and Nucl. Physics* **59**, 147 (2007)
17. K.-L. Kratz, et al., *ApJ* **662**, 39 (2007)
18. K.-L. Kratz, *AIP Conf. Proc. "Carpathian Summer School on Exotic Nuclei & Nuclear/Particle Astrophysics"* **972**, in print (2008)
19. K. Farouqi, K.-L. Kratz, F.-K. Thielemann, et al., to be subm. to ApJ
20. Y.-Z. Qian and G.J. Wasserburg, *Physics Reports* **442**, 237 (2007), and refs. therein
21. C. Fröhlich, et al., *Phys. Rev. Lett.* **96**, 142502 (2006)
22. J. Pruet, et al., *ApJ* **644**, 1028 (2006)
23. B. Pfeiffer, K.-L. Kratz, and P. Möller, *Prog. Nucl. Energy* **41**, 39 (2002)
24. K.-L. Kratz, et al., *European J. Phys. A Supp.* **25**, 633 (2005)
25. C.Travaglio, et al., *ApJ* **601**, 864 (2004)
26. A. Burrows, et al., *ApJ* **655**, 416 (2007)
27. C. Arlandini, et al., *ApJ* **525**, 886 (1999)
28. C.L. Fryer, et al., *ApJ* **646**, L131 (2006)
29. S. Wanajo, *arXiv:0706.4360.v2 [astro-ph]* (2007); and refs. therein
30. T. Kuroda, et al., *arXiv:0707.225v1 [astro-ph]* (2007)
31. J. Beun, et al., *arXiv:0707.4498v1 [astro-ph]* (2007); and refs. therein
32. H. Ning, et al., *arXiv:0708.1748v1 [astro-ph]* (2007)
33. J. Westin, et al., *ApJ* **530**, 783 (2000)
34. C. Sneden, et al., *ApJ* **591**, 936 (2003)
35. P.S. Barklem, et al., *A&A* **439**, 129 (2005)
36. J.J. Cowan, et al., *ApJ* **627**, 238 (2005); and refs. therein
37. I. Ivans, et al., *ApJ* **645**, 613 (2006)
38. S. Honda, et al., *ApJ* **666**, 1189 (2007); and refs. therein
39. L.I. Mashonkina, et al., *Astron. Rep.* **51**, 903 (2007); and refs. therein
40. M. Liebendoerfer, et al., *arXiv:0708.4296* (2007)

Presolar Silicates In Meteorites And Interplanetary Dust Particles

Peter Hoppe[†] and Christian Vollmer[†]

[†]Max-Planck-Institute for Chemistry, Particle Chemistry Department, P.O. Box 3060, 55020 Mainz, Germany

Abstract. Presolar silicates remained unrecognized in primitive solar system materials for a long time and only the recent development of advanced ion probe imaging techniques led to their discovery. Silicates turned out to be the most abundant presolar mineral. To date more than 200 presolar silicates have been found. Most grains (>70 %) exhibit enrichments in ^{17}O with $^{17}O/^{16}O$ ratios of up to 25x the solar ratio, close-to-solar or slightly lower than solar $^{18}O/^{16}O$ ratios, and Si-isotopic ratios close to the SiC mainstream line, although slightly shifted to the ^{30}Si-poor side. These grains most likely formed in the winds of 1-2.5 $M_{\odot}$ RGB/AGB stars. $^{17}O/^{16}O$ ratios of $>3x10^{-3}$ may be the result of mass transfer in binary systems as was recently proposed for presolar oxide grains. This scenario is also supported by the Si-isotopic ratios. Some (a few %) silicate grains have enrichments in ^{17}O and strong depletions in ^{18}O, probably the result of cool bottom processing during the RGB or AGB phase. Presolar silicate grains with moderate to strong excesses in ^{18}O (~10 % of all silicates) most likely have a supernova origin.

Keywords: Meteorites – IDPs – Presolar Grains – Silicates – Isotopic Compositions.
PACS: 96.30.Za, 96.50.Dj, 26.20.Kn, 26.20.Cd, 26.30.-k

1. INTRODUCTION

Primitive meteorites, interplanetary dust particles (IDPs), and Antarctic micrometeorites (AMMs) contain small quantities of presolar dust grains that formed in the winds of evolved stars or in the ejecta of nova and supernova (SN) explosions [1]. Individual presolar grains can be studied in the laboratory for their isotopic compositions, chemistry, and mineralogy. These studies have provided a wealth of information on stellar nucleosynthesis and evolution, mixing in nova and SN ejecta, dust formation in stellar environments, dust chemistry in the interstellar medium (ISM), and on the inventory of stars that contributed dust to the Solar System.

While most of the presolar minerals (e.g., SiC, graphite, corundum, spinel) can be chemically separated from meteorites this does not hold for presolar silicates. For this reason the presolar silicates remained unrecognized for a long time and only the application of advanced ion probe imaging techniques led to their discovery [2]. Of special importance in this respect is the invention of the Cameca NanoSIMS 50 ion probe [3] which permits isotope mapping of thin sections of meteorites and IDPs with a spatial resolution of better than 100 nm. Circumstellar O-rich dust is characterized by large O-isotopic anomalies and O isotope ion imaging with the NanoSIMS allows to identify presolar silicates with sizes >100-200 nm in-situ. Presolar silicates turned

CP1001, *9th Torino Workshop on Evolution and Nucleosynthesis in AGB Stars and the 2nd Perugia Workshop on Nuclear Astrophysics*, edited by R. Guandalini, S. Palmerini, and M. Busso

out to be the most abundant presolar mineral type (disregarding the nanodiamonds, most of which might not represent circumstellar dust). The first presolar silicates were found in an IDP [2]. Later studies have constrained the abundance of presolar silicates in isotopically primitive IDPs to ~800 ppm [4]. Following the discovery in IDPs presolar silicates were also identified in meteorites [5-7]. To date more than 200 presolar silicates were found in primitive solar system materials. Most of them were identified in the Acfer 094 meteorite, which turned out to be exceptionally rich in presolar grains, and in which the presolar silicates represent about 200 ppm of the matrix material [8].

Silicates are important constituents of circumstellar and interstellar dust. Features of amorphous and crystalline silicates are seen in the spectra of young stars, C- and O-rich asymptotic giant branch (AGB) stars, post-AGB stars, planetary nebulae, and massive stars [9]. The most prolific suppliers of O-rich (mostly silicate) dust to the ISM are AGB stars during early stages in their evolution [10]. Other important dust sources are red supergiants and novae; their contribution to the presolar grain inventory, however, is small. Not much known is about the nature and efficiency of dust production in SNe but as shown below there are grains with likely SN origin among the presolar silicates.

Although most abundant there is only limited information on the isotopic compositions and mineralogy of presolar silicates to date. Precise in-situ measurements of isotopic compositions are difficult due to the small size of presolar silicates (see below) and dilution of secondary ion signals by neighboring isotopically normal material. Here, we will briefly review the available isotope data (O, Si, Mg-Al; NanoSIMS data only) and mineralogical properties and will discuss the putative stellar sources of presolar silicate grains.

2. PHYSICAL PROPERTIES

Sizes of presolar silicates vary from 150 to 800 nm with most grains having sizes between 200 and 400 nm (Fig. 1). Detailed information on the mineralogy, obtained by transmission electron microscopy (TEM), was reported for 12 grains. Many of these TEM investigations have been performed on already ultramicrotomed IDP sections, for which no further preparation step between NanoSIMS and TEM work is necessary. However, most presolar silicates have been detected in meteorite thin sections or samples pressed onto gold foils which are not accessible for direct TEM analysis. These grains have to be cut-out from the samples in which they were discovered using the focused ion beam (FIB) technique [11]. This is a difficult and time-consuming task which explains the small number of grains which were analyzed so far. Among the studied silicates are nine amorphous grains, two olivines, and a $MgSiO_3$ with a perovskite (high-pressure) like structure [2,4,8,12-16]. The structure of the latter has been explained to be likely the result of shock-transformation of a silicate precursor either in the ISM or induced by the grain's parent star [14]. The chemistry of additional grains was inferred from NanoSIMS and electron-microscopy (SEM/EDX, Auger) data. Two of these grains have a complex, i.e., heterogeneous chemistry. One of them consists of an Al-rich core, probably corundum (Al_2O_3), surrounded by a silicate mantle [17]. The other grain comprises at least three different Ca-Al-rich minerals and can thus be regarded as a presolar CAI [18].

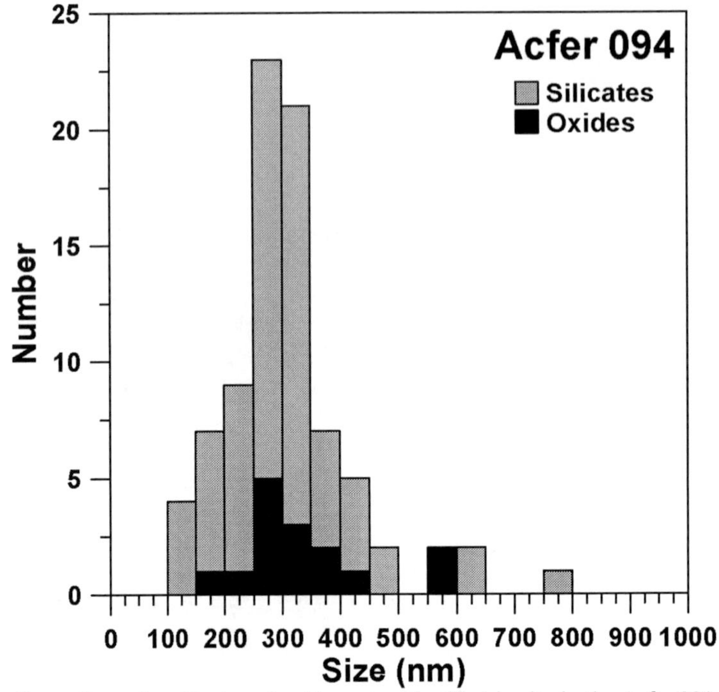

FIGURE 1. Sizes of presolar silicate and oxide grains identified in-situ in the Acfer 094 meteorite. Data sources: [5,8,19].

3. ISOTOPIC COMPOSITIONS AND STELLAR SOURCES

Oxygen isotope data exist for more than 200 presolar silicate grains (Fig. 2). Large variations are evident, with $^{17}O/^{16}O$ ratios ranging from 0.3 to 25x the solar ratio, and $^{18}O/^{16}O$ ratios from 0.1 to 15x the solar ratio. Based on the O-isotopic signature Nittler et al. [20] divided presolar oxide (corundum, spinel, hibonite) grains into 4 distinct groups (grey ellipses in Fig. 2). The distribution of the O-isotopic compositions of presolar silicate grains is roughly compatible with that of presolar oxide grains, i.e., they are likely to come from the same types of stellar sources. Most silicate grains (>70 %) exhibit enrichments in ^{17}O and close-to-solar or slightly lower than solar $^{18}O/^{16}O$ ratios (Group 1). These grains most likely formed in the winds of 1-2.5 $M_{\odot}$ red giant branch (RGB) and AGB stars as can be inferred from a comparison with the O-isotopic ratios predicted for the 1. and 2. dredge-up in red giant stars [21]. $^{17}O/^{16}O$ ratios > $\sim 3 \times 10^{-3}$ (6 grains in Fig. 2) can not be explained by models for the 1. and 2. dredge-up. However, mass transfer in a binary system has been proposed as a mechanism to account for the highest $^{17}O/^{16}O$ ratios [22]. Some (a few %) of the presolar silicate grains show high $^{17}O/^{16}O$ ratios and strong depletions in ^{18}O (Group 2 grains). The latter signature indicates that the parent stars experienced partial H burning, probably by cool bottom processing (CBP) [23,24] during the RGB or AGB phase.

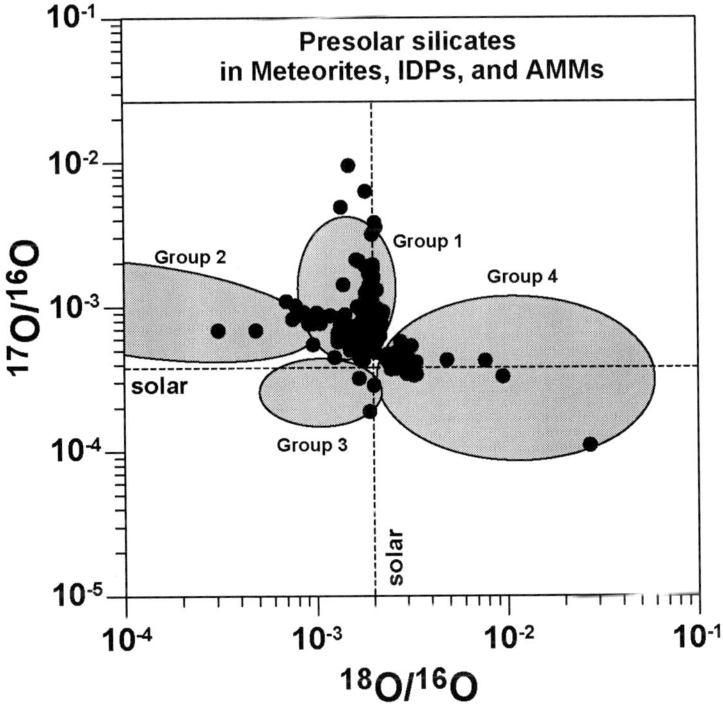

FIGURE 2. O-isotopic compositions of 204 presolar silicate grains from primitive meteorites, IDPs, and AMMs measured with the NanoSIMS [2,4,5,7,8,12-14,17,19,25-29]. The 4 groups defined by Nittler et al. [20] for presolar oxides are indicated by the grey-shaded ellipses. Note that the Group 4 has been modified here such that all grains with moderate to strong enrichments in ^{18}O are included.

The Group 4 grains are characterized by moderate to strong enrichments in ^{18}O and a limited range for $^{17}O/^{16}O$ ratios around solar (both above and below) for all except one grain (Fig. 2). These grains comprise about 10 % of all presolar silicate grains. One olivine grain exhibits a strong depletion in ^{17}O and a Type II SN has been considered as the most likely stellar source [12]. The origin of the common Group 4 grains, on the other hand, has been a matter of debate for a long time and both Type II SNe and high-metallicity AGB stars have been proposed as stellar sources [5,20,30]. A detailed comparison between O-, Mg-, and Ca-isotopic ratios of Group 4 oxide grains with predictions from mixing calculations for Type II SN ejecta, however, has shown that the grain data can be well reproduced by these models [22,31]. The Si isotope data of Group 4 silicate grains support this view (see below), and Type II SNe are clearly favoured as stellar sources for this sub-population of presolar silicate grains.

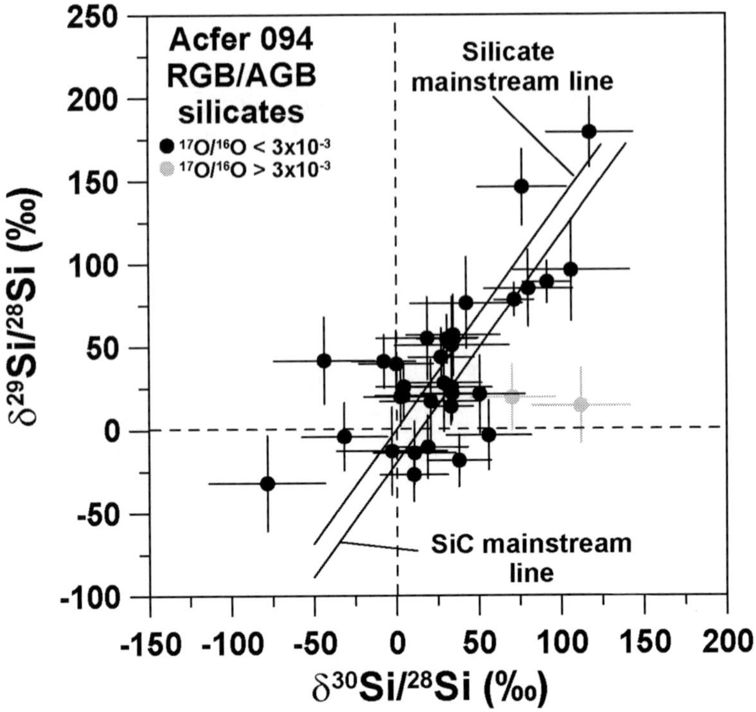

FIGURE 3. Si-isotopic compositions of presolar silicate grains of likely RGB/AGB star origin (Groups 1-3) from the Acfer 094 meteorite measured with the NanoSIMS [5,13,28]. Errors are 1σ. Only data with error in δ^{30}Si of <35 ‰ are shown.

Si-isotopic compositions were measured with the NanoSIMS in 55 presolar silicate grains. In a Si-three-isotope representation the Group 1 and 2 grains plot along the SiC mainstream line ($\delta^{29}Si/^{28}Si = 1.37 \times \delta^{29}Si/^{28}Si - 20$ [32]), although most of them are slightly shifted to the ^{30}Si-poor side of this line. The SiC mainstream line is believed to reflect mainly the Galactic chemical evolution of the Si isotopes, both in time and space, but also to a small extent the dredge-up of s-process Si in 1-3 $M_{\odot}$ AGB stars of roughly solar metallicity [32]. Based on the available data for Group 1 and 2 silicate grains, Vollmer et al. [28] have constructed a silicate mainstream line, assuming this line to be parallel to the SiC mainstream line. This line passes through the origin in Fig. 3. The shift between the two lines is a measure for the different fractions of s-process Si dredged-up from the He intershell to the envelope of the silicate and SiC parent stars. The dredge-up of s-process Si is expected to shift the Si-isotopic ratios to the ^{30}Si-rich side along lines of slope 0.2-0.7 in a Si-three-isotope representation [32]. Since silicates form when C/O < 1, i.e. earlier than SiC, the imprint of s-process Si dredge-up should be smaller, in agreement with the shift of the silicate mainstream line to the ^{30}Si-poor side. Shifts of 3-21 ‰ in $^{29}Si/^{28}Si$ and, respectively, of 17-30 ‰ in $^{30}Si/^{28}Si$ are inferred [28], in good agreement with predictions for 1.5-3 $M_{\odot}$ AGB stars of solar metallicity [32].

Three of the extremely ^{17}O-rich Group 1 grains have been measured for their Si-isotopic composition. The Si data of one grain are subject to large errors, making an interpretation difficult. Interestingly, the remaining two grains plot clearly to the ^{30}Si-rich side of the silicate mainstream line (Fig. 3). Following the idea of. [22], Vollmer et al. [28] have proposed mass transfer from a ONe nova to a main sequence star which could account at least qualitatively for the observed O- and Si-isotopic signatures.

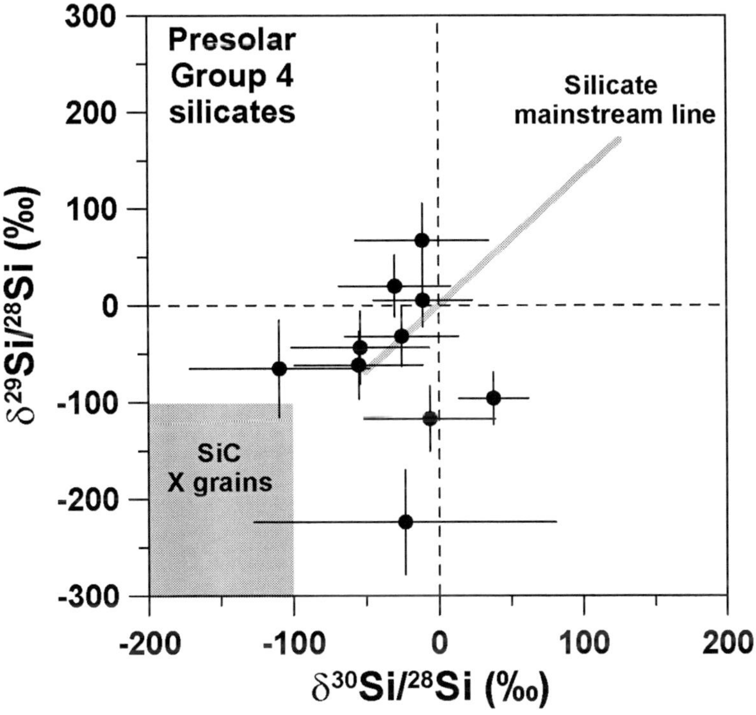

FIGURE 4. Si-isotopic compositions of presolar silicate grains from Group 4, which likely formed in SN ejecta [5,13,25,28]. Errors are 1σ. The silicate mainstream line is shown for comparison.

The Si-isotopic signatures of the Group 4 silicate grains are clearly distinct from those of most Group 1 and 2 grains (Fig. 4). All but one grain show lower than solar ^{30}Si/^{28}Si ratios and 7 out of 10 grains have lower than solar ^{29}Si/^{28}Si ratios. This is qualitatively the signature of the SiC X grains which are from Type II SNe [1]. In order to evaluate the Type II SN origin more quantitatively, Vollmer et al. [28] performed mixing calculations following the procedure applied by Nittler et al. [22] for Group 4 oxide grains and by Messenger et al. [12] for the extremely ^{17}O-poor Group 4 olivine grain. Close to solar ^{17}O/^{16}O ratios, enrichments in ^{18}O, and slightly lower than solar ^{29}Si/^{28}Si and ^{30}Si/^{28}Si ratios can be obtained by mixing matter from the Si/S, O/Ne, O/C, He/C, He/N, and H zones in appropriate proportions. The alternative scenario of an origin of Group 4 grains in high metallicity AGB stars, on the other hand, has severe problems in view of the O and Si isotope data. Silicon-isotopic ratios close to the low metallicity end of the SiC/silicate mainstream line argue against a high-metallicity RGB/AGB star origin as do the observed lower than solar $^{17}O/^{18}O$ ratios in most Group 4 grains. Clearly, a Type II SN origin is favoured for the Group 4 silicate grains.

The Mg-isotopic composition was measured in one Group 2 silicate grain [7]. A strong ^{26}Mg enrichment was found, most likely due to in-situ decay of ^{26}Al. The inferred $^{26}Al/^{27}Al$ ratio of about 0.12, which is among the highest ratios in O-rich presolar dust, together with the low $^{18}O/^{16}O$ ratio of 0.3x the solar ratio point to an origin in a low-mass thermally pulsing AGB star undergoing cool bottom processing. The $^{26}Al/^{27}Al$ ratio is indicative of very deep mixing as temperatures of > 57 million K are required. The $^{18}O/^{16}O$ ratio, on the other hand, constrains the mass circulation rate to $>10^{-6}$ $M_{\odot}$/yr. Combined O- and Mg-isotopic measurements on presolar silicates can thus provide detailed information on mixing processes in the envelopes of RGB/AGB stars.

ACKNOWLEDGMENTS

We thank Maurizio Busso for organizing the IXth Torino Workshop on Evolution and Nucleosynthesis in AGB stars where this paper was presented.

REFERENCES

1. E. Zinner, "Presolar Grains", in *Meteorites, Comets, and Planets,* edited by A. M. Davis (Elsevier, Amsterdam, 2004), p. 17
2. S. Messenger, L.P. Keller, F. Stadermann, R.M. Walker, and E. Zinner, *Science* **300**, 105 (2003)
3. F. Hillion, B. Daigne, F. Girard, G. Slodzian, and M. Schumacher, "A new high performance SIMS instrument: The Cameca "Nanosims 50", in *Secondary Ion Mass Spectrometry SIMS IX*, edited by A. Benninghoven, Y. Nihei, R. Shimizu et al. (John Wiley & Sons Ltd., Chichester, 1994), p. 254
4. C. Floss, F.J. Stadermann, J.P. Bradley, Z.R. Dai, S. Bajt, G. Graham, and A.S. Lea, *Geochim. Cosmochim. Acta* **70**, 2371 (2006)
5. S. Mostefaoui and P. Hoppe, *ApJ* **613**, L149 (2004)
6. K. Nagashima, A.N. Krot, and H. Yurimoto, *Nature* **428**, 921 (2004)
7. A.N. Nguyen and E. Zinner, *Science* **303**, 1496 (2004)
8. C. Vollmer, P. Hoppe, F. Brenker, and C. Holzapfel, *Lunar Planet. Sci.* **38**, abstr. #1262 (CD ROM) (2007)
9. F.J. Molster and L.B.F.M. Waters, "The mineralogy of interstellar and circumstellar dust", in *Astromineralogy*, edited by Th. Henning (Springer, New York, 2002), p. 121
10. H.-P. Gail, "Formation and evolution of minerals in accretion disks and stellar outflows", in *Astromineralogy*, edited by Th. Henning (Springer, New York, 2002), p. 55
11. T.J. Zega, L.R. Nittler, H. Busemann, P. Hoppe, and R.M. Stroud, *Meteorit. Planet. Sci.* **42**, 1373 (2007)
12. S. Messenger, L.P. Keller, and D.S. Lauretta, *Science* **309**, 737 (2005)
13. A.N. Nguyen, F.J. Stadermann, E. Zinner, R.M. Stroud, C.M. O'D. Alexander, and L.R. Nittler, *ApJ* **656**, 1223 (2007)
14. C. Vollmer, P. Hoppe, F. Brenker, and C. Holzapfel, *ApJ.* **666**, L49 (2007)
15. T. Yada, F. J. Stadermann, C. Floss, E. Zinner, C.T. Olinger, G.A. Graham, J.P. Bradley, Z.R. Dai, T. Nakamura, T. Noguchi, and M. Bernas, *Lunar Planet. Sci.* **36**, abstr. #1227 (2005)
16. L.P. Keller and S. Messenger, *Meteorit. Planet. Sci.* **42**, A81 (2007)

17. C. Vollmer, P. Hoppe, F. Brenker, and H. Palme, *Lunar Planet. Sci.* **37**, abstr. #1284 (2006)
18. C. Vollmer, F.J. Stadermann, M. Bose, C. Floss, P. Hoppe, and F. Brenker, *Meteorit. Planet. Sci.* **42**, A158 (2007)
19. P. Hoppe, S. Mostefaoui, and T. Stephan, *Lunar Planet. Sci.* **36**, abstr. #1301 (2005)
20. L.R. Nittler, C.M. O'D. Alexander, X. Gao, R.M. Walker, and E. Zinner, *ApJ* **483**, 475 (1997)
21. A.I. Boothroyd and I.-J. Sackmann, *ApJ* **510**, 232 (1999)
22. L.R. Nittler, C.M. O`D Alexander, R. Gallino, P. Hoppe, A.N. Nguyen, F.J. Stadermann, and E.K. Zinner, *ApJ, submitted*, (2007)
23. K.M. Nollett, M. Busso, and G.J. Wasserburg, *ApJ* **582**, 1036 (2003)
24. G.J. Wasserburg, A.I. Boothroyd, and I.-J. Sackmann, *ApJ* **447**, L37 (1995)
25. P.A. Bland, F.J. Stadermann, C. Floss, D. Rost, E.P. Vicenzi, A.T. Kearsley, and G.K. Benedix, *Meteorit. Planet. Sci.* **42**, 1417 (2007)
26. K.K. Marhas, P. Hoppe, F.J. Stadermann, C. Floss, and A. S. Lea, *Lunar Planet. Sci.* **37**, abstr. #1959 (2006)
27. S. Mostefaoui and P. Hoppe, *Lunar Planet. Sci.* **35**, abstr. #1593 (2004)
28. C. Vollmer, P. Hoppe, and F. Brenker, *ApJ, submitted*, (2007)
29. T. Yada, F.J. Stadermann, C. Floss, E. Zinner, T. Nakamura, T. Noguchi, and A.S. Lea, *Lunar Planet. Sci.* **37**, abstr. #1470 (2006)
30. B.-G. Choi, G.R. Huss, G.J. Wasserburg, and R. Gallino, *Science* **282**, 1284 (1998)
31. L.R. Nittler, *Workshop on Chronology of Meteorites*, abstr. #4016 (2007)
32. E. Zinner, L.R. Nittler, R. Gallino, A.I. Karakas, M. Lugaro, O. Straniero, and J.C. Lattanzio, *ApJ* **650**, 350 (2006)

Thermal And Dynamical Histories Of Saturn's Satellites: Evidence For The Presence Of Short Lived Radioactive Isotopes

Torrence V. Johnson†, Julie C. Castillo-Rogez†, and Dennis L. Matson†

† Jet Propulsion Laboratory, California Institute of Technology, 4800 Oak Grove Drive, Pasadena, CA 91109, USA

Abstract. Recent observations of Saturn's satellite system from the Cassini/Huygens mission present serious challenges to understanding the current dynamical states and thermal histories of these icy bodies using conventional thermal models that use long lived radioactive isotopes (LLRI) as the primary heat sources. In particular, the most distant of the regular satellites, Iapetus, is in synchronous rotation about Saturn, implying relatively high levels of dissipation of tidal energy. However, it retains a highly non-equilibrium, oblate spheroid, shape, implying a thick, cold, mechanically rigid outer layer or lithosphere. Thermal history models of Iapetus that successfully explain these apparently contradictory characteristics require significant heating early in the satellite's history from short lived radioactive isotopes (SRLI), particularly ^{26}Al, implying a formation time for Iapetus of between 2.5 and 5 Myr after the formation of Calcium Aluminum Inclusions (CAIs) [1]. The characteristics of the other icy satellites in the system are consistent with this formation time, and the current thermal geyser activity on the more silicate-rich satellite Enceladus may be related to such an early heating event. A consequence of these early formation time models is that the early crust of Iapetus is too thin and weak to retain large impact basin topography until about 100 Myr after formation, and despinning to synchronous rotation might have occurred from 200-900 Myr after formation. This chronology is consistent with the formation of the large impact basins observed on Iapetus' surface by the 'late heavy bombardment' or 'lunar cataclysm' event recorded in the dating of samples from the Moon at 3900 Ma.

Keywords: ^{26}Al, Iapetus, Satellites, Chronology, Saturn, Cassini
PACS: 96.30.N-, 91.80.Vm, 95.55.Pe

1. BACKGROUND

Saturn's satellite Iapetus was the first planetary satellite beyond the Earth's Moon known to be in synchronous rotation (*i.e,* rotation period equal to the orbital period about its primary planet). When Iapetus was discovered in 1671 by the Italian/French astronomer Giovanni Cassini, he noted that the moon was visible when on one side of Saturn but invisible on the other. He correctly deduced that the object must be much darker on one hemisphere than the other and keep the same face toward Saturn as it orbits, in the same manner as the Earth's Moon.

In the modern era, the development of tidal theory and improved astronomical data on the characteristics of the planetary satellites combined to suggest that Iapetus' synchronous rotation state is a significant anomaly. Eq. 1 gives the basic relation be-

CP1001, *9th Torino Workshop on Evolution and Nucleosynthesis in AGB Stars and the 2nd Perugia Workshop on Nuclear Astrophysics,* edited by R. Guandalini, S. Palmerini, and M. Busso

tween the change of the angular rotation velocity due to tidal forces and a satellite's physical and orbital properties.

$$\frac{d\omega}{dt} = -\frac{3GM_p^2 a^5 k_2}{CD^6 Q} \tag{1}$$

Where ω is the spin angular velocity, G the gravitational constant, M_p the mass of the planet, a the satellite's mean equatorial radius, C the maximum moment of inertia about the center of mass, D the semi-major axis of the satellite's orbit, k_2 the dynamical tidal Love number, and Q the energy dissipation factor for tidal frequencies.

In a review of satellite rotation states published in 1977, a few years prior to the first spacecraft observations of the Saturn system, Stanton Peale, using some simplifying assumptions (*e.g.* assuming a fluid k_2 of 3/2), derived the time to achieve complete tidal retardation to the synchronous rate for all the satellites known at that time [2]. The dimensionless dissipation factor, Q, was the primary unknown in Peale's formulation and it was estimated to range from 100 to 1000 based on observations of the Earth and Moon. In his discussion of the results he noted that "The tidal decay times for Iapetus and Hyperion exceed the age for the solar system for moderate values of Q ...". Iapetus stands out from the other regular satellites primarily due to its large distance from Saturn (it is the most distant synchronous satellite known), which has a strong effect on de-spinning per Eq. 1. All other regular satellites, except Hyperion, should easily achieve synchronous rotation in comparatively very short times. This is illustrated in Fig. 1 with data (ca 1977) taken from Peale's review.

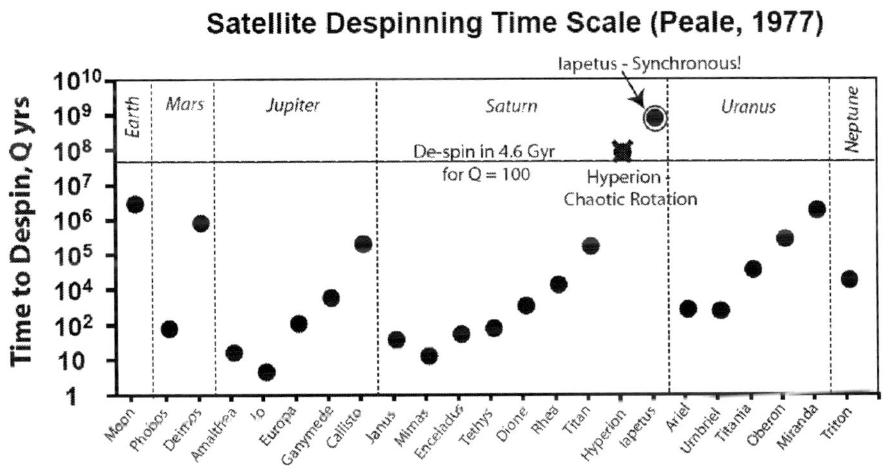

FIGURE 1. Comparison of the times to de-spin the major satellites for a constant Q. Data from Peale [2].

When the Voyager spacecraft explored the Saturn system in 1980 and 1981, images of Iapetus showed details of its dramatic hemispheric brightness dichotomy (approximately a factor of 10), but due to relatively limited spatial resolution, provided few details about its geologic history that might help explain its anomalous rotation state [3]. Hyperion was found to be not only non-synchronous but in chaotic rotation due to resonant interactions with other satellites, primarily Titan [4].

2. NEW DATA AND THERMAL MODELS

2.1 Cassini Observations

The Cassini spacecraft has been studying Saturn, its satellites, rings and magnetospheric environment since entering orbit in July of 2004. The data it has returned concerning Iapetus has made it even more difficult to understand the history of this strange satellite. First, high resolution imaging has revealed an ancient, heavily cratered surface with no evidence of endogenic activity or heating events that might be associated with creating the high dissipation required to de-spin the satellite [5]. Second, accurate measurements of its mass and shape yield a very low bulk density of 1083 kg/m^3 [6], implying a very ice-rich composition and low levels of radiogenic heat from its silicate fraction, which is about twenty weight percent. Finally, its shape is significantly different from the hydrostatic equilibrium shape expected for its current, slow rotation period of 79 days. Its oblate figure has a difference between the equatorial and polar axes of ~35 km, compared with an expected value of only ~ 10 m. This implies a thick lithosphere with significant strength.

2.2 Iapetus Thermal Models

When the currently known physical properties of Iapetus are included in a conventional thermal history model using heat production from long lived radioactive isotopes (*i.e.* U, Th, K) [7], Castillo-Rogez et al. [1] find that Iapetus' interior remains very cold for its entire history, with most of the interior remaining far below the water ice creep temperature required to produce significant dissipation. In these models there is little or no tidal de-spinning and Iapetus should have remained a cold, rapidly spinning object, very different from its currently observed state.

Explaining Iapetus' evolution requires models with both more heat and a different time dependence of heating than conventional models. The satellite must not only de-spin to synchronous rotation but at the same time also cool rapidly to prevent relaxation to a synchronous, near-spherical shape. Short lived radioactive isotopes (SLRI) are attractive candidates for such additional heating. SLRI have been recognized for some time as potentially important for heating planetesimals accreting in the early solar system [8]. Most past studies have focused on their role in melting the parent bodies of igneous meteorites [9], but their possible inclusion in bodies forming in the outer solar system is of considerable interest, both for understanding the thermal evolution of the bodies and for relating the time of formation of giant planets to the meteoritic record related to the inner solar system.

Castillo-Rogez et al. [1] have explored a wide range of possible thermal models for Iapetus, incorporating porosity in the water ice, both LLRI and SLRI heat sources, and using temperature-dependent parameterization. The dissipative behavior of the mostly icy interior is modeled using Maxwell rheology, which includes both temperature and frequency dependence so that the heating history and tidal de-spinning processes are properly coupled. The SLRI included in the thermal calculation are ^{26}Al, ^{60}Fe, and ^{53}Mn. The amount of heat supplied as a function of time is set in the model by varying the assumed time of Iapetus' formation after CAIs (whose age is set by the Pb-Pb age of 4567.2 ± 0.6 Ma [10]), combined with the SLRI half-lives and initial abundances. Iapetus' heat budget from the SLRI is dominated by ^{26}Al, with an uncertainty in the total of the SLRI heat being set by the current uncertainty in the initial abundance ratio of ^{60}Fe/^{56}Fe. The lower values reported for the initial ^{60}Fe/^{56}Fe, ~ 10^{-7}, yield a negligible heat contribution from this isotope, while higher reported initial values, ~ 10^{-6}, lead to about a factor of two greater total heat production in the time range covered.

A major result of this modeling effort is determining the range of heat required from SLRI necessary to both de-spin the satellite while allowing cooling of the outer layers to the rigidity needed to sustain its non-hydrostatic shape. Figure 2 illustrates the relationships between heat available from SLRI and heat required by models that successfully explain Iapetus' current state, as a function of time of formation after CAIs. The range in heat required is based on varying the amount of initial accretional heating and therefore initial temperature assumed for Iapetus. Combining the uncertainties in available and required heat leads to the authors' estimate that Iapetus formed between 2.5 and 5.0 Myr after CAIs. In models with earlier formation times, Iapetus becomes warm and dissipative very quickly and de-spins, but relaxes to a hydrostatic shape very quickly. Models formed after the suggested period either never have enough heating to de-spin (similar to the conventional models) or produce interior warming much later in the satellite's history and do not retain the non-hydrostatic shape because of their significant remnant porosity and warm temperatures at relatively shallow depths.

2.3 Other Satellites

If all the regular satellites formed at the same time in the environment around Saturn, they should also have experienced heating from SLRI during their early history. The effects of this heating depend strongly on the satellites' size and their silicate rock fraction, as reflected in their bulk densities. Looking at the general characteristics of the other satellites, there are no obvious features that are inconsistent with the early formation inferred for Iapetus. As noted previously, the other satellites (except chaotic Hyperion) are in synchronous rotation as expected, whether SLRI heat was available or not. Small satellites with estimated rock fractions similar to Iapetus or lower (e.g. Mimas and Tethys) appear to also have ancient highly cratered surfaces, although some tectonic features such as Ithaca Chasma on Tethys [11] might be related to early heating. The larger satellites with higher rock fractions (Rhea and Dione) would be expected to have had a period of early SLRI heating also, but direct evidence of this is not obvious on their surfaces today.

Tiny Enceladus (only about one third the size of Iapetus {1/9 of Iapetus' volume} but with a high, ~60%, rock fraction) is a special case. This satellite's spectacular water geysers and thermal anomalies [12,13,14] are a clear indication of an unusual thermal history for such a small body. The direct heating effects of SLRI are, of course, limited to the first few million years following satellite formation, after which the SLRI are essentially extinct due to their short half-lives. On the other hand, early heating in this satellite could produce early differentiation and a hot silicate core, which may be relevant to understanding the composition of the gases in the plumes [15], and Enceladus' unusual thermal state today.

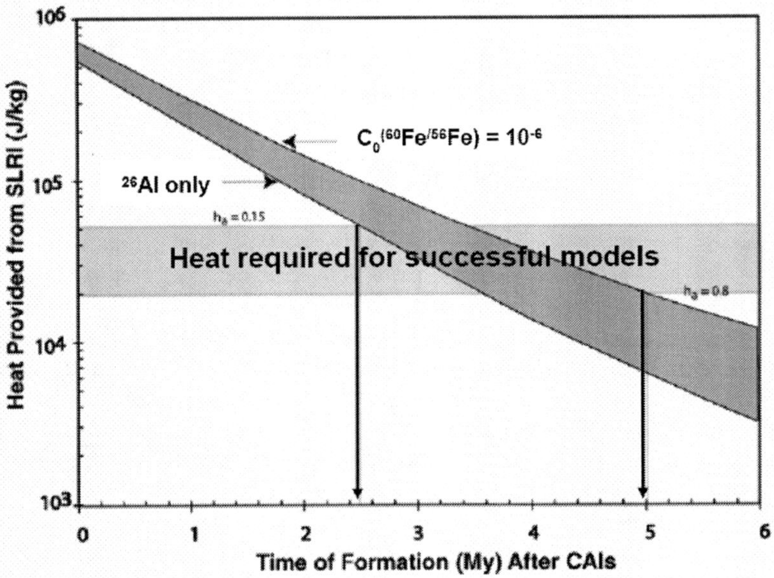

FIGURE 2. Heat from SLRI required for successful Iapetus models compared with available heat as a function of time of formation after CAIs. h_a refers to the fraction of heat of accretion initially retained by the satellite. From Castillo-Rogez *et al.* [1], reprinted by permission from Elsevier.

3. DISCUSSION: IMPLICATIONS FOR CHRONOLOGY

3.1 Formation Time of Iapetus

The Iapetus formation time reported by Castillo-Rogez et al. is tied directly to the early solar system chronometer supplied by ^{26}Al from the study of meteorites. Since we do not yet have samples from outer solar system satellites available for isotopic dating, this provides the first reasonably high precision estimate of the age of the giant planets and their satellites that can be related directly to the inner solar system record.

Figure 3 shows a comparison of the Iapetus formation time estimate with a compilation by McKeegan and Davis [16] of relative ^{26}Al ages tied to absolute ages from Pb-Pb dating [9]. This suggests that Iapetus, and Saturn, formed in the first few million years after the CAIs at about the time that the differentiated parent bodies of eucritic meteorites were present in the inner solar system, but before the final formation of the inner terrestrial planets, which occurred possibly tens of Myr afterward [16,17].

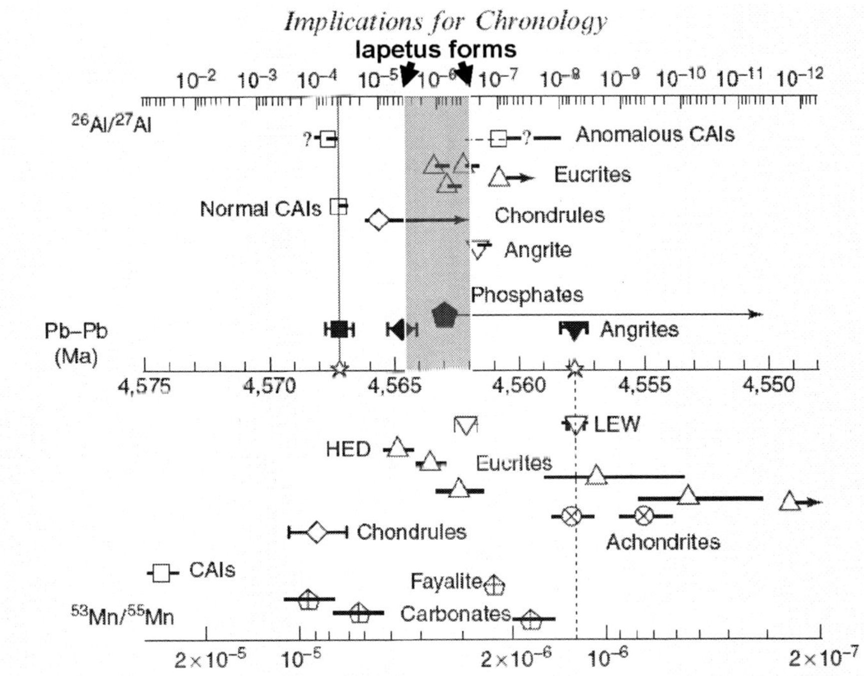

FIGURE 3. Short lived radioisotope chronometry compared with absolute ages from Pb-Pb dating. Modified after McKeegan and Davis [16], reprinted by permission from Elsevier, and compared with the estimated time of formation of Iapetus from Castillo-Rogez *et al.* [1].

3.2 Late Heavy Bombardment

Another important dated event in the inner solar system that may be related to outer solar system events is the "late heavy bombardment" or "lunar cataclysm", inferred from the dating of lunar samples to have occurred about 3900 Ma. Current models of the dynamical evolution of the early solar system suggest that this event, involving a large "spike" in the impact rate on the Moon, may have been triggered by Jupiter and Saturn evolving into a temporary resonance, having the effect of pushing the orbits of Uranus and Neptune into the outer protoplanetary disk and gravitationally scattering large numbers of large icy planetesimals throughout the solar system. This model, referred to as the "Nice model" after the location of most of the developers, predicts that the "lunar cataclysm" was not confined to the terrestrial planet zone or the inner solar system, but should have been a solar system-wide phenomena, distinct from the last stages of accretion of the planets and satellites themselves [18].

Iapetus' cratered surface has multiple large basins, resembling, on a smaller scale, the lunar record [5]. There are two clues to establishing at least limits on the age of Iapetus' impact basins. First, the thermal history models that heat the satellite enough to accomplish de-spinning do not result in a lithosphere rigid enough to retain large basins for about 100 Myr after Iapetus forms. Second, at least one of the basins disrupts and therefore postdates a portion of the prominent equatorial ridge on Iapetus. The mechanism for forming this ridge is still not well understood, but Castillo-Rogez et al. suggest that it may have been related to the surface area decrease and stresses associated with the de-spinning event, which occurs from 200 to 900 Myr after Iapetus' formation. While not establishing the exact age of the basins, these constraints suggest that these basins formed significantly later than the end of the satellite's formation and are consistent with formation during the Nice model's system-wide, late heavy bombardment at 3900 Ma.

ACKNOWLEDGMENTS

The research described in this paper was carried out at the Jet Propulsion Laboratory, California Institute of Technology, under a contract with the National Aeronautics and Space Administration.

REFERENCES

1. J.C. Castillo-Rogez, et al., *Icarus* **190**, 179 (2007)
2. S.J. Peale, in *Planetary Satellites*, edited by J. A. Burns (University of Arizona Press, Tucson, AZ, 1977), p. 87
3. B.A. Smith, et al., *Science* **215** , 504 (1982)
4. J. Wisdom, S.J. Peale, and F. Mignard, *Icarus* **58**, 137 (1984)
5. C.C. Porco, et al., *Science* **307** , 1237 (2005)
6. P.C. Thomas, et al., *Icarus* **190**, 573 (2007)
7. K. Ellsworth and G. Schubert, *Icarus* **54**, 490 (1983)
8. H.C. Urey, *Proc. Natl. Acad. Sci. USA* **41**, 127 (1955)
9. H.Y. McSween and R.E. Grimm, *Meteoritics* **29**, 500 (1994)
10. Y. Amelin and A.N. Krot, *Meteoritics & Planetary Science* **37**, A12 (2002)
11. B. Giese, et al., *Geophysical Research Letters* **34**, L21203 (2007)
12. C.J. Hansen, et al., *Science* **311**, 1422 (2006)
13. J.R. Spencer, et al., *Science* **311**, 1401 (2006)
14. C.C. Porco, et al., *Science* **311**, 1393 (2006)
15. D.L. Matson, et al., *Icarus* **187**, 569 (2007)
16. K.D. McKeegan and A.M. Davies, in *Treatise on Geochemistry: Vol1. Meteorites, Comets, and Planets*, edited by A. Davis (Elsevier, 2007)
17. A.N. Halliday, in *Treatise on Geochemistry: Vol1. Meteorites, Comets, and Planets*, edited by A. Davis (Elsevier, 2007)
18. R. Gomes, et al., *Nature* **435**, 466 (2005)

A Massive AGB Star as Source of Short-Lived Nuclei in the Early Solar System

M. Lugaro*, A.I. Karakas†, M. van Raai*, D.A. García-Hernández** and J.M. Trigo-Rodríguez‡

*Sterrekundig Instituut, University of Utrecht, Postbus 80000, 3508 TA Utrecht, The Netherlands
†Research School of Astronomy and Astrophysics, Mt. Stromlo Observatory, Cotter Rd., Weston, ACT 2611, Australia
**Instituto de Astrofísica de Canarias, La Laguna (Tenerife), España
‡Institute of Space Sciences (CSIC-IEEC), Campus UAB, Facultat de Ciencies, Torre C5 pares, 08193 Bellaterra (Barcelona), España

Abstract. The origin of short-lived radioactive nuclei at the time of the formation of the Sun is a much debated mystery in modern astronomy. We show that abundance predictions for massive AGB stars (roughly 6 $M_\odot$) of solar metallicity can self-consistently match the abundances of ^{26}Al, ^{41}Ca, ^{60}Fe, and ^{107}Pd in the early Solar System, as derived from meteorite analysis. On the other hand our models do not produce ^{182}Hf, thus confirming previous results. Recent measurements of excesses in ^{62}Ni and ^{96}Zr, correlated to ^{60}Fe, can be accounted for by a massive AGB source. Moreover, these stars could be an important source of ^{87}Rb, as recently observed, and thus induce the fluctuations in the Rb/Sr ratio observed in primitive chondritic materials. Finally, we discuss recent observations of decoupling between the presence of ^{26}Al and ^{60}Fe in differentiated planetesimals and their implications on candidate stellar sources.

Keywords: AGB Stars, Nucleosynthesis, Meteorites, Solar System
PACS: 97.10.Cv

1. WHO SALTED THE SOUP?

To introduce the problem of radioactive nuclei in the early Solar System we first present the text of a short performance that may be helpful to student and non-experts to understand the topic. This piece was inspired by the title of the Torino workshop held in 2000 in celebration of Gerald J. Wasserburg's Laurea Honoris Causa in Physics at the University of Torino: "Salting the early soup: trace nuclei from stars to the solar system". Equipment needed: a transparent container with soup inside (preferably a not too liquid soup), a spoon, a salt shaker, and a few little plates or bowls. The slides accompanying this performance can be found on the Perugia-Torino workshop website at http://www.fisica.unipg.it/~torino/i1.html

Part one. *(Holding the soup container and the spoon, stirring and walking around the room - Slide 2).* I want you to stretch your imagination and visualize that this soup is actually the material from which the Solar System formed 4.6 billion years ago. It is made of gas and dust, which have been hanging around in a big cloud. Since I have been stirring it for some hundred million years it is quite well mixed. As for the exact ingredients of the recipe, the soup contains carbon, oxygen, iron, and all the other elements, just in the right amount as we observe them today in the Solar System.

CP1001, *9th Torino Workshop on Evolution and Nucleosynthesis in AGB Stars and the 2nd Perugia Workshop on Nuclear Astrophysics*, edited by R. Guandalini, S. Palmerini, and M. Busso

FIGURE 1. M. Lugaro salting the soup in Perugia.

Part two. *(Standing at the presentation desk with the salt shaker nearby.)* Now, a big cook suddenly comes around my back and tells me: "Come on, what are you waiting for? Put some salt in and serve it up!" - "Oh, gosh! That's right, I did forget to put the salt in". *(Put the salt in the soup - Fig. 1)* At this point, perhaps because I perturbed the soup by suddenly putting the salt in, the material start collapsing, mainly going to form the Sun. The little bit that is left I can serve it up. *(Spooning the soup into the small plates - Slide 3).* In this way I start making the first solid bodies in the Solar System, the proto-planets. Now, one of the big questions coming out of this picture is if, in fact, the shock of salting the soup triggered the collapse of the proto-solar nebula. So, when studying this salt we also perhaps shed some light on our own origin.

Part three. *(Addressing the audience).* At this point some of you may have wondered about an important question: since there is so little salt that gets mixed with a lot of soup, how do we know, today, that someone put some salt in the soup 4.6 billion years ago? The answer is that salt tastes salty, so one is able to tell if a soup was salted or not. The special taste of the salt that was put in the early Solar System was radioactivity. It was radioactive salt *(Slide 4).* This means that we can recognize it today by analysing the composition of primitive meteorites, which are some of the first rocks that formed in the Solar System. For example, two of the most famous types of the "radioactive salt" are ^{26}Al and ^{60}Fe. While these nuclei are still part of elemental Al and Fe you would not be able to recognize them in a rock full of Al and Fe: they are as "grey" as the rest of the rock. But, in about one million year, they suffer radioactive decay and transform into ^{26}Mg and ^{60}Ni, which are nuclei of different elements, Mg and Ni, respectively: they become like "pink" on a "grey" rock and thus are easy to see and measure. Finally, it should be mentioned that, while decaying, these nuclei released lots of energy in the form of high-energy photons - γ rays. These were adsorbed by the rock, which consequently melted. Hence, the radioactive salt was probably also responsible for the early melting of the proto-planets in the Solar System (see also T.V. Johnson, these proceedings).

TABLE 1. Short-lived radioactive isotopes found to be present in the early Solar System, in order of increasing mean life τ. The ratios observed in the early Solar System are also indicated for each radionuclide.

	τ(Myr)	reference ratio	observed
^{41}Ca	0.15	^{41}Ca/^{40}Ca	1.5×10^{-8}
^{36}Cl	0.43	^{36}Cl/^{35}Cl	5×10^{-6}
^{26}Al	1.03	^{26}Al/^{26}Al	5×10^{-5}
^{60}Fe	2.2	^{60}Fe/^{56}Fe	2×10^{-6}-2×10^{-7}
^{10}Be	2.3	^{10}Be/^{10}Be	1×10^{-3}
^{53}Mn	5.3	^{53}Mn/^{55}Mn	6×10^{-5}-5×10^{-6}
^{107}Pd	9.4	^{107}Pd/^{108}Pd	2×10^{-5}
^{182}Hf	13	^{182}Hf/^{180}Hf	2×10^{-4}

1.1. Radioactivity in the Early Solar System

In Table 1 we list the eight most-short-lived isotopes that have been found to be present in the early Solar System, in order of increasing mean life up to 13 Myr, together with the ratio of their abundances to the abundances of the reference stable isotopes, as observed in the early Solar system (see [1] for a thorough discussion of these values). Note that also isotopes with mean lives higher than 13 Myr, such as ^{244}Pu (τ=115 Myr) and ^{235}U (τ=1020 Myr), were present in the early Solar System and some of them are still today.

Different origins can be attributed to the radioactive isotopes listed in Table 1. As discussed at length by [1], the abundances of most radioactive nuclei heavier than iron, such as ^{182}Hf and ^{244}Pu, can be explained by Galactic uniform production, in other words, they were already in the soup. On the other hand, the first isotopes listed in Table 1 have such a short mean life that they must have been put in the soup at the very last minute before the collapse of the proto-solar nebula. The presence of ^{10}Be requires irradiation in the early Solar System since this isotope cannot be produced in stars, and this may be the same for ^{36}Cl, since the detection of these two isotopes in primitive materials appears to be coupled. Of the remaining isotopes, ^{26}Al and ^{41}Ca could have originated either from irradiation in the early Solar System (see, e.g, [2]), or as part of the material expelled by a nearby dying star, such as a supernova explosion or the winds of a giant star. ^{60}Fe and ^{107}Pd must have come from a stellar site, and, finally, ^{53}Mn could have been produced by a supernova explosion or via irradiation in the early Solar System. Its relatively long mean life allows for the possible parent supernova to have occurred up to 10 Myr before the formation of the Sun.

2. RADIOACTIVE NUCLEI FROM MASSIVE AGB STARS

We now discuss the possibility of finding a self-consistent solution for the origin of the abundances of radioactive nuclei in the Solar System using models of evolved intermediate-mass stars. After core H and He burning, stars of masses between approximately 1 and 8 $M_\odot$ climb up the asymptotic giant branch (AGB, see [3] for a review).

There, they become very cool and luminous while progressively shedding all the material in their convective envelope into the interstellar medium via strong stellar winds. Here, we focus on massive AGB stars, $\simeq 6$ $M_\odot$. The nucleosynthesis that occurs in these stars leads to the production of several radioactive nuclei, which are expelled into the interstellar medium by the stellar winds. As done by [1], we can restrict our first analysis of the match with AGB models to four ratios involving radioactive nuclei in the early Solar System: ^{26}Al/^{27}Al, ^{41}Ca/^{40}Ca, ^{60}Fe/^{56}Fe and ^{107}Pd/^{108}Pd.

We use a stellar model of 6.5 $M_\odot$ and Z = 0.02 (solar metallicity). The structure was calculated with the Monash version of the Mount Stromlo Stellar Structure Program where the numerical method used to compute the stellar models has been previously described in detail in [4, 5, 6]. Mass loss on the AGB was included using the formulation provided by Vassiliadis and Wood [7], and initial solar abundances from the compilation of Andres and Grevesse [8]. This model experienced 39 thermal pulses with the third dredge-up episode occurring 36 times. The temperature in the thermal pulses reached 372 $\times 10^6$ K, thus efficiently activating the ^{22}Ne(α,n)^{25}Mg reaction. The neutrons released by this reaction led to the production of ^{41}Ca, ^{60}Fe, and ^{107}Pd via neutron captures on ^{40}Ca, ^{58,59}Fe, and ^{106}Pd, respectively. These isotopes were carried to the stellar surface via the third dredge-up, with the total amount of matter dredged into the envelope during the AGB phase equal to 0.048 $M_\odot$. This model also experienced proton captures at the base of the convective envelope (hot bottom burning), where the maximum temperature at the base of the convective envelope was 86 $\times 10^6$ K. For this reason the star retained an O-rich composition, and produced a large amount of ^{26}Al via proton captures on ^{25}Mg.

Our reaction library includes neutron captures from the Bao et al. [9] compilation as well as the latest ^{41}Ca(n,α)^{38}Ar rate [10], by far the main channel for the destruction of ^{41}Ca in neutron-rich conditions. For electron captures on ^{41}Ca we took the terrestrial mean life of 0.15 Myr, even though in fully ionized stellar conditions this half life is longer if the density is less than a few 10^4 gr/cm^3 [11].

The ratios of interest predicted by our model at the stellar surface are displayed in Fig. 2. The final values are: ^{26}Al/^{27}Al = 0.015, ^{41}Ca/^{40}Ca = 1.6 $\times 10^{-4}$, ^{60}Fe/^{56}Fe = 0.001, and ^{107}Pd/^{108}Pd = 0.012. We have also estimated the ^{182}Hf/^{180}Hf ratio for this model and found it to be extremely low, of the order of 10^{-5}, thus confirming previous results that ^{182}Hf is not produced in AGB stars [12]. We have also computed a final ^{36}Cl/^{35}Cl ratio of 9.8 $\times 10^{-5}$, which is too low to match the observed value.

2.1. Comparison to Early Solar System Abundances

We consider the final values listed above and follow the same procedure described in detail by [1]. We obtain from ^{107}Pd an allowed range for the dilution factor $f_0 = M_{ENV}/M_0$, where M_{ENV} is the mass of injected stellar envelope and M_0 is the mass of the cloud, of $1.65 \times 10^{-3} < f_0 < 3.48 \times 10^{-3}$. We chose $f_0 = 3.3 \times 10^{-3}$ and then use ^{41}Ca to derive a Δ_1, i.e., the time interval before ejection of the radionuclides and formation of the solar system, equal to 0.53 Myr, which is a reasonable value. Finally, we determine Δ_2, which is the time interval between the initial state and the time of

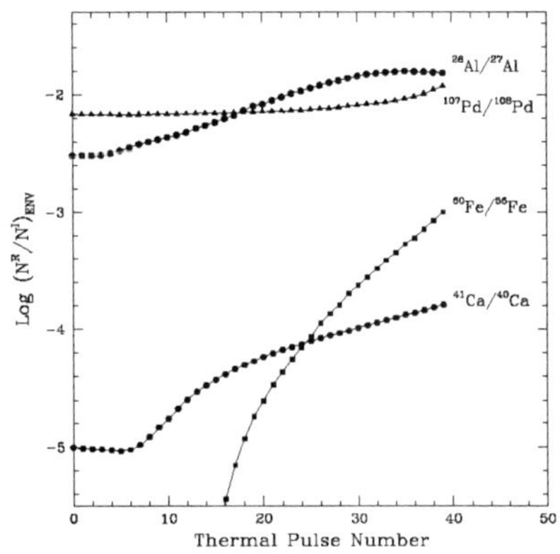

FIGURE 2. Evolution of the number abundance ratios, as indicated by the labels, at the surface of our 6.5 $M_\odot$ and Z=0.02 AGB model as function of the number of thermal pulses.

formation of differentiated objects, to be equal to 6.0 Myr assuming that at such time $^{107}Pd/^{108}Pd = 2 \times 10^{-5}$, as observed. With this parameters, we obtain $^{26}Al/^{27}Al = 3.2 \times 10^{-5}$ and $^{60}Fe/^{56}Fe = 2.6 \times 10^{-6}$ at Δ_1. Within nuclear reaction uncertainties, such as the 50% uncertainty related to the $^{25}Mg(p,\gamma)^{26}Al$ reaction rate [13], these numbers match the ratios observed in the early Solar System. On the other hand, considering the large uncertainty associated to the $^{26}Al(p,\gamma)^{27}Si$ reaction rate, values for ^{26}Al two orders of magnitude times lower than reported here are also allowed [13].

3. OTHER CONSTRAINTS

[14] measured small anomalies (of the order of one part per ten thousand) in calcium-aluminium inclusions from the Allende and Efremovka meteorites and found a correlation between the ^{60}Ni anomalies - which are also due to the decay of ^{60}Fe - and the ^{62}Ni anomalies. [15] further found a possible correlation between ^{62}Ni and ^{96}Zr anomalies in the same materials. Since massive AGB stars can also make ^{62}Ni and ^{96}Zr via neutron captures (see Fig. 3), we check if these isotopes are produced in our model in the same relative abundances as observed. We find that the predicted ^{60}Ni anomaly is 0.3 of the predicted ^{62}Ni anomaly, while observations indicate a ratio of 0.53. We should point out that in this model we have used the value from [9] for the neutron capture cross section of ^{62}Ni of 12 mbarn at 30 keV. Two recent experiments [16, 17] have instead obtained values of 28 mbarn, at 25 keV, and 37 mbarn, at 39 keV and further experiments are currently under consideration (F. Käppeler and M. Heil, private communication). Using a higher ^{62}Ni+n rate we would obtain a smaller abundance of ^{62}Ni, thus improving the match with the observations.

As for the ^{96}Zr, the sparse data do not allow to obtain a reliable number for the correlation with ^{62}Ni. Our models produces an anomaly in ^{96}Zr of the same order of that of ^{62}Ni (Fig. 3), which is consistent with the observations. This point should be tested

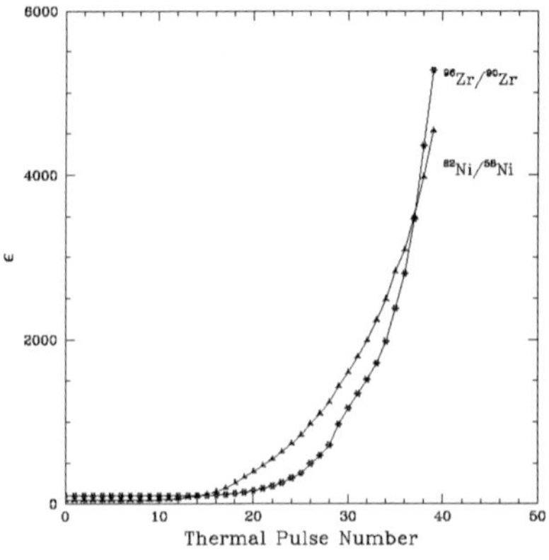

FIGURE 3. Evolution of the $^{62}Ni/^{58}Ni$ and $^{96}Zr/^{90}Zr$, in form of variations with respect to the solar ratios per ten thousand (ε-values), at the surface of our 6.5 $M_\odot$ and Z=0.02 AGB model as function of the number of thermal pulses.

in the light of new estimates for the ^{62}Ni and the ^{96}Zr neutron capture cross sections.

Another point of interest regards the very long-living heavy isotope ^{87}Rb (τ=3.3$\times 10^{10}$ yr). Our massive AGB model produces this isotope by almost a factor of three (see M. van Raai et al., these proceedings). [18] suggested possible heterogeneity at the level of parts per ten thousands in Sr in the early Solar System to interpret small variations observed in the Sr/Rb ratios in different calcium-aluminium inclusions. This heterogeneity could have been produced by the injection of ^{87}Rb-rich massive AGB material in the early Solar System.

4. POSSIBLE DECOUPLING OF ^{26}AL AND ^{60}FE?

Recently, Bizzarro et al. [19] have presented measurements of bulk meteorites which seems to indicate that: "^{60}Fe was injected into the protoplanetary disk $\sim$ 1 Myr after Solar System formation, when ^{26}Al was already homogeneously distributed." While the validity of such measurements themselves is much debated, with several other laboratories not being able to obtain the same results, here, we will only comment on their possible interpretation. Bizzarro et al. [19] state that: "The observed decoupling is a predicted feature of [...] $>$ 30 $M_\odot$ stars, where ^{26}Al is expelled [...] during the Wolf-Rayet phase, [...] ^{60}Fe is predicted to be ejected during the following supernova of type Ib/c event, where no ^{26}Al is co-ejected [20]." However, this interpretation cannot be applied because, looking at Fig 7 of [20], one can see that most ^{26}Al from these stars is expelled during the supernova explosion. The fraction of ^{26}Al from the winds accounts only for 10%, 20%, and 33% of the total ^{26}Al yield in the 35, 60, and 80 $M_\odot$ stellar models, respectively. In the massive stars models of Limongi and Chieffi [20], actually, there is no case where ^{60}Fe is produced and ^{26}Al is not (see Fig. 4). On the other hand, AGB models of masses 4 to 5 $M_\odot$, can produce up to 20 times more ^{60}Fe than ^{26}Al, as

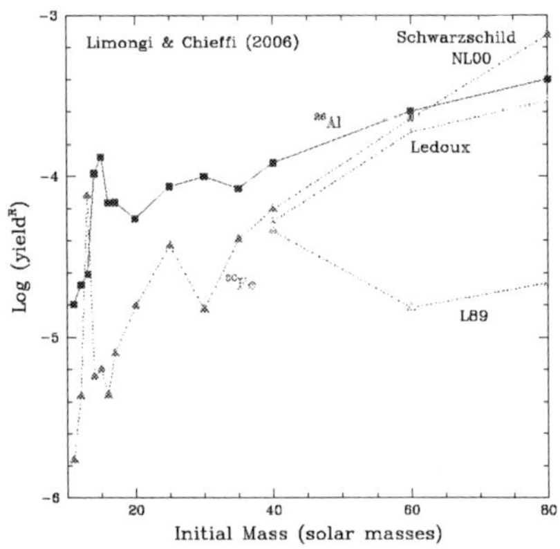

FIGURE 4. Yields of ^{26}Al and ^{60}Fe integrated over the whole stellar lifetime (in $M_\odot$) for massive star models of different masses and solar metallicity from [20]. The different lines for ^{60}Fe and masses from 40 $M_\odot$ up correspond to the following cases: Schwarzschild NL00 $\rightarrow$ Schwarzschild criterion for convective borders and mass loss rate from [21]; Ledoux $\rightarrow$ Ledoux criterion for convective borders; and L89 $\rightarrow$ mass loss rate from [22]. See [20] for a full discussion.

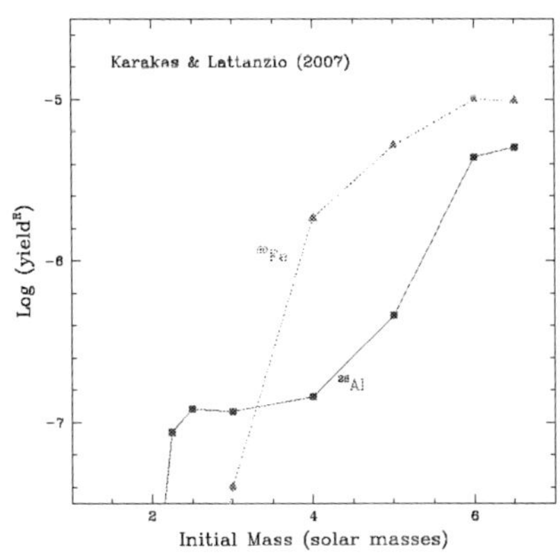

FIGURE 5. Yields of ^{26}Al and ^{60}Fe integrated over the whole stellar lifetime (in $M_\odot$) for AGB models of different masses and solar metallicity from [6].

neutron captures and the third dredge-up are efficient in these models, while hot bottom burning is not (see Fig. 5). In any case, more laboratory data are needed to assess the reliability of the measurements of [19].

5. CONCLUSION

There is evidence that massive AGB stars may have contributed to salting the early soup: ^{26}Al, ^{41}Ca, ^{60}Fe, ^{107}Pd, ^{87}Rb, ^{62}Ni, and ^{96}Zr are all made in these stars. A more detailed study on the topic can be found in Trigo-Rodríguez et al. (2007, submitted to Meteoritics & Planetary Science). In particular, it is interesting to point out that, from comparison of Fig. 4 and 5, it appears that massive AGB models produce as much ^{60}Fe as supernova models. Further efforts in the modeling of nucleosynthesis in massive AGB stars are justified in order to address these issues.

ACKNOWLEDGMENTS

ML is supported by NWO (VENI fellow). JMTR thanks MEC for a JdC research grant.

REFERENCES

1. G. J. Wasserburg, M. Busso, R. Gallino, and K. M. Nollett, *NucPhysA* **777**, 5 (2006)
2. T. Lee, F. H. Shu, H. Shang, A. E. Glassgold, and K. E. Rehm, *ApJ* **506**, 898 (1998)
3. F. Herwig, *Ann.Rev.A&A* **43**, 435 (2005)
4. M. Lugaro, C. Ugalde, A. I. Karakas, J. Görres, M. Wiescher, J. C. Lattanzio, and R. C. Cannon, *ApJ* **615**, 934 (2004)
5. A. I. Karakas, M. Lugaro, M. Wiescher, J. Goerres, and C. Ugalde, *ApJ* **643**, 471 (2006)
6. A. I. Karakas and J. C. Lattanzio, *PASA* **24**, 103 (2007).
7. E. Vassiliadis and P. R. Wood, *ApJ* **413**, 641 (1993)
8. E. Anders and N. Grevesse, *Geochim. Cosmochim. Acta* **53**, 197 (1989)
9. Z. Y. Bao, H. Beer, F. Käppeler, F. Voss, K. Wisshak, and T. Rauscher, *Atomic Data and Nuclear Data Tables* **76**, 70 (2000)
10. L. de Smet, C. Wagemans, J. Heyse, S. Vermote, and J. Van Gils, *International Symposium on Nuclear Astrophysics - Nuclei in the Cosmos, Proceedings of Science, PoS(NIC-IX)020* (2006)
11. G. M. Fuller, W. A. Fowler, and M. J. Newman, *ApJSS* **48**, 279 (1982)
12. G. J. Wasserburg, M. Busso, R. Gallino, and C. M. Raiteri, *ApJ* **424**, 412 (1994)
13. R. G. Izzard, M. Lugaro, A. I. Karakas, C. Iliadis, and M. van Raai, *A&A* **466**, 641 (2007)
14. G. Quitté, A. N. Halliday, B. S. Meyer, A. Markowski, C. Latkoczy, and D. Günther, *ApJ* **655**, 678 (2007)
15. M. Schönbächler, D.-C. Lee, M. Rehkämper, A. N. Halliday, M. A. Fehr, B. Hattendorf, and D. Günther, *Earth and Planetary Science Letters* **216**, 467 (2003)
16. H. Nassar, M. Paul, I. Ahmad, D. Berkovits, M. Bettan, P. Collon, S. Dababneh, S. Ghelberg, J. P. Greene, A. Heger, M. Heil, D. J. Henderson, C. L. Jiang, F. Käppeler, H. Koivisto, S. O'Brien, R. C. Pardo, N. Patronis, T. Pennington, R. Plag, K. E. Rehm, R. Reifarth, R. Scott, S. Sinha, X. Tang, and R. Vondrasek, *Physical Review Letters* **94**, 092504–+ (2005)
17. A. Tomyo, Y. Temma, M. Segawa, Y. Nagai, H. Makii, T. Shima, T. Ohsaki, and M. Igashira, *ApJL* **623**, L153 (2005)
18. F. A. Podosek, E. K. Zinner, G. J. MacPherson, L. L. Lundberg, J. C. Brannon, and A. J. Fahey, *Geochim. Cosmochim. Acta* **55**, 1083 (1991)
19. M. Bizzarro, D. Ulfbeck, A. Trinquier, K. Thrane, J. Connelly, and B. Meyer, *Science* **316**, 1178 (2007)
20. M. Limongi and A. Chieffi, *ApJ* **647**, 483 (2006)
21. T. Nugis and H. J. G. L. M. Lamers, *A&A* **360**, 227 (2000)
22. N. Langer, *A&A* **220**, 135 (1989)

SESSION VIII:

PROBLEMS OF STELLAR MIXING

News on Deep Mixing

John C. Lattanzio*, Davis S. P. Dearborn† and Peter P. Eggleton†

*Centre for Stellar and Planetary Astrophysics, Monash University, Australia
†Lawrence Livermore National Laboratory, California, USA

Abstract. We briefly summarize the abundant observational evidence for the need of a "deep mixing" mechanism in first-ascent red-giant stars, and probably in AGB stars as well. By the term "deep mixing" we mean some mixing mechanism which operates in the radiative zone below the convective envelope, and which transports material from the convective region to hotter regions, near the top of the hydrogen shell, where nuclear burning may take place. We then discuss a recent discovery of deep-mixing caused by the burning of ^{3}He following first dredge-up in low-mass stars. This is expected to be a thermohaline process and preliminary calculations show that it has many of the properties required to explain the observations.

Keywords: Mixing; Giant Stars, Abundances
PACS: 97.10.Cv. 97.10.Tk, 97.20.Li, 26.20.f, 26.20.Cd, 26.20.Np

1. INTRODUCTION

For many years now we have been making fairly successful models of stars using computers. The standard view of stellar interiors is that in the case of non-rotating, non-magnetic stars, the only mass motions are associated with turbulent convection. In regions where the energy transport is radiative, there is assumed to be no mixing. The failure of these standard models to reproduce some observed abundances is now well known. Some form of "extra-mixing" is required to transport material from the bottom of the convective envelope to hotter regions, presumably the top of the hydrogen burning shell, where nuclear burning can take place.

Here we will review the evidence for this required mixing, see that rotationally driven mixing is not the answer, and then discuss a recently discovered mixing mechanism that might solve the problems, or at least some of them.

2. ABUNDANCE ANOMALIES

We begin by apologizing to the many researchers who have contributed to the summary we give below. Space and time do not permit us to provide all the deserved references.

We have for a long time known (dating back to Iben's papers in the 60s) that when a low mass star ascends the giant branch for the first time, the convective envelope extends inwards into regions that have seen some hydrogen burning. This mixes to the surface the products of such burning, and alters (mostly) the $^{12}C/^{13}C$ ratio as well as the C/N ratio. This is called the "First Dredge-Up" and observations of red-giants show that the predicted abundances are not bad, dropping from an initial $^{12}C/^{13}C \sim 90$ to ~ 20. As the star ascends the giant branch a little further, when it reaches the position of the bump in

CP1001, 9th Torino Workshop on Evolution and Nucleosynthesis in AGB Stars and the 2nd Perugia Workshop on Nuclear Astrophysics, edited by R. Guandalini, S. Palmerini, and M. Busso

the luminosity function, we start to see a further drop in $^{12}C/^{13}C$, typically to $\sim 10-15$ for Population I stars and maybe as low as the equilibrium value of 4 for Population II stars.

Work by Charbonnel and co-workers (see especially [1]) showed that the onset of this extra-mixing coincided with the hydrogen shell reaching the maximum inward extent of the convective envelope during the first dredge-up; it is this same event that causes the bump in the luminosity function. The standard assumption since [2] has been that rotation is the missing ingredient, and serves to provide the required mixing. The first dredge-up event leaves behind a substantial discontinuity in the molecular weight, and this is believed to inhibit mixing (by rotation). When the hydrogen shell reaches this region the resulting burning removes the discontinuity, and hence mixing across this region may begin. But a recent investigation of rotation showed that it fell well short of providing the required mixing (see [3]).

Although we have discussed only the C-isotope ratios above, we note that other species are also implicated from observations. Specifically, the C/N ratio as well as oxygen isotope ratios do not match the predictions of standard theory. There is also the observation of enhanced Li seen in some giants when they reach the position of the bump in the luminosity function [4]. All of these data resist explanation from standard models and seem to require some extra-mixing.

3. NEW MIXING MECHANISM

While investigating the core-flash using the 3D hydrodynamics code *Djehuty*, we discovered an unusual mixing event occurring just above the hydrogen shell in our red-giant models (see [5]; hereafter Paper I). In that initial paper we identified the mixing with a Rayleigh-Taylor instability. We now believe this was wrong, for the reasons we describe below. It is likely a thermohaline instability, although we prefer to identify it by the name "$\delta\mu$ mixing" since it is driven by a decrease in the local molecular weight, as we describe below.

3.1. The Role of ^{3}He

Low-mass stars produce ^{3}He by the PPI chain and then destroy it again through PPII and PPIII. It is well known that standard models predict a region in the Sun that is enriched in ^{3}He. The outer envelope has none, due to the temperature being too low for PPI reactions. Further in there is some ^{3}He produced, and further inward still the PPII and PPIII reactions destroy ^{3}He so that there is a region in between that shows enhanced ^{3}He. This is shown in the left panel of Fig. 1.

Once the star reaches the giant branch and the convective envelope extends inwards, it homogenizes the abundances. This results in a discontinuity at the deepest part of the convective penetration, as shown in the right had panel of Fig. 1. Now, as the hydrogen shell advances on this discontinuity, one of the first reactions to start burning is the destruction of ^{3}He:

$$^3\mathrm{He}+^3\mathrm{He}\rightarrow^4\mathrm{He} + 2\mathrm{p}.$$

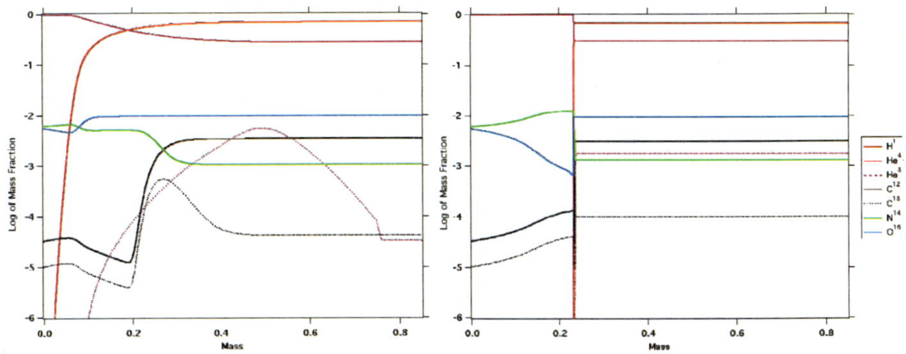

FIGURE 1. Composition structure of a $0.85 M_{\odot}$ Pop I model with $X = 0.7, Z = 0.02$. Left panel – near the end of its main-sequence life. Right panel – near the deepest penetration of the surface convective region on the giant branch. H – vermilion; ^{4}He – magenta; ^{12}C – black; ^{16}O – blue; ^{14}N – green; ^{3}He – red dots; ^{13}C – black dots.

This is an unusual fusion reaction, in that it actually decreases the mean molecular weight by virtue of the three particles resulting from the fusion of the two ^{3}He nuclei. One can show ([6]; hereafter Paper II) that

$$\delta\left(\frac{1}{\mu}\right) = \frac{\delta X_3}{6}$$

where X_3 is the mass fraction of ^{3}He. Clearly, the maximum extent of the inversion occurs if one burns all of the ^{3}He. We discuss this below, as this is the origin of the high speeds seen in Paper I.

We observed a typical change in $\delta\mu/\mu$ of about 10^{-4} resulting from this reaction. Normally this is not seen against a background of varying H and ^{4}He, but in this case the homogenized composition allows the μ inversion to develop.

3.2. The Resulting Mixing

In Paper I we took one of these red-giant models and ran it with the 3D hydrodynamical code *Djehuty*. We saw rapid motion develop in the area of the μ inversion and identified this with a Rayleigh-Taylor instability. In fact we now realize that such rapid motion is unlikely to occur. Once the μ inversion begins to develop it will drive mixing which will move the burned material upwards, replacing it with unburnt material from above. Thus the composition of the material, and hence the actual extent of the μ inversion, depends on the competition between the mixing (which acts to increase the ^{3}He content, reducing the size of the inversion) and the burning (which decreases the ^{3}He abundance and hence increases the inversion).

The model used to start the *Djehuty* calculation was a 1D model which had not been told to mix if a molecular weight inversion developed. So the material stayed in place until all of the ^{3}He was burnt, thus producing the maximum possible μ inversion as per the equation above. For this reason we saw rapid motion in the *Djehuty* calculation - the initial model had an inversion that is larger than would really develop in a star. To jump ahead slightly, when we added an approximate algorithm to the 1D code to allow for mixing, the maximum extent of the μ inversion was reduced to a value more like 10^{-5}.

In fact we now believe that the $\delta\mu$ mixing is more likely a form of double diffusive mixing, with the speed determined by the competition between the diffusion of heat, on the one hand, and the ^{3}He mixing on the other hand. Such double diffusive schemes are usually called "thermohaline" mixing, by analogy with the case of salty sea-water. Of course, the mixing of the material is not a diffusion process, although it is common to approximate it as such in stellar evolutionary calculations. But it is wise to remember that this is not truly the case, it is just a convenient way to calculate the composition.

4. CONSEQUENCES OF THE NEW MIXING MECHANISM

In Paper I we investigated the consequences of the new mixing mechanism for the galactic abundance of ^{3}He. Briefly, observations show an amount of ^{3}He which is consistent with what is predicted from Big Bang Nucleosynthesis. Yet low mass stars produce ^{3}He and as they ascend the giant branch, as shown in Fig. 1, they should return this ^{3}He to the interstellar medium unless it is destroyed somehow. Again, one usually invokes rotation to somehow provide the required mixing. We showed in Paper I that the $\delta\mu$ mechanism can efficiently destroy all the ^{3}He in the star's envelope. This means that low-mass stars become neutral to the galactic inventory of ^{3}He.

In Paper II we investigate the effect of $\delta\mu$ mixing on the carbon isotope ratios. Later work will extend this to oxygen isotopes and lithium. Here we provide a summary of the main results from Paper II.

Rather than use any of the existing theories for thermohaline mixing, such as [7] or [8], we decided to use a phenomenological approach. There are many reasons for this, among the most prominent being that the two theories described above differ substantially.

Rather, we use the diffusion equation with a diffusion co-efficient that we constrain from estimated limits to the mixing velocity (see Paper II). We then implement a simple algorithm that mixes when there is a negative gradient in the molecular weight, and with a speed proportional to that gradient, and constrained to be about the values we estimated from first principles. As discussed in Paper II, we expect this mixing to be of order 1 cm/sec. We found that varying this speed by factors of 100 either way produced only small changes in the results, this is because the key point is that the mixing is slow compared to the advance of the hydrogen shell.

In Fig. 2 we show the carbon isotopic ratios for the evolution of a 1 $M_\odot$ model with $Z = 0.02$. The initial value of about 90 drops to 29.5 after first dredge-up. When the $\delta\mu$ mixing begins, this ratio rapidly drops further to near 15, reaching 14.3 by the tip of the giant branch. During this evolution 93% of the ^{3}He was destroyed.

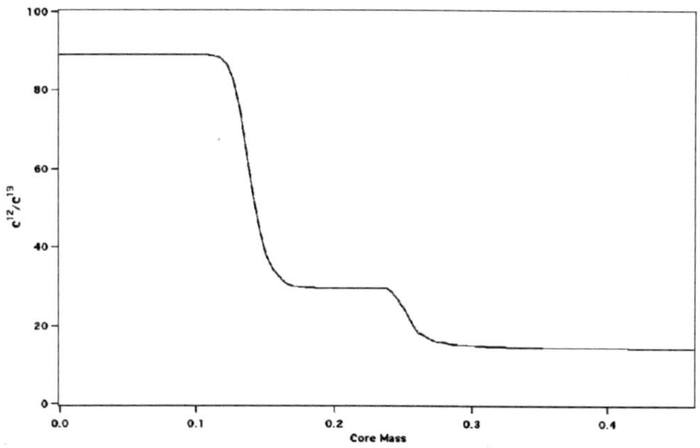

FIGURE 2. Variation of carbon isotopic ratios as a function of evolution for stars of different masses.

4.1. Dependence on Mass

We next investigated how the final carbon isotopic ratio depends on the stellar mass. Figure 3 shows values of the isotopic ratio in the Hertzsprung-Russell diagram for low mass stars of the same composition, namely $Z = 0.02$ and $X = 0.70$. The onset of the first dredge-up is clearly seen, and results in values of $^{12}C/^{13}C \simeq 20 - 30$ as expected. The nearly vertical black line shows where the $\delta\mu$ mixing begins, and we see that the final ratio is then reduced to something like 15, exactly as required by the observations.

4.2. Dependence on Metallicity

We next investigated how the final carbon isotopic ratio depends on the stellar metallicity. It is well known that in Population II stars the final carbon isotopic ratio can drop as low as the CN cycle equilibrium value of about 3-4. Figure 4 shows the change in this ratio with evolution for stars of $0.9M_\odot$ with varying metallicities Z. We see that the final ratio drops from about 15 for Population I to as low as 3 for Z=0.0001, typical of globular clusters. Again, this is exactly as is required by the observations.

Table 1 shows results for seven different masses and three different compositions. We present the values at the end of first dredge-up as well as the value at the tip of the giant branch. This shows nicely how the extra-mixing becomes less effective at higher masses, closer to the limit for the core-flash. It also shows how the mixing becomes more effective at reducing the carbon isotope ratios as the metallicity decreases.

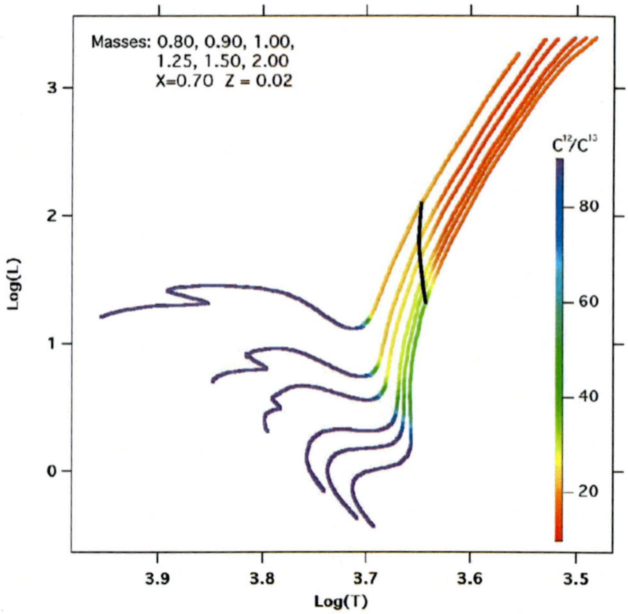

FIGURE 3. Variation of carbon isotopic ratios as a function of evolution for stars of different masses.

TABLE 1. Carbon isotope ratios at the end of first dredge- up (FDU) and at the tip of the giant branch (Tip) for stars of seven different masses and three different compositions.

	X=0.70, Z=0.02		**X=0.738, Z=0.001**		**X=0.74, Z=0.0004**	
Mass	FDU	Tip	FDU	Tip	FDU	Tip
0.80	36.9	15.9	34.1	5.3	35.0	4.2
0.85	34.0	15.3	31.5	5.0	31.8	4.0
0.90	32.2	14.5	29.6	4.9	30.0	4.0
1.00	29.5	13.4	27.3	4.9	27.4	4.0
1.25	25.6	13.0	24.3	5.0	24.3	4.1
1.50	23.6	13.7	24.3	5.2	22.7	4.6
2.00	22.3	17.0	21.2	14.2	21.0	21.0

4.3. Concerning the Galactic Inventory of ^{3}He

We have explained briefly the problem with ^{3}He, which was discussed in Paper I. The dependence on mass and composition that is revealed in Paper II may explain why some planetary nebulae show large amounts of ^{3}He despite the fact that low mass stars must destroy ^{3}He for consistency with Big Bang Nucleosynthesis. Table 2 shows the destruction of ^{3}He as a function of mass and composition. Just as was the case for the carbon isotopes, the mixing has more effect for the lower masses and the lower metallicities. Hence it can be seen that stars with moderate masses, perhaps even as low

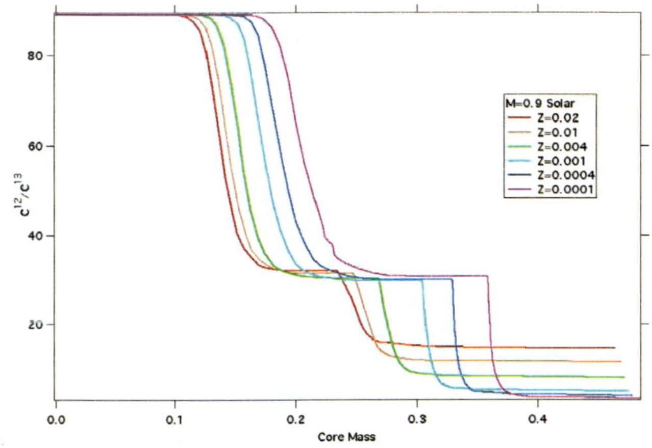

FIGURE 4. The $^{12}C/^{13}C$ ratio as a function of core-mass for a range of metallicities, at $0.9M_{\odot}$.

TABLE 2. The destruction of ^3He for various masses and compositions. We give values for the ratio of the envelope ^3He abundance to the initial value. The "Peak" value is the maximum, prior to the start of the $\delta\mu$ mixing. The "Mixed" value is the final value at the tip of the giant branch.

	X=0.70, Z=0.02		X=0.738, Z=0.001		X=0.74, Z=0.0004	
Mass	Peak	Mixed	Peak	Mixed	Peak	Mixed
0.80	76.6	2.7	54.7	0.86	53.0	0.62
0.85	68.2	2.9	46.5	0.90	45.1	0.66
0.90	61.1	3.1	40.0	0.95	38.7	0.73
1.00	49.7	3.7	32.9	1.12	30.0	0.83
1.25	31.7	4.5	21.8	1.60	19.9	1.21
1.50	21.8	5.5	15.7	2.08	14.4	1.91
2.00	12.8	7.0	9.5	6.70	8.7	8.39

as $2M_{\odot}$, can produce nearly an order of magnitude more ^3He than they started with, even with the $\delta\mu$ mechanism operating. However, given the IMF's weighting toward low mass stars, and the increased efficiency of the ^3He destruction at lower metallicities, we are still able to solve the problem of over-production of ^3He while allowing some planetary nebulae to still show ^3He enhancements, provided they are the more metal-rich and/or massive stars.

4.4. Other Species

We will address other species in detail in a later paper. Here we note that this mechanism may have problems with the observed abundance of lithium seen in stars near the bump in the luminosity function. To produce lithium requires something like the

Cameron-Fowler mechanism to operate, which requires the freshly synthesized ^{7}Be to be removed from the hot region (where it can capture an electron to form ^{7}Li) before it burns further, to complete the PP chains. If the mixing is as slow as we have estimated, then the production of lithium may be difficult for our mechanism.

5. CONCLUSIONS

We have given a brief summary of the observational reasons for believing that some mixing is required to join material from the convective envelope, through the radiative region, to the top of the hydrogen burning shell. Recent calculations indicate that rotation is not the answer. We propose that burning of ^{3}He following first dredge-up creates a molecular weight inversion which drives mixing from just above the hydrogen shell to the bottom of the convective envelope. This is similar to a double-diffusive or "thermohaline" mixing process, but the details are not well understood. With our estimates of the mixing speed, the model seems to nicely account for the required destruction of ^{3}He and the observed variation in the ratio of the carbon isotopes. It remains to be seen if it can also contribute to problems associated with other species, such as lithium and oxygen isotopes.

Finally, we note that there is growing evidence for some form of extra-mixing on the AGB that also joins the convective envelope to the top of the hydrogen shell (e.g. [9]). It remains to be seen if the $\delta\mu$ mechanism operates on the AGB.

ACKNOWLEDGMENTS

This study has been carried out under the auspices of the U.S. Department of Energy, National Nuclear Security Administration, by the University of California, Lawrence Livermore National Laboratory, under contract No. W-7405-Eng-48. JCL was partially supported by the Australian Research Council. We are indebted to R. Palasek for assistance with the code. JCL also thanks the organizers of the Torino workshop for helping with practical details enabling his attendance. A special thanks to Sarah for all her help, and to Alessandro for transport to and from Roma.

We are all indebted to Roberto Gallino for all of the Torino workshops!

REFERENCES

1. C. Charbonnel, *A&A* **282**, 811 (1994)
2. A.V. Sweigart and J.G. Mengel, *ApJ* **229**, 624 (1979)
3. A. Palacios, C. Charbonnel, S. Talon, and L. Siess, *A&A* **453**, 261 (2006)
4. A. Palacios, C. Charbonnel, and M. Forestini, *A&A* **375**, L9 (2001)
5. P.P. Eggleton, D.S.P. Dearborn, and J.C. Lattanzio, *Science* **314**, 1580 (2006) [Paper I]
6. P.P. Eggleton, D.S.P. Dearborn, and J.C. Lattanzio, *ApJ*, in press (2008) [Paper II]
7. R.K. Ulrich, *ApJ* **172**, 165 (1972)
8. R. Kippenhahn, G. Ruschenplatt, and H.-C. Thomas, *A&A* **91**, 175 (1980)
9. K. Nollett, M. Busso, and G.J. Wasserburg, *ApJ* **582**, 1036 (2003)

Theory and Numerics: New Results on Convection in Stars

David Arnett*,†, Casey Meakin*,**, Sumner Starrfield‡, Frank Timmes‡ and Patrick Young‡

*Steward Observatory, 933 N. Cherry Avenue, Tucson AZ 85721 USA
†International Center for Relativistic Astrophysics (ICRA), Roma, It
**FLASH Center, University of Chicago, Chicago IL USA
‡School of Earth and Space Exploration, Arizona State University, Tempe AZ USA

Abstract. We have begun a new program to improve the treatment of turbulent convection in stars; it features Convection Algorithms Based on Simulations (CABS). We analyze stellar convection with the aid of 3D hydrodynamic simulations, introducing the turbulent cascade into our theoretical analysis. We devise closures of the Reynolds-decomposed mean field equations by simple physical modeling of the simulations. The procedure is testable and allows systematic improvement. The interpretation of mixing-length theory (MLT) as generally used in astrophysics is incorrect. Direct tests show that the damping associated with the turbulent cascade is that suggested by Kolmogorov [1]($\varepsilon_K \approx (u')^3_{rms}/\ell_{CZ}$), where ℓ_{CZ} is the depth of the convection zone. This implies that the mixing-length parameter α is a simple function of depth of the convection zone, and not freely adjustable. Appropriate adjustment of another parameter of MLT, the "geometric factor", can leave solar models almost unchanged, except for their velocity scale. This parameter adjusts the super adiabatic region just below the photosphere, and represents a crude interpolation function for characterizing 3D hydrodynamic atmospheres.

Keywords: Turbulence - Stars - Stellar Evolution - Hydrodynamics
PACS: 26.20.-f, 26.30.-k, 47.27.E-, 47.27.ep, 97.10.Cv, 97.10.Ex, 97.20.-w

1. INTRODUCTION

Treatment of turbulent convection is the most popular choice for the weakest aspect of stellar evolution theory. The mixing-length theory (MLT), as now used, was invented more than fifty years ago [2, 3]. It is now possible to directly simulate three-dimensional (3D) turbulent convection numerically, and further computer development will enhance such an approach. However, even if we assume Moore's law of computer improvement (doubling every 18 months), a direct simulation of the whole Sun, with resolution down to the Kolmogorov scale, would not be possible before 2080 or so. We must resort to a hybrid approach, with direct simulations of smaller volumes of spacetime directly simulated, and tied together with a fundamental theory. This discussion is a preview of material to be published elsewhere (e.g. [4]).

We have begun such a project; [5] have simulated a sector of the oxygen burning shell in a star of 23 $M_\odot$. The rapid rate of neutrino cooling in such a star makes the thermal relaxation time much closer to the sound travel time for the convective region, and consequently reduces the computation requirements (by a factor of 10^5 with respect to a simulation of the solar convection zone, top to bottom).

CP1001, 5th Torino Workshop on Evolution and Nucleosynthesis in AGB Stars and the 2nd Perugia Workshop on Nuclear Astrophysics,
edited by R. Guandalini, S. Palmerini, and M. Busso

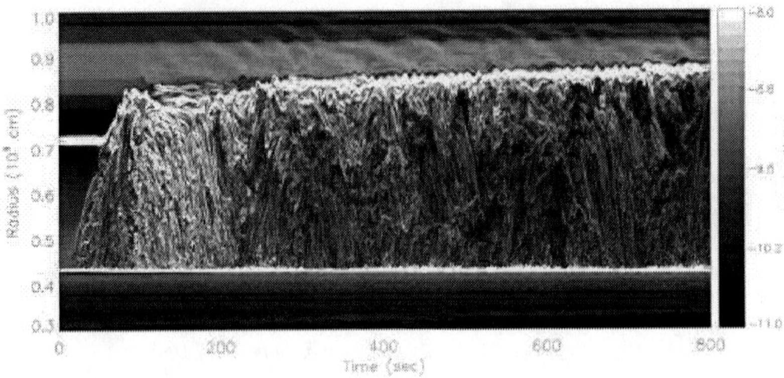

FIGURE 1. Composition gradients (white) in convection zone, radius versus time.

Boundary Conditions. Figure 1 shows the composition gradients in the oxygen burning shell; white regions have large gradients in O^{16}. The vertical axis is radius and the horizontal axis is time. The initial model was constructed with a one-dimensional stellar evolution code, and evolved into silicon burning. A sector of the overlying oxygen burning shell was mapped onto the computational grid. At the left of the figure, we see this initial model being modified by 3D flow. In particular, the white band which represented the outer edge of the convective region is overtaken by turbulent mixing. The Schwarzschild criterion for convective motion only tells us where buoyant acceleration is zero, not if flow can erode stably stratified layers. The bulk Richardson criterion [6, 7] does this, and does define the convective boundary seen in the simulations. After this initial transient, there is a slow entrainment of new matter into the convective region, from both the top and bottom [5].

Figure 2 shows the kinetic energy density in the same radius-time graphical context. There are strong bursts of kinetic energy production, at an interval of about one transit time $\tau = \ell_{CZ}/u_{rms} \approx 20$ s. The burst are driven by thermonuclear burning, and recur in an approximate steady state. These bursts must be characterized by an average behavior if long term evolution is to be investigated.

2. NUMERICS

2.1. Kolmogorov Cascade

A key element in understanding the simulation is the physical effect of the turbulent cascade from the largest scale eddys to the final smallest eddys at which molecular diffusion converts macroscopic motion into heat [1, 8]. This may be sketched as Large Eddys $\rightarrow$ Turbulent cascade $\rightarrow$ Tiny Eddys $\rightarrow$ Molecular dissipation. A numerical hydrodynamic simulation also has a minimum scale, that of the grid size; below this

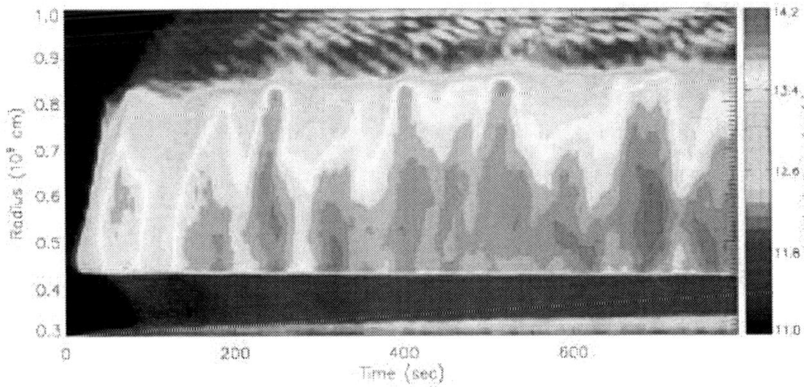

FIGURE 2. Kinetic energy in convection zone, radius versus time. Note bursty behavior, which must be averaged over.

information is reduced to average properties. This is equivalent to entropy increase [9] and viscosity [10].

Implicit Large Eddy Simulations (ILES). Some hydrodynamics methods automatically approximate the dissipation at the subgrid scale by a Kolmogorov-like cascade [11]: PPM and other monotone shock-capturing algorithms. We are lucky:

1. Most of the energy is in the large eddys.
2. Most of the time is spent in the large eddys.
3. The rate of dissipation is controlled by the large eddys.

Thus, if we are able to resolve the large eddys well, we may capture the physics without needing to resolve the whole cascade. The ILES approach seems to be the most efficient use of computer resources to deal with turbulent flow; it explicitly calculates the large eddys and at the grid scale dissipates like the Kolmogorov spectrum does.

3. THEORY

3.1. Taming Turbulent Fluctuations

As we have seen in Fig. 1 and Fig. 2, turbulent convection has a rich time behavior. The Reynolds decomposition splits each variable into an average and a fluctuating part, $p = \langle p \rangle + p'$, where to *first* order, the average of the fluctuations is zero by construction ($\langle p' \rangle = 0$). Here $\langle p \rangle$ denotes an average over angle. Example: the thermal flux is $F = \rho C_P \langle u'T' \rangle$. For times greater than the transit time, L_{CZ}/u_{rms}, the time averages become smooth. (e.g., $\overline{p}$).

We may derive a kinetic energy (KE) equation [5, 4],

$$\partial_t \langle \overline{\rho E_K} \rangle \nabla \cdot \langle \overline{\rho E_K \mathbf{u_0}} \rangle = -\nabla \cdot \langle \overline{\mathbf{F_p} + \mathbf{F_K}} \rangle + \langle \overline{p' \nabla \cdot u'} \rangle + \langle \overline{\rho' \mathbf{g} \cdot \mathbf{u}'} \rangle - \varepsilon_K. \tag{1}$$

where ε_K is the Kolmogorov dissipation per unit volume.

4. COMPARISON WITH MLT

While we choose the KE equation (Eq. 1) to determine the convective velocity, MLT uses

$$u_{rms}^2 = g H_P \delta \Delta \nabla \ (\alpha^2/8), \tag{2}$$

where $\Delta\nabla = \nabla - \nabla_a$, $\nabla = \partial \log T / \partial \log P$, H_P is the pressure scale height, and $\alpha = \ell_{ML}/H_P$. In MLT α is taken to be freely adjustable, can be calibrated (is a universal constant?), and chosen by fitting the present-day solar radius in a solar model.

Are these assumptions valid? If we assume a quasi-steady state, and neglect small terms in Eq. 1,

$$\langle \overline{\rho' g u'} \rangle = \varepsilon_K. \tag{3}$$

Integrating over the convective volume, $\int \langle \overline{\rho' g u'} \rangle d\mathscr{V} = -M_{CZ} u_{rms}^3 / \ell_D$. Assuming low Mach-number flow, we have $\rho'/\rho_0 = -\delta T'/T_0$. In the simulation we find $(T'/T_0)_{rms} = \alpha_T \Delta\nabla$, and $\langle \overline{T'u'} \rangle = \alpha_E (T')_{rms} (u')_{rms}$, with $\alpha_E \approx 0.7$. Collecting terms, and rudely treating integration (pulling u'_{rms} out of the integral),

$$u_{rms}^2 \approx g H_P \delta \Delta\nabla \ (\alpha_E \alpha_T \ell_D / H_P), \tag{4}$$

which is to be compared to Eq. 2, which are equal if

$$\alpha^2/8 = \alpha_E \alpha_T \ell_D / H_P. \tag{5}$$

On further examination it appears that α is a function of the depth of the connection zone and of the dissipation length for Kolmogorov damping (which itself tends to scale with the depth of the convection zone); α is not a universal constant.

5. CONVECTIVE ALGORITHMS BASED ON SIMULATIONS (CABS)

We may summarize the CABS procedure:

1. Use Reynolds-decomposed average equations for Kinetic and for Internal Energy.
2. Explicitly include Kolmogorov dissipation in theory.
3. Use 3D, time-dependent simulations to design physically motivated closure schemes.

If we add differential rotation and magnetic fields, we get the master equations used by [12] for the magnetic resonance instability in accretion disks.

TABLE 1. Mixing-Length Parameters for Solar Models.

Model	α	g_{ML}	v_{max} km/s
A	1.643	1.0	2.25
B	2.323	42.0	2.80
C	3.286	280.0	3.05
D	5.190	1540.0	3.20

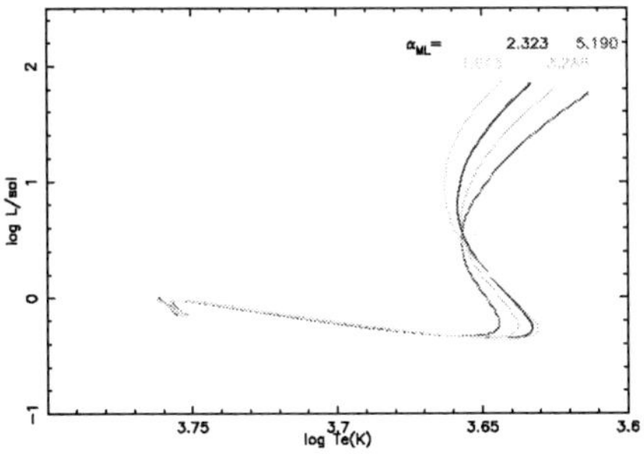

FIGURE 3. Trajectories in the HR diagram for Solar models with different mixing lengths *and* geometric factors.

CABS is *general* (it is has a theoretical component), and allows us to derive particular effects in a coherent way.

CABS is *improvable*, either by new simulations or experiment. It is a procedure more than a result.

CABS makes contact with modern research in fluid mechanics, meteorology, oceanography, etc., and can be *terrestrially* defined.

6. THE SOLAR EXAMPLE

To be consistent with the KE equation (Eq. 1), larger values of α are required than are now used in stellar evolution. Entrainment of new matter, acoustic and kinetic energy fluxes, and mechanical heating by the turbulent cascade may also have significant effects. We examine only one new aspect here, that of larger values of α.

We note that MLT has other parameters which are left at their historical values. An important one is the "geometric form factor"; see [13]. In Table 1 are given the values of α and the scaling for the "geometric form factor" ($g_{ML} = 1$ is the historical value, larger values imply more efficient radiative cooling, as would be expected for smaller length scales in a turbulent cascade. Each of these pairs produce an almost identical solar model, except that the velocity scale increases for A to D.

Figure 3 shows the trajectories for models A-D (i.e., using these pairs of parameters) in the HR diagram. Below the birth line at $\log L/L_{\odot} \approx 1$, the curves are very similar. Above this line the models are inappropriate; they do not contain accretion disks. We conclude that there is no problem with larger values of α for this stage of stellar

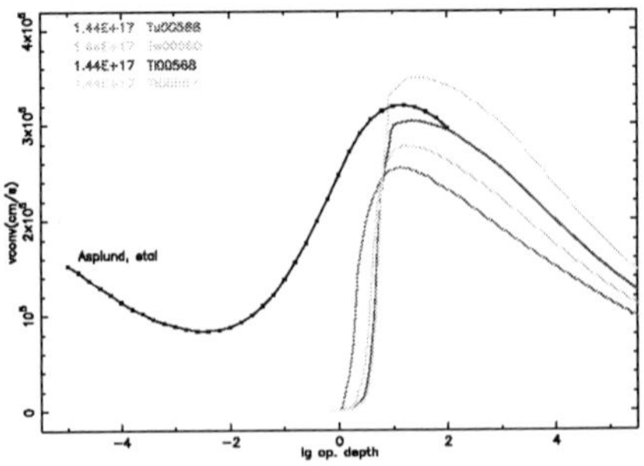

FIGURE 4. Surface layers of Solar models with different mixing lengths *and* geometric factors, compared to 3D solar hydrodynamic atmospheres [14].

evolution.

Figure 4 shows the surface layers of the same models, compared to the 3D hydrodynamic atmospheres of [14]. The convective velocity is plotted versus the optical depth. There is a dramatic difference between the 3D atmospheres and the stellar models because the "atmospheres" algorithm used the Schwarzschild criterion for the convective boundary; the Richardson criterion gives finite velocities well into the visible regions (as do the 3D atmospheres and the observed doppler shifts of spectral lines) because the stable layers are too weakly stable to resist the vigorous convective motion. This is a very old problem, and is "solved" by the introduction of "microturbulent" and "macroturbulent" velocities to get the width of spectral lines as large as observed. These velocities are related to convective flow, with a velocity scale of about 3 km/s, comparable to the larger values in Table 1. Below the fictional "stable" region, the 3D atmospheres merge with models using the larger α values needed to satisfy the KE equation.

6.1. Why Did MLT Work?

For stars like the Sun, the new effects beyond MLT are small, and discrepancies can be absorbed in the free choice of an adjustable parameter, the mixing length. Only challenging constraints, such as those found in helioseismology, show the flaws.

In standard algorithms for stellar evolution, the convective regions are constrained to have a particular temperature gradient relative to the pressure gradient, $\nabla = d\ln T/d\ln P$ which is set equal to the adiabatic value (constant entropy S, $\nabla_a = (d\ln T/d\ln P)_S$).

The total convective energy flux is taken to be proportional to $(\Delta\nabla)^{\frac{3}{2}}$, and is a local quantity. Correlation lengths for temperature and velocity are much less than a pressure scale height H_P, 0.3 and 0.8 H_P respectively (Fig. 19, [5]). This is why the local MLT approximation works; most of the enthalpy flux is carried by the chaotic, slow moving regions, not the vigorous downdrafts [15].

The absolute value of the kinetic energy flux is smaller than the enthalpy flux in most regions, so that its neglect does not usually cause large errors.

MLT lumps many errors together and compensates by adjusting the mixing-length parameter α_{ML} to insure the entropy jump between the interior and the photosphere gives the observed solar radius.

Effects of boundary physics are small, and would only be seen in precision tests (e.g., solar helioseismology).

Errors are absorbed by other effects which are not easily included with precision (e.g., rotation and MHD).

Standard stellar evolution, which has been successful, only uses velocity, if at all, as a measure of mixing speed. Except in the super-adiabatic region, the key is to have an adiabatic run of T and P. This ignores physics of boundaries with stably stratified regions, but these effects are not large for solar mass stars near the main sequence.

6.2. What MLT Missed

Strictly speaking, MLT has no boundary physics because the development does not maintain derivatives, but replaces them with algebraic factors [16]. Entrainment is ignored and an incorrect definition of the boundary is used (Schwarzschild instead of bulk Richardson).

MLT systematically underestimates velocity scale, which has impact of estimates of wave-induced mixing, entrainment, wave generation, coronal heating, and mass loss.

MLT assumes no net flux of KE, which is generally false.

MLT ignores the dissipation of KE by the Kolmogorov cascade.

Replacing the KE equation in MLT by an adjustable parameter is bad theory, and makes difficult the use of developments from related fields.

6.3. Goals

There are several lines of research that are needed to go beyond these interesting speculations:

- Systematic simulations to test and strengthen the theoretical ideas relating to the Reynolds- ecomposion of the KE and internal energy equations.
- Characterization of the bursty behavior to connect it with chaos theory (the Lorenz model of convection).
- A general formulation of these new effects in a way they may be easily used in stellar evolutionary calculations.
- Exploitation of the time dependent nature of the KE equation for nova and late stages of stellar evolution.

ACKNOWLEDGMENTS

This work was supported in part by NSF AST-0708871. DA thanks the International Center for Relativistic Astrophysics (ICRA) and Professor R. Ruffini for support in Italy, and Martin Asplund for providing details of the 3D hydrodynamic atmospheres.

REFERENCES

1. A.N. Kolmogorov, *Dokl.Akad.NaukSSSR* **30**, 299 (1941)
2. L. Biermann, *Zeit.f.Ap.* **28**, 305 (1951)
3. E. Böhm-Vitense,*Zeit.f.Ap.* **46**, 108 (1958)
4. D. Arnett, C. Meakin, and P.A. Young, *in preparation* (2008)
5. C. Meakin and D. Arnett, *ApJ* **667**, 448 (2007)
6. J.A. Dutton, *The Ceaseless Wind, Dover Publications, Inc., Mineola, NY* , (1986)
7. O.M. Phillips, *The Dynamics of the Upper Ocean, Cambridge University Press, Cambridge*, (1966)
8. A.N. Kolmogorov, *J. Fluid Mechanics* **13**, 82 (1962).
9. C. Shannon, *Bell System Technical Journal* **27**, 379 (1948)
10. L.D. Landau and E.M. Lifshitz, *Fluid Mechanics, Pergamon Press, London*, (1957)
11. J. Boris, *Implicit Large Eddy Simulations, ed. F. F. Grinstein, L. G. Margolin, & W. J. Rider, Cambridge University Press, Cambridge*, 9 (2007)
12. S.A. Balbus and J.F. Hawley, *RevModPhys* **70**, 1 (1998).
13. R. Kippenhahn and A. Weigert, *Stellar Structure and Evolution, Stringer-Verlag, Berlin*, 50 (1996)
14. M. Asplund, N. Grevesse, A.J. Sauval, C. Allende Prieto, and R. Blomme, *A&A* **431**, 693 (2005)
15. F. Cattaneo, N. Brummel, J. Toomre, A. Malagoli, and N.E. Hurlburt, *ApJ.* **370**, 282 (1991).
16. E. Spiegel, *AnnRevA&A* **10**, 261 (1972)

Magnetic Bubbles and Extramixing in Stars

Gerald J. Wasserburg [a] and Maurizio Busso [b]

[a] The Lunatic Asylum, Divisionof Geological and Planetary Sciences, California Institute of Technology, Pasadena, CA 91125, USA, gjw@gps.caltech.edu
[b] Department of Physics, University of Perugia, via A. Pascoli, I-06123 Perugia, Italy busso@fisica.unipg.it

Abstract. The possible role of magnetic flux tubes in transporting matter from near the H shell through the radiative zone and into the convective envelope is explored. It is shown that the required rates of mass transport necessary to provide nuclear processed material to the envelope can be achieved if large magnetic fields are present just above the H shell in AGB and RGB stars. The required fields in this zone reach 5 MG for the AGB case and 0.5-0.05 MG for the RGB case. It may thus be possible that magnetic bouyancy play a major role in providing the extra mixing needed for these stars.

Keywords: Stars: Red Giants; Stars: Evolution; Extended Mixing; MHD.

PACS: 97.20.Li, 97.30.Hk, 97.10.Cv, 95.30Qd

1. INTRODUCTION

It has been widely recognized that, during Red Giant Branch (RGB) and Asymptotic Giant Branch (AGB) phases, low mass stars (~1-3 $M_\odot$) must have a weak transport mechanism that moves matter from the vicinity of the H burning shell, through the stable, non-convecting radiative zone, and injects it into the massive, overlying convective envelope. Evidence for this comes, e.g., from the following observations: 1) The ^{12}C/ ^{13}C ratios observed in such stars are lower than predicted by purely convective dredge up; 2) High enrichments of Li (which is usually destroyed in stellar interiors) are sometimes observed and this requires nuclear processing at depth and rapid transport into the envelope; 3) Anomalous abundances of the ^{18}O and ^{17}O isotopes, plus high abundances of ^{26}Al are present in grains of presolar, circumstellar condensates recovered from meteorites. These grains are predominantly from AGB and RGB stars (see Fig. 1) and require some "extra nuclear" processing that is not the result of standard stellar models and is not understood from well defined, dynamical calculations, but requires long-lasting circulation of partially processed material in the radiative layers below the envelope. For reviews of these topics see [1,2].

The amounts of material that need to be transferred are large, being ~1 $M_\odot$ over the appropriate stellar time scale (RGB or AGB), but with low circulation rates. This process must not disturb the stellar structure or the basic thermonuclear sources.

A variety of mechanisms have been brought forward to address some or all of the above issues: cf. [3,4,5]. Most of these suggestions require some form of *ad hoc* parameterization. A more recent study [6], using a detailed model of a low mass star,

CP1001, *9th Torino Workshop on Evolution and Nucleosynthesis in AGB Stars and the 2nd Perugia Workshop on Nuclear Astrophysics*, edited by R. Guandalini, S. Palmerini, and M. Busso

showed that, at the boundary of the H burning shell, an instability is naturally produced due to the combustion of ^{3}He.

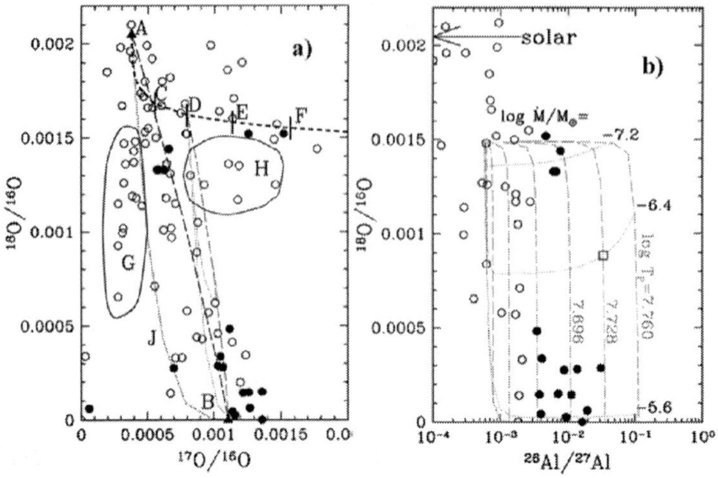

FIGURE 1. a) Diagram, showing the evolution of the oxygen isotopes starting with solar abundances for stars ranging in mass from 1-3 $M_\odot$, for 1st dredge up (trajectory A, C, D, E and F). It is not much changed by 2nd dredge up. High precision data on circumstellar dust grains show that many of them lie far outside of the possible range for evolutionary models without CBP. Note: point B is the equilibrium value in the H burning shell where almost complete destruction of ^{18}O occurs (see [7] and references therein). These results unambiguously demonstrate the need for added nuclear processing near the H shell and injection into the convective envelope. **b)** The ^{18}O/^{16}O – ^{26}Al/^{27}Al diagram, showing the observed grains and the trajectories for a stellar envelope with CBP as a function of $\dot{M}$ and T_P. Note that, after a drop of ^{18}O/^{16}O from a Solar ratio to ~ 0.015, due to 1st dredge up, the curves for a fixed $\dot{M}$ are rather flat and not a steep function of temperature. In contrast, the ^{26}Al/^{27}Al is strongly dependent on T_P and requires that it be close to the temperature of the H burning shell. Thus the ^{18}O/^{16}O is mostly a measure of $\dot{M}$, and ^{26}Al/^{27}Al is a thermometer.

This nuclear process causes a molecular weight inversion, resulting in matter transport into the radiative zone, and decreases the abundance of ^{3}He in the convective envelope. This has been recognized as a form of thermohaline mixing [8], especially relevant in stars of about one solar mass. As the efficiency of this mechanism is reduced for increasing mass, it cannot apparently account for the observations outlined above, which concern the most common AGB stars in the galactic disc, whose masses are in the range 1.5 to 2.5 $M_\odot$ [9].

2. IS EXTRAMIXING DRIVEN BY MAGNETIC BUOYANCY?

Investigation of possible stellar stages including 1st and 2nd dredge up showed that many grains lie in a region in the ^{18}O/ ^{16}O -^{17}O/ ^{16}O diagram that is forbidden by either normal AGB or RGB evolution (or from Hot Bottom Burning, hereafter HBB) and thus requires additional nuclear processing: such a processing was called "Cool Bottom Processing", or CBP [10]. This term was invented by GJW, on his patio in Pasadena, at the request of Juliana Sackman-Christy, who strongly felt that a name was required: it *distinguishes between ongoing small levels of nuclear processing from efficient envelope burning (HBB).* CBP is amongst the phenomenological transport models under consideration to explain the extra nuclear burning. This model has been extensively explored by [11] and by [7]. It was shown that CBP may be treated in a simple "Box Car" model, in which matter circulates from the base of the convective envelope to a depth where the temperature is T_P and then moves upward into the envelope. This produces a small amount of nuclear processing at each pass and only depends on the path integral of the nuclear reaction rates and the mass flow rate. The processing does not significantly disturb the overall stellar structure.

The only parameters that define the results are the mass flow rate ($\dot{M}$) and the maximum temperature T_P reached (for a star of a given mass and metallicity). The schematic Box Car flow pattern was simply assumed and no mechanism was proposed to drive the flow. The results of this approach are that the observed range of isotopic shifts can be very plausibly explained by CBP and that its conditions $\dot{M}$ and T_P can be easily defined. These results permit a clear interpretation of most observational data. However, the question remains as to the physical mechanism that causes the transport and what the velocity field must be. In this report, we will present the possibility that buoyancy of magnetic bubbles might be the cause of the transport. E. N. Parker first proposed [12] that "magnetic buoyancy" might be a cause of Sunspots. Gurm and Wentzel then studied the conditions of convection in a zone of radiative equilibrium due to B fields [13]. In a very insightful paper, Hubbard and Dearborn proposed [14] that magnetic mixing might be a mechanism to explain the $^{12}C/^{13}C$ data and would have direct effects on ^{18}O,^{17}O and C/N (there was no data available at that time).

In 2005, the senior author of this paper received an inquiry from Maurizio Busso, who said "What do you think about magnetic bubbles?" (see Fig. 2). He responded, "That is a great idea, why don't you do it". After some time and with considerable work with Andrea Calandra, Maurizio, upon visiting with the Wasserburg family in Florence (Oregon), showed the results of their preliminary work and asked GJW if he would join them. GJW responded that he had little knowledge of the subject as the last intimate contact that he had with magneto-hydrodynamics (MHD) was when he took Chandrasekhar's course more than 56 years ago. This did not dissuade Maurizio and it was agreed to make a joint effort. The idea was that if material were magnetized near the H burning shell, then the magnetic pressure ($B^2/8\pi$) might produce sufficient buoyancy to transport matter up to and into the convective zone. This subject had been already considered by many authors in various scenarios, after the early suggestions by E. N. Parker for the case of the Sun [15]. In particular, the interested reader is referred to [16]. McDonald and Mullan studied shear instability in rotating RGB stars [5] and showed that it could result in B fields between the H shell and the convective envelope.

The fundamental issues were: 1) the basic MHD mechanisms that must provide the fields; 2) the consequences that would follow if magnetic bubbles were to transport sufficient material to address the circulation that CBP required. After consideration of the great complexity associated with the MHD problem, it was chosen to ignore the first basic question (n. 1 above) and focus on problem number 2. We now have the special privilege to give the first public presentation of this work [17], that is just now published (here it will be referred to as BWNC). The senior author guesses that this is a birthday present for him.

FIGURE 2. A message from MB to GJW in 2005.

3. A PHENOMENOLOGICAL MODEL FOR MAGNETIC CBP

In the matter at hand, the question is: "What do we know and how far can it take us?". We know the stellar structure of low mass stars from extensive and intensive investigations. Hence, we know the density, pressure, composition and structure of stars over their evolution with a high degree of certainty. Using this basis, as provided by several of you in the audience, we can determine, using the observational data and the CPB model, the rate of mass transfer ($\dot{M}$) and the temperature T_P (or depth) to which the circulation must reach. This is all that is available to us now.

We show the data for a star of initial mass $1.5\,M_\odot$ during the RGB phase, taking into account CBP (see Fig. 3). It can be seen that to achieve the required value of $^{12}C/^{13}C$ ~10-12 over the time spent on the RGB, this would require $\dot{M}$ between $4\text{x}10^{-7}$ and $4\text{x}10^{-8}$ $M_\odot$/yr. This result is not very dependent on the temperature T_P (values of log $T_P/T_H \cong -0.2$ to -0.25 are sufficient).

For the AGB phase it can be seen from Fig. 1 that many observed data points require major depletion of ^{18}O. As seen in Fig. 1b), the $^{18}O/^{16}O$ versus $^{26}Al/^{27}Al$ trajectories of CBP as a function of $\dot{M}$ and T_P show that $^{26}Al/^{27}Al$ is essentially a thermometer, while $^{18}O/^{16}O$ is most strongly dependent on $\dot{M}$. From an analysis of the CBP results [7] one derives that, in order to account for the general effects on the AGB, we require $\dot{M}$ values between $4\text{x}10^{-7}$ and $2.5\text{x}10^{-6}$ $M_\odot$/yr and temperature values such that log $T_P/T_H \cong -0.1$.

From the above considerations, we therefore have clear guides to the values of $\dot{M}$ and T_P required for a magnetic bubble transport model. As typical rates, we take from the above considerations $\dot{M}=10^{-6}$ and $4\text{x}10^{-8}$ $M_\odot$/yr for the AGB and RGB, respectively.

We shall consider, as a reference, the case of a star of initial mass 1.5 $M_\odot$ with a metallicity half of the Solar value and an internal structure at the RGB and AGB phases as computed previously [18,19], including mass loss with the parameterization presented in [20].

TABLE 1.

	AGB	RGB
$\dot{M}$ ($M_\odot$/yr)	10^{-6}	$4\mathrm{x}10^{-8}$
$f_{MT}\, x\, f_{vs}$	$2\mathrm{x}10^{-7}$	$8.7\mathrm{x}10^{-10}$
v_{TRANS}	*1 m/s*	*1 m/s*
f_{MT}	$7\mathrm{x}10^{-2}$	$2\mathrm{x}10^{-4}$

The first issue is whether the transport across the boundary between the upper part of the radiative zone and the base of the convective layer will cause any difficulties. If v_{TRANS} is the average velocity of transport across this interface and f_{MT} the fraction of the mass at the interface that came from CBP, then the relationship with $\dot{M}$ is:

$$\dot{M} = f_{MT}\, v_{TRANS}\ \mathrm{x}\ 4\,\pi\, R^2\, \rho = f_{MT}\, f_{VS}\, v_{SOUND}\ \mathrm{x}\ 4\,\pi\, R^2\, \rho \qquad (1)$$

where v_{SOUND} is the sound speed, and both f_{MT} and f_{VS} must be <<1 (see again BWNC for details). The results for the product ($f_{MT}\, \mathrm{x}\, f_{VS}$) are shown in Table 1 for

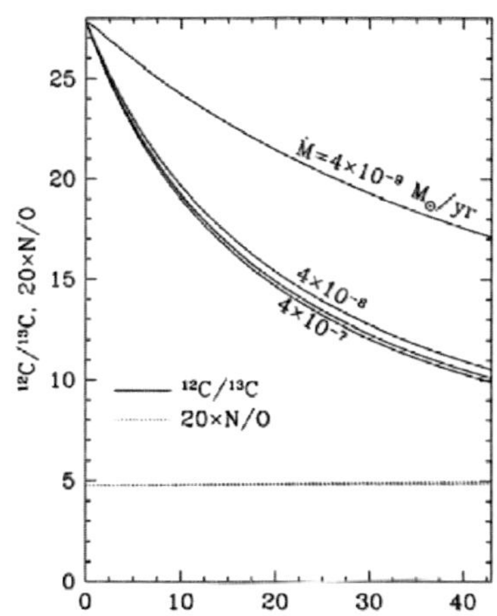

FIGURE 3. Evolution of the ^{12}C/ ^{13}C ratio in the envelope of a 1.5 $M_\odot$ star during the RGB phase. Evolution curves for different values of $\dot{M}$ are shown. To achieve ^{12}C/ ^{13}C ~ 10 at the end of the RGB phase, the requirements on $\dot{M}$ are shown (see BWNC).

both cases. If we assume that $v_{TRANS} \sim 1$m/s, as is inferred for the Solar case, then values of f_{MT} can be obtained. It is clear for the RGB case that CBP only requires a very small part of the matter at the upper interface to be from the extra transport. For the case of the AGB, it is small but substantial. As we will show later, the velocity of the ascending magnetic bubbles on the AGB are much more rapid (1km/s) and it is reasonable to consider that they will readily intrude into the convective layer at speeds much greater than 1m/s. We therefore do not consider that the flow across this upper boundary be a problem.

We also note that the frequency with which the magnetic bubbles must cross the upper boundary is: $\dot{M}/M_{bubble} = \hat{V}/\Delta R$. Here $M_{\rm bubble}$ is the mass of the bubble, $\hat{V}$ is its average velocity over the trajectory and ΔR is the distance traveled. Note also that, if only one "event" occurs per month, then this would require a bubble of ~1000km in scale for the RGB case.

Now we turn to the matter of the transport from the processing zone through the radiative zone. This requires that we define the geometry of the "bubble" and the dynamics of motion. We assume here that the bubble is a torus, – a toroidal magnetic flux tube (Fig. 4a) and is transported radially from the CBP processing zone (temperature T_P) to the base of the convective envelope with a velocity $v(R)$ (see Fig. 4b). The expression for the average velocity is given in the figure. The equation relating the density shift in the magnetized material of the torus to the magnetic field is $(\rho-\rho')/\rho(R) = B^2/8\pi P(R)$. The buoyant force per unit length on the torus is: $\pi a^2(R)g(R)(\rho - \rho')$, where $a(R)$ is the cross sectional radius of the torus and $g(R)$ is the local acceleration of gravity. The drag force per unit length on the torus is $1/2 C_{DRAG}\rho a(R)v^2(r)$. We assume mass conservation and magnetic flux conservation for the flux tube giving: $Ra^2(R)\rho(R) = const.$ and $B(R)a^2(R) = const.$ The mass of the toroidal flux tube is $2\pi^2\rho_0 R_0 a_0^2$.

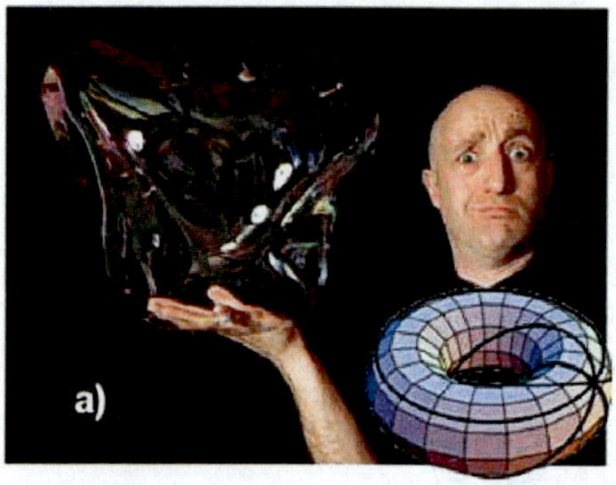

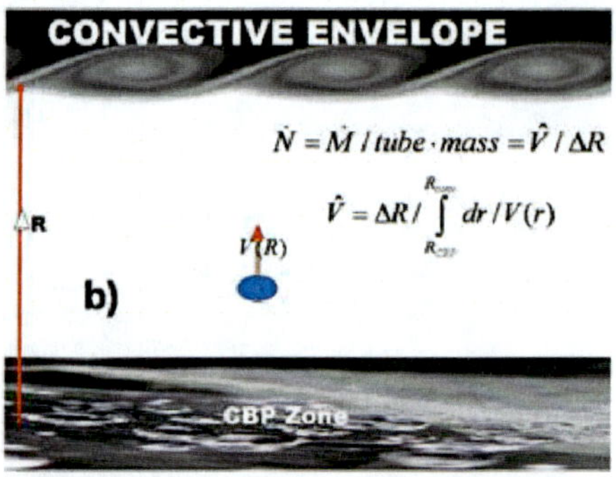

FIGURE 4. **a)** The problem of bubble geometry. A simple torus was chosen representing an undeformed magnetic flux tube; **b)** Cartoon showing the transport between the CPB processing region and the convective envelope .

This treatment follows original works on flux tubes in the Sun [15, 21]. Using these relationships, we obtain for the velocity as a function of radius:

$$v(r) = 1/2(\rho(r)/\rho_0)^{3/4}(r/r_0)^{-1/4}(g_0 a_0/C_D)^{1/2}\{B_0/\sqrt{P(r)}\} \tag{2}$$

The values with subscript 0 correspond to the deepest processing zone (T_P). The drag coefficient C_D is typically estimated at 1, but much lower values, down to 0.04 are possible. Note that C_D occurs as $B_0/\sqrt{C_D}$, thus even this wide uncertainly range does not grossly affect B_0. We have taken $g(R)/g_0=(R_0/R)^2$ as there is very little range in mass across the radiative zone. In calculating $v(R)$ we have used the values of ρ and P determined in the stellar model. We have inferred the values of a_0 from Solar observations. The resulting curves for $v(R)$ and $\hat{v}$ are shown in Fig. 5. Table 2 shows the required magnetic fields (B_0) at the processing zone, the initial velocities and average velocities to obtain the required flows for each case. The value of B_{CONV} at the base of the convective envelope is also given. The frequencies for a bubble event for the parameters used are $\dot{N} = 2.6x10^{-6}/s$ (7 per month) for the AGB, and $\dot{N} = 1.6x10^{-7}/s$ (5 per year) for the RGB.

TABLE 2.

	AGB	RGB I	RGB II
a_0 *(km)*	65	65	150
$\hat{v}$ *(km/s)*	1.45	0.098	0.018
v_0*(km/s)*	6.8	0.336	0.062
B_0*(G)*	5x10^6	3.8x10^5	4.8x10^4
B_{CONV}*(G)*	8.6x10^3	3.4x10^3	4.4x10^2

From the results given in Table 2, it can be seen that:

1) For the RGB case the fields at the processing zone are substantial, but are much lower at the base of the convective envelope;
2) For the AGB case the required fields at the processing zone are quite large (~5 Megagauss), but again they are much reduced at the base of the envelope.
3) The size of the hypothecated flux tubes is justifiable in terms of the Solar observations.
4) The velocity of upward transport by the bubbles is quite rapid so that very little nuclear processing will result from this part of the trajectory.
5) The hypothecated velocity of trajectory in most of the treatments [7, 11] assumed a rather low velocity field (≅ 1m/s) for both downward and upward motion. Thus, in order to accomplish the nuclear processing, the magnetic bubble model requires that the complimentary downward flow (D) be broad and slow (f_D≅0.5, as in WBS).

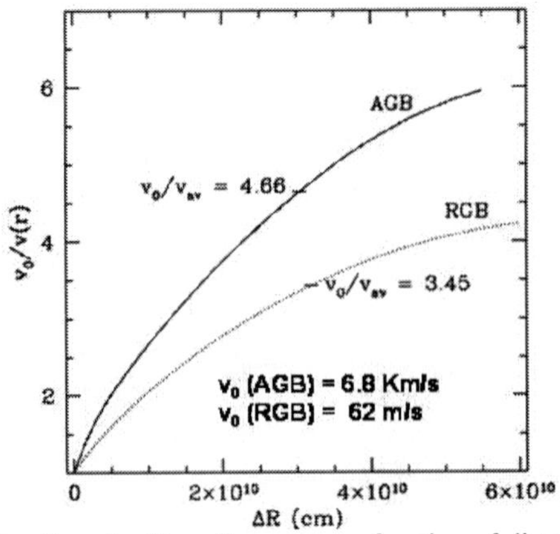

FIGURE 5. Curves showing the velocities of ascent as a function of distance for the AGB and RGB cases (relative to the initial velocities at T_P).

We must emphasize that the model outlined here was guided by the results from SOHO using Helioseismology, which proved the differential rotation of the sub-envelope zones of the Sun relative to the rather rigid interior. This motion penetrates to a depth ~0.04 $R_\odot$ and 0.01 $M_\odot$ thick into the radiative zone. This results in a weakly convective velocity field that penetrates into the radiative zone immediately below the envelope, a region called the *Tachocline.* It is further inferred that the fields deep in the radiative tachocline must exceed ~10^5 Gauss [22, 23]. For a thorough presentation of the structure and flow dynamics of the Sun the reader is referred to the site: http://solarphysics.livingreviews.org/About/copyright.html, which contains an excellent review [24].

The fundamental issue of the basic MHD mechanisms that must provide these observed fields remains to be resolved and is the subject of intense investigation in the Solar MHD community. We have simply assumed that they also exist for RGB and AGB stars.

There is then the issue of whether the available observations on RGB or AGB stars are in conflict with the proposed model. There is no *direct* observational evidence of strong magnetic fields in either RGB or AGB stars. A rough estimate for AGB stars, using flux conservation extrapolated to the surface, gives surface fields of up to 20G, in line with values inferred (from observational estimates in the circumstellar envelope) by [25]. This is compatible with the fact that AGB stars are not observed to have X-ray emitting coronae. For the RGB case the surface fields would again be small, in the range from ~ 1 to a few G. However, it is quite plausible that the fields we find at the base of the envelope would be greatly subdued, far below the values assuming flux conservation, when the flux tubes are engulfed in the very massive overlying convecting stellar envelope.

4. DISCUSSION AND PRELIMINARY CONCLUSIONS

In this note we have reported on the possibility that magnetic buoyancy be responsible for driving extramixing processes in red giants, during their first and second ascent on the giant branch. In our original work, published in the Astrophysical Journal, several important earlier papers were overlooked. We here have noted and recognized these critical contributions.

With regard to the high fields found to be necessary for the AGB case, we note that plasma jets of magnetic origin have been observed for many planetary nebulae [26]. A planetary nebula is the result of the end of AGB evolution, when the envelope is blown off, leaving a white dwarf remnant. Blackman et al. considered the development of dynamos in AGB stars as the origin of magnetic fields shaping planetary nebulae [27]. In their calculations they inferred fields of 5×10^4G at the base of the AGB convective envelope (at a radius of 1.5 $R_\odot$) in order to explain the collimation. B values of up to a few $\times10^6$ G were inferred in the case of low filling factors (which is in fact our case, due to the low values of f_{MT}). There is some problem in sustaining such fields [28]. In general, there is some independent evidence for significant internal magnetic fields in AGB stars. Magnetic fields in planetary nebulae are considered to be the cause of the bi-polar outflows and are the subject of intense investigation. In particular Nordhaus and Blackman are seeking possible scenarios that might provide the ongoing (or frequently recurring) high fields as presented here -- see their report in this volume [29].

It is further of note that white dwarfs, which are the end product of AGB evolution, sometimes show very high fields. Recent studies have revived the old hypothesis [30] that these fields might be fossil remnants of the stellar ones; for modern studies on this subject see [31, 32, 33]. About 10 − 20 % of the available white dwarf sample show fields larger than 1 MG (see Fig. 6), and up to 2 Gigagauss [34]. These observations suggest that possibly, the high fields of such white dwarfs might be inherited from the AGB precursors, with high enhancement factors ensuing upon the expulsion of the envelope.

From the arguments presented in this report we conclude that magnetic buoyancy is a very plausible mechanism for transporting material upward from near the H shell into the convective envelope, for both RGB and AGB stars. This requires very high internal fields for those AGB stars with extensive CBP. We have only given phenomenological arguments regarding the mass transport, assuming that the stars provide the required magnetic fields. The fundamental issue as to how the fields are generated by current flows is not addressed in our work.

Gene Parker, in commenting on this work stated "I feel that your paper makes a fundamental point, that deep mixing is an essential part of the evolution on the RGB and the AGB. What causes the convection? I understand from your paper that the only theoretical possibility that cannot be ruled out at the present time is the magnetic buoyancy of Megagauss fields. It is not possible, so far as I am aware, to explain the presence of a Million Gauss in the radiative zone, or anywhere else. We should recognize that a similar problem exists in the Sun. [.....] Nowhere is there enough convective muscle to manipulate so strong a field. So one turns to schemes whereby a weaker field, of say, 5 KiloGauss, is greatly intensified by the upward bulging Ω-shaped loops that exist at intervals along the flux bundle. The upward convection, when an Ω-loop is formed, sucks the gas out of the portion of the flux bundle that remains deep down in the convective zone. [...] If there is some hope, you can find it written up in [21]." An analysis of the magnetic instabilities suitable to enhance otherwise weaker fields was later presented in [34].

It is our hope that future investigations will be found to provide the "convective muscle" required by many stars, including the Sun, to yield the apparently necessary magnetic fields.

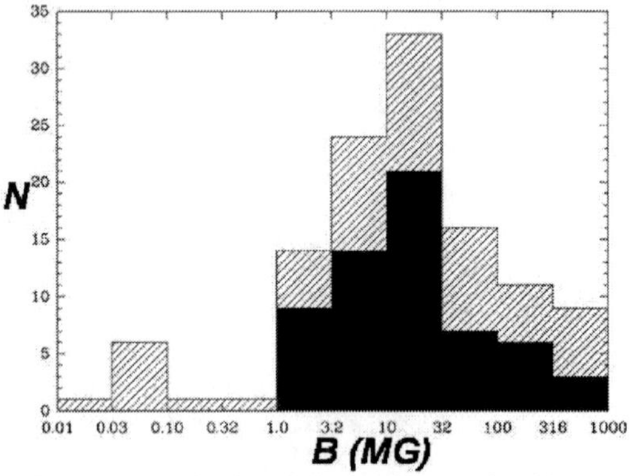

FIGURE 6. Histogram of magnetic white dwarfs in equal intervals of log B (MG). SDSS discoveries are shaded black; the distribution for all known magnetic white dwarfs is hatched [34].

ACKNOWLEDGEMENTS

GJW acknowledges the support of DOE-FG03-88ER13851, the generosity of the Epsilon foundation, and Caltech's contribution 9003(1124). M.B. thanks support from the Italian Ministry of research, through the contract PRIN2006-022731.

REFERENCES

1. C. Charbonnel, *In "Origin and Evolution of the Elements", ed. A. McWilliam and M. Rauch, (Pasadena: Carnegie Observatories),* p. 59, (2004)
2. G.J. Wasserburg, M. Busso, R. Gallino, and K.M. Nollett, *Nuc.Phys. A* **777**, 5 (2006)
3. C. Charbonnel and J.D.Jr. Do Nascimiento, *A&A* **336**, 915 (1998)
4. P.A. Denissenkov, G.S. Da Costa, J.E. Norris, and A. Weiss, *A&A* **333**, 926 (1998)
5. J. McDonald and D.J. Mullan, *ApJ* **598**, 560 (2003)
6. P.P. Eggleton, D.S.P. Dearborn, and J.C. Lattanzio, *Science* **314**, 1580 (2006)

7. K.M. Nollett, M. Busso, and G.J. Wasserburg, *ApJ* **582**, 1036 (2003)
8. C. Charbonnel and J.P. Zahn, *A&A* **467**, L15 (2007)
9. M. Cantiello, H. Hoekstra, N. Langer, and A.J.T. Poelarends *in "Unsolved Problems in Stellar Physics", AIP Conf. Proc. 948,* p. 73 (2007)
10. A.I. Boothrpyd, I.-J. Sackman, and G.J. Wasserburg, *ApJ* **430**, L77 (1995)
11. G.J. Wasserburg, A.I. Boothroyd, and I.-J. Sackmann, *ApJ* **447**, L37 (1995)
12. E.N. Parker, *ApJ* **121**, 491 (1955)
13. H.S. Gurm and D.G. Wentzel, *ApJ* **149**, 139 (1967)
14. E.N. Hubbard and D.S.P. Dearborn, *ApJ* **239**, 248 (1980)
15. E.N. Parker, *Ap. Space Sci.* **31**, 261(1974)
16. H.C. Spruit, *A&A* **381**, 923 (2002)
17. M. Busso, G.J. Wasserburg, K.M. Nollett, and A. Calandra, *ApJ* **671,** 802 (2007)
18. O. Straniero, A. Chieffi, M. Limongi, et al., *ApJ* **478**, 332 (1997)
19. M. Busso, R. Gallino, and. G.J. Wasserburg, *PASA* **20**, 356 (2003)
20. D. Reimers, *in "Problems in Stellar Atmospheres and Envelopes". Ed. B. Bascheck, W.H. Kegel, G. Traving (New York: Springer-Verlag Inc.),* p. 229 (1975)
21. E.N. Parker, *ApJ* **433**, 867 (1994)
22. M. Rempel, M. Schuessler, and G. Toth, *A&A* **363**, 789 (2000)
23. M. Schuessler and M. Rempel, in *"From Solar Min to Max: Half a Solar Cycle with SOHO", ed. A. Wilson, ESA SP-508 (Nordwjik:ESA Publ. Division),* p. 499 (2002)
24. M.S. Miesch, *Living Rev. Solar Phys.* **2**, 1 (2005)
25. F. Herpin, A. Baudry, C. Thum, et al., *A&A* **450**, 667 (2006)
26. J.H. Kastner, B. Balick, E.G. Blackman, et al., *ApJ* **591**, L37 (2003)
27. E.G. Blackman, A. Frank, J.A. Markiel, et al., *Nature* **409**, 485 (2001)
28. J. Nordhaus, E.G. Blackman, and A. Franck, *MNRAS* **376**, 599 (2007)
29. J. Nordhaus and E.G. Blackman, this volume (2007)
30. E.N. Hubbard and D.S.P. Dearborn, *ApJ* **247**, 236 (1981)
31. C.A. Tout, D.T. Wickramasinge, and L. Ferrario, *MNRAS* **355**, L13 (2004)
32. D.T. Wickramasinge and L. Ferrario, *MNRAS* **356**, 1576 (2005)
33. G. Valyavin, S. Bagnulo, S. Fabrika, et al., *ApJ* **648**, 559 (2006)
34. G.D. Schmidt, H.C. Harris, J. Liebert, et al., *ApJ* **595**, 1101 (2003)
35. H.C. Spruit, *A&A* **349**, 189 (1999)

Dynamos and Chemical Mixing in Evolved Stars

J. Nordhaus[*,†] and E. G. Blackman[*,†]

[*]Dept. of Physics and Astronomy, University of Rochester, Rochester, NY 14627
[†]Laboratory for Laser Energetics, University of Rochester, Rochester, NY 14623

Abstract. In low-mass Red Giant Branch (RGB) and Asymptotic Giant Branch (AGB) stars, anomalous mixing must transport material near the hydrogen-burning shell to the convective envelope. Recently, it was suggested that buoyant magnetic flux tubes could supply the necessary transport rate [1]. The fields are assumed to originate from a dynamo operating in the stellar interior. Here, we show what is required of an $\alpha - \Omega$ dynamo in the envelope of an AGB star to maintain these fields. Differential rotation and rotation drain via turbulent dissipation and Poynting flux, so if shear can be resupplied by convection, then large-scale toroidal field strengths of $\langle B_\phi \rangle \simeq 3 \times 10^4$ G can be sustained at the base of the convection zone.

Keywords: Stars: AGB and Post-AGB – Stars: Magnetic Fields – Stellar Dynamos – Stars: Mixing
PACS: 95.30.Qd

1. INTRODUCTION

In low-mass Red Giant Branch (RGB) and Asymptotic Giant Branch (AGB) stars, an anomalous transport mechanism must cycle material from the convective envelope to the stable radiative region where nuclear processing can occur [2, 3]. In particular, material must be pumped downward from the convection zone, exposed to partial H burning (so-called Cool Bottom Processing; CBP) and returned to the convective envelope. Isotopic abundances provide constraints on the required processing temperatures and mass transfer rates. Recently, it was proposed that buoyant magnetic flux tubes can provide the necessary transport [1].

In evolved stars, magnetically mediated outflows have been proposed as the origin of bipolarity in post-Asympotitc Giant Branch stars (post-AGB) and planetary nebula (PN) [4, 5, 6, 7]. The needed large-scale field likely results from a dynamo operating during the RGB or AGB. In both phases, a convective envelope coupled with differential rotation in the stellar interior can amplify the field via an $\alpha - \Omega$ dynamo [7]. However, as the field amplifies, differential rotation energy is drained and rapidly terminates the dynamo ($\leq$ 100 yrs) [4, 5]. This would favor binary shaping mechanisms for post-AGB/PNe and a binary induced dynamo (see [7, 8] for detail). Only if convection resupplies differential rotation (analogous to the λ-effect in the Sun [9]) could a sustained dynamo operate throughout the RGB or AGB phases without a binary [7]. While sustained fields are required to shape post-AGB outflows, the dynamo required for mixing may be weaker and need not influence the outflow. It may be that a weaker dynamo (supplying sufficient mixing) is sustained throughout the AGB phase, but that a binary companion is required

[1] nordhaus@pas.rochester.edu

CP1001, 9th Torino Workshop on Evolution and Nucleosynthesis in AGB Stars and the 2nd Perugia Workshop on Nuclear Astrophysics, edited by R. Guandalini, S. Palmerini, and M. Busso

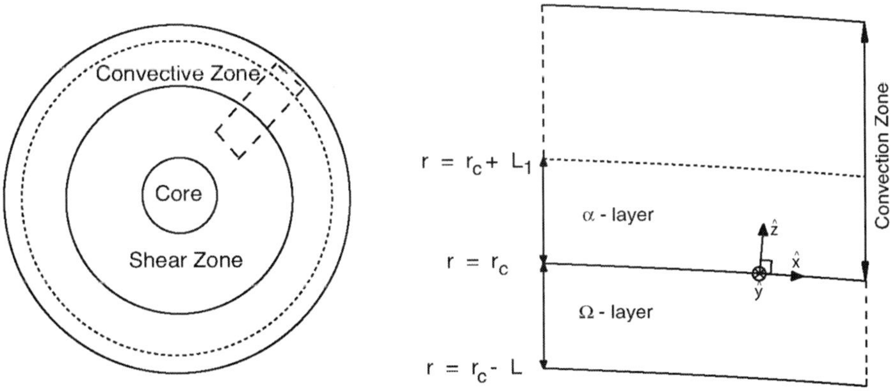

FIGURE 1. Figure from [7]. A meridional slice of dynamo geometry. The left figure shows the global geometry of the AGB star. The right figure is a close-up view of the dashed region on the left. The α-effect is driven by convection and occurs in layer of thickness L_1 above the differential rotation zone. The poloidal component of the field is pumped downwards into the differential zone, where it is wrapped torodially due to the Ω-effect.

to power post-AGB bipolarity.

Here, we present a dynamical $\alpha-\Omega$ dynamo operating at the base of the convection zone in a 3 $M_{\odot}$ AGB star. We consider two models: (i.) shear is not resupplied, and a transient dynamo results (ii.) convection resupplies shear, sustaining the dynamo. For both models, the field penetrates the shear zone and may transport material to the H burning shell. We compare our results to the field strengths and penetration depths needed for viable magnetic mixing in low-mass RGB and AGB stars [1]. We comment on future directions.

2. AN EVOLVED STAR DYNAMO

A 2-D schematic of our 1-D, 3 $M_{\odot}$ AGB star[2] is presented in Fig. 1. In such a star, the convection zone extends from the stellar surface to the interface between the convective and radiative zones. In this layer, convective twisting motions convert buoyant toroidal field into poloidal field through the α-effect. Below the convection layer, the differential rotation zone extends to the core and shears poloidal field back into toroidal field through the Ω-effect.

The differential rotation profile in the AGB interior is unknown. In order to employ a reasonable initial profile, we assume that the main sequence surface rotation velocity is 200 km s^{-1}. A radial shear profile is then obtained by conserving angular momentum on spherical shells during post-MS evolution [6].

We solve for the time evolution of dynamic quantities at the interface between the

[2] Stellar model courtesy of S. Kawaler – personal communication.

convective region and shear zone ($r = r_c$). The rotation profile across the shear layer varies from Ω at the interface to $\Omega + \Delta\Omega$ at $r = r_c - L$. Thus, $\Delta\Omega$ is a measure of shear in the differential rotation zone. If $\Delta\Omega = 0$, the system exhibits solid body rotation. The average poloidal field, $\langle B_p \rangle$, and average toroidal field, $\langle B_\phi \rangle$ amplify from 1 G seed values until they are quenched through a drain of the available differential rotation energy. We refer to reader to [7] for a derivation of the precise 1-D, mean-field equations solved.

2.1. Shear Zone Penetration Depth

To capture aspects of 2-D, mean-field geometry within the framework of our simple, 1-D time-dependent model, we employ two turbulent diffusion coefficients: β_p, corresponding to diffusion of the poloidal field (which grows primarily in the convective region) and β_ϕ, corresponding to diffusion of the toroidal field (which is amplified in the differential rotation zone; see Fig. 1). We also use β_ϕ as the turbulent diffusion coefficient for the toroidal velocity. Since the convective region is highly turbulent and the differential rotation zone is more weakly turbulent, we take $\beta_\phi \ll \beta_p$. The value of β_ϕ governs how far the poloidal component of the field can diffuse into the shear zone in a cycle period. The further into the shear zone the poloidal field can penetrate, the greater the shear energy that can be extracted and utilized by the dynamo. The coefficient, β_ϕ also governs how much heat is generated by turbulent dissipation in the shear layer.

To quantify this, we define δ as the depth at which the poloidal field can diffuse into the shear layer in a cycle period, τ. We have

$$\delta \simeq (\beta_\phi \tau)^{\frac{1}{2}}. \tag{1}$$

If $\delta/L = 1$, the poloidal component of the field reaches the core boundary and extracts all available shear energy. If $\delta/L < 1$, then the poloidal field extracts a fraction of the total shear energy and can not penetrate as deeply. Thus for a given β_ϕ, we can determine the depth to which the field diffuses. As an aside, we note that for the maximally growing dynamo mode, τ is weakly dependent on β_ϕ and hence, would change the power index slightly. Although we do not expect this effect to be significant, future work should use the maximally growing mode.

2.2. The Need For Convection Resupplying Shear

To investigate the back-reaction of field growth on the shear, we apply our dynamical model using the initial rotation profile previously described. Our results are shown in Fig. 2. The dynamo terminates rapidly as the shear energy is drained after ~ 20 yrs. In addition, the loss of Poynting flux spins down the interface by a factor of ~ 3. In order to sustain the dynamo through an AGB lifetime, a constant differential rotation profile must be established. This occurs in the Sun as convection re-seeds shear through the λ-effect [9]. Although it remains to be established if a similar effect occurs in evolved stars, by analogy to the solar case, we allow a fraction of the turbulent energy

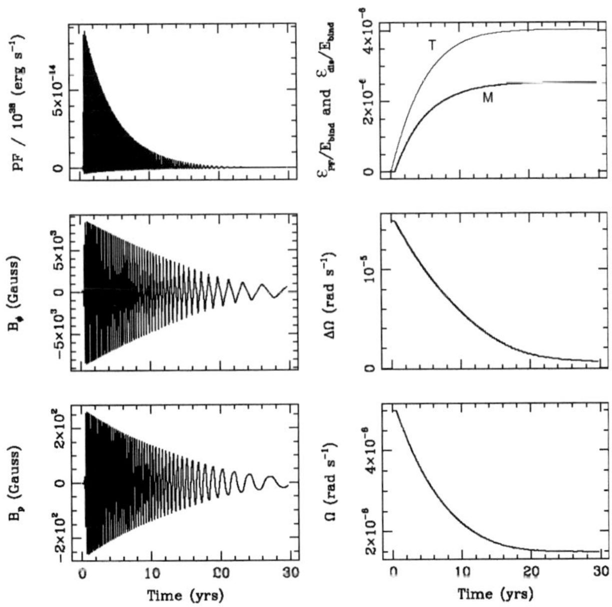

FIGURE 2. Figure from [7]. The differential rotation energy is allowed to drain through field amplification and turbulent dissipation. We define, $\varepsilon_{[PF,dis]} \equiv \int_0^t E_{[PF,dis]}(t')\,dt'$ and label M (ε_{PF}) and T (ε_{dis}) on the top right plot to distinguish between the thermal and magnetic contributions. If $\varepsilon_{[PF,dis]}/E_{bind} \geq 1$, then the AGB envelope was unbound via heat from turbulent dissipation or time-integrated Poynting flux. Peak field strengths are a factor of $\sim 5-10$ less then those obtained in [5]. Differential rotation energy is drained in < 20 yrs. Lowering the turbulent dissipation coefficient in the shear region results in the differential rotation energy draining at a slower rate, allowing the field to sustain for longer periods of time ($\sim 40-50$ yrs). However, peak field strengths remain the same and are a function of $\Delta\Omega$ and Ω.

cascade to resupply shear. Additionally, we keep the rotation at the interface fixed. This is physically equivalent to storing Poynting flux in the interface region. If the field is trapped, Poynting flux does not emerge from the layer and thus, does not spin down the envelope. An AGB dynamo will thus sustain when the following two conditions are met: (i.) convection resupplies shear (ii.) Poynting flux is stored inside the envelope [7].

For our 3 $M_\odot$ AGB star, under those conditions, a steady-state dynamo is established when $\sim 1\%$ of the turbulent cascade energy per unit time reinforces the shear (see Fig. 3).

3. REQUIREMENTS FOR MAGNETIC MIXING

For a given bulk transport rate, Cool Bottom Processing predicts the chemical and isotopic evolution of the convective envelope. Without appealing to a specific driving mechanism, CBP yields explicit temperature requirements and nuclear processing times needed to match observations for a specific stellar model. To investigate the physical origin of CBP, hydrodynamical models, such as rotationally induced meridional mixing

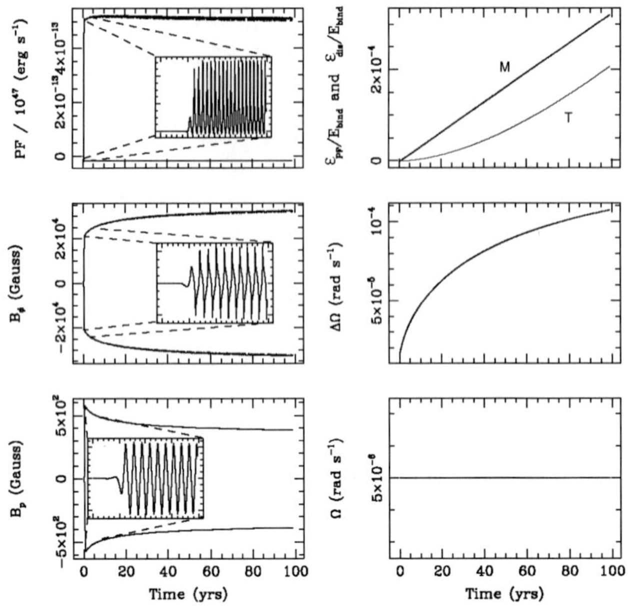

FIGURE 3. Figure from [7]. Convective resupply results in a steady-state differential rotation profile. For the left column, the envelope of the poloidal, toroidal and Poynting flux is plotted. The Poynting flux is sustained at $\sim 5\times10^{34}$ erg/s. The sustained Poynting flux supplies enough energy to unbind the envelope of our 3 $M_\odot$ model at the end of the AGB phase ($\sim 10^5$ yrs). In this figure, $\sim 0.1\%$ of the cascade power must be used to sustain the differential rotation to supply the requisite Poynting flux. This choice of parameters corresponds to a magnetically dominated explosion at the end of the AGB.

were suggested. However, these models do not supply sufficient mixing to explain observations of low-mass RGB field and globular cluster stars [10].

Recently, as an alternative to hydrodynamic models, magnetically induced mixing was proposed [1]. Buoyant toroidal fields in the radiative zone carry material to the convective envelope and drive mass circulation. By requiring CBP transport rates, it is possible to constrain magnetic field strengths and buoyant velocities necessary for a given stellar model.

For an initial 1.5 $M_\odot$ star with metallicity one-half solar [11], constraints are calculated for the subsequent RGB and AGB phases. In both phases, subadiabatic zones similar to the solar tachocline exist [12]. Since our previous dynamo calculations apply to an AGB star, we focus on that phase.

For the AGB phase, the total mass has decreased to $M = 1.2\ M_\odot$. The base of the convection zone is located at $r_c = 5.7\times10^{10}$ cm. For a mass flux of $\dot{M} = 10^{-6}\ M_\odot/yr$, material must mix to a radius of $r_{CBP} = 2\times10^9$ cm which is slightly above the location of the H burning shell ($r_H = 1.5\times10^9$ cm) [1].

Given a mass transfer rate, mixing depth and buoyant rise velocity, the magnetic field strength at r_H and r_c can be constrained. If the field is purely toroidal and in flux tubes that rise individually, then $B_t = B\left(r_{CBP}\right) = 5\times10^6$ G [1]. At the base of the convection zone, this corresponds to a field of $B_c = B(r_c) \simeq 9 \times 10^3$ G for a dipole geometry.

Direct comparison to the dynamo calculation is difficult as our AGB star is 2.5 times more massive. In addition, our 3 $M_\odot$ structure corresponds to a star that has just reached the AGB. Future work should investigate a dynamo operating in a lower mass star, during the interpulse periods. Nevertheless, for plausibility arguments, a general comparison with our sustained dynamo is useful.

From our 3 $M_\odot$ AGB star, the toroidal field strength at the base of the convection zone is $\langle B_{\phi,c} \rangle \sim 3 \times 10^4$ G, suggesting that it may be possible to produce sufficiently high average fields. However, since we do not solve for the radial structure, we cannot determine the field strength where CBP nucleosynthesis occurs.

We can, however, estimate the depth to which the magnetic field penetrates the shear layer. Based on the initial conditions used in Fig. 3, Eq. 1 gives that the field reaches a radius of $\sim 3 \times 10^{10}$ cm with approximately 15 cycles per year. This is encouraging and suggests that significant penetration into the shear zone may occur.

3.1. Flux Tubes versus Average Fields

Our dynamo calculations are mean-field and correspond to volume averages. When the ratio of average thermal to magnetic pressure $\beta = \frac{\langle P \rangle}{\langle B^2 \rangle / 8\pi} >> 1$ it is possible for the magnetic field to be concentrated in flux tubes with an overall volume filling fraction of the magnetic field $f << 1$, to be small, such that $f \sim \frac{1}{\beta}$ [13]. When flux tubes are in pressure balance with their exterior, the mean-field strength is then related to a flux tube field, B_t by $\langle B^2 \rangle \sim f \langle B_t^2 \rangle$ [13]. Since the field constraints on magnetic mixing from [1] constrain the magnitude of flux tube fields with total $f << 1$, the mean field can be much lower than the flux tube field, relaxing the demands on the strength of the dynamo.

Further investigation of all of the above is warranted for relevant stellar models. Determining the radial diffusion coefficients will constrain the amount of shear energy available to the dynamo and thus, determine the penetration depth and field strengths. For a given stellar model, constraints on the fraction of convective energy needed to sustain a steady shear profile should be calculated in both the RGB and AGB phases. In the RGB phase, weaker fields are required. Realistic rotation profiles should be utilized when known. Finally, $\geq$ 2-D model calculations incorporating Ω-quenching will provide radial information on magnetic field strengths.

4. CONCLUSIONS

It was recently suggested that buoyant toroidal fields, amplified in the stellar interior, may supply the necessary chemical transport rates from the hydrogen burning shell to the convective envelope required from Cool Bottom Processing in low-mass RGB and AGB stars [1].

In this paper, we have presented results of a dynamic $\alpha - \Omega$ dynamo operating in an AGB star. Amplification of the large-scale field drains shear energy rapidly (≤ 100 yrs). For an isolated star, stringent conditions (re-supply of shear + storage of Poynting flux) must be met if a dynamo is to sustain through the RGB or AGB phases. This can be accomplished if $\sim 1\%$ of the turbulent energy cascade reinforces shear. In this case, our AGB model produces strong toroidal fields ($B_\phi \simeq 3 \times 10^4$ G) at the base of the convection zone that can diffuse into the shear layer. This suggests that magnetic fields may be a plausible mechanism for mixing in low-mass RGB and AGB stars, however future work is needed before this can be firmly established.

ACKNOWLEDGMENTS

JN thanks M. Busso and G. J. Wasserburg for their gracious invitation and ensuing discussions. This research was carried out under the financial support of a Horton Fellowship from the Laboratory for Laser Energetics through the U. S. Department of Energy and HST grant AR-10972.

REFERENCES

1. M. Busso, G.J. Wasserburg, K.M. Nollett, and A. Calandra, *ArXiv e-prints* **708** (2007), 0708.2949
2. K.K. Gilroy and J.A. Brown, *ApJ* **371**, 578 (1991)
3. G.J. Wasserburg, A.I. Boothroyd, and I.J. Sackmann, *ApJL* **447**, L37 (1995)
4. G. Pascoli, *ApJ* **489**, 946 (1997)
5. E.G. Blackman, A. Frank, J.A. Markiel, J.H. Thomas, and H.M. Van Horn, *Nature* **409**, 485 (2001)
6. J. Nordhaus and E.G. Blackman, *MNRAS* **370**, 2004 (2006)
7. J. Nordhaus, E.G. Blackman, and A. Frank, *MNRAS* **376**, 599 (2007)
8. J. Nordhaus and E.G. Blackman, *ArXiv e-prints* **707** (2007) 0707.3792
9. G. Rüdiger and R. Hollerbach, *The magnetic universe : geophysical and astrophysical dynamo theory*, The Magnetic Universe: Geophysical and Astrophysical Dynamo Theory, by Günther Rüdiger, Rainer Hollerbach, pp. 343. ISBN 3-527-40409-0. Wiley-VCH , August 2004., (2004)
10. A. Palacios, C. Charbonnel, S. Talon, and L. Siess, *A&A* **453**, 261 (2006)
11. O. Straniero, A. Chieffi, M. Limongi, M. Busso, R. Gallino, and C. Arlandini, *ApJ* **478**, 332 (1997)
12. K.M. Nollett, M. Busso, and G.J. Wasserburg, *ApJ* **582**, 1036 (2003)
13. E.G. Blackman, *Physical Review Letters* **77**, 2694 (1996)

Testing Evolutionary Models with Observations of Galactic Bulge AGB Stars

Stefan Uttenthaler[*,†], Thomas Lebzelter[*], Josef Hron[*], Bernhard Aringer[*], Michael T. Lederer[*], Sara Palmerini[**], Maurizio Busso[**] and Hans Ulrich Käufl[‡]

[*]*Institute for Astronomy, Türkenschanzstraße 17, A–1180 Vienna, Austria*
[†]*Instituut voor Sterrenkunde, K. U. Leuven, Celestijnenlaan 200D, B–3001 Leuven, Belgium*
[**]*Department of Physics, University of Perugia, Via A. Pascoli 1, I–06123 Perugia, Italy*
[‡]*ESO, Karl Schwarzschild Straße 2, D–85748 Garching near Munich, Germany*

Abstract. We performed spectroscopic observations of a sample of oxygen-rich long-period AGB variables in the outer Galactic Bulge and use these data to test evolutionary models of the AGB phase in various respects.

The detection of the radioactively unstable element technetium (Tc) can be used to test model predictions for the onset of the third dredge-up (3DUP) mixing episode. No Tc-rich stars below the predicted lower luminosity limit for 3DUP are found, confirming this limit. From the luminosity of the least luminous Tc-rich star the minimum core mass required for 3DUP is estimated from our observations.

Surprisingly, lithium (Li) was detected in some of our sample stars. The occurrence of this fragile element is interpreted as the result of an "extra mixing" process going on between the convective envelope and the H-burning shell. We propose the name "Extended Envelope Mixing" for this phenomenon. An estimate for the mass circulation rate for the most Li-rich sample star is given, and considerations about extra mixing and 3DUP are made.

Finally, we were also able to measure the abundance of fluorine (F) in one of the sample stars, and conclude that F is produced only in small quantities in low-mass AGB stars, as predicted by theoretical models.

Keywords: Stars: AGB and Post-AGB – Stars: Abundances – Stars: Evolution
PACS: 01.30.Cc

1. INTRODUCTION

Evolutionary models of the AGB need constraints and fine tuning by comparison with observations. Aiming at such constraints, we obtained high-resolution optical spectra of a sample of 27 oxygen-rich long-period AGB variables in the outer Galactic Bulge (Palomar-Groningen field no. 3, $b = -10^\circ$) with the UVES spectrograph at ESO's VLT. The observations, data reduction and analysis with respect to the occurrence of technetium are described in more detail in [1], referred to as Tc-paper in the following. Four stars with the signatures of line absorption by Tc were identified in our sample. Technetium has only radioactively unstable isotopes and thus serves well as 3DUP indicator.

CP1001, *9th Torino Workshop on Evolution and Nucleosynthesis in AGB Stars and the 2nd Perugia Workshop on Nuclear Astrophysics,* edited by R. Guandalini, S. Palmerini, and M. Busso

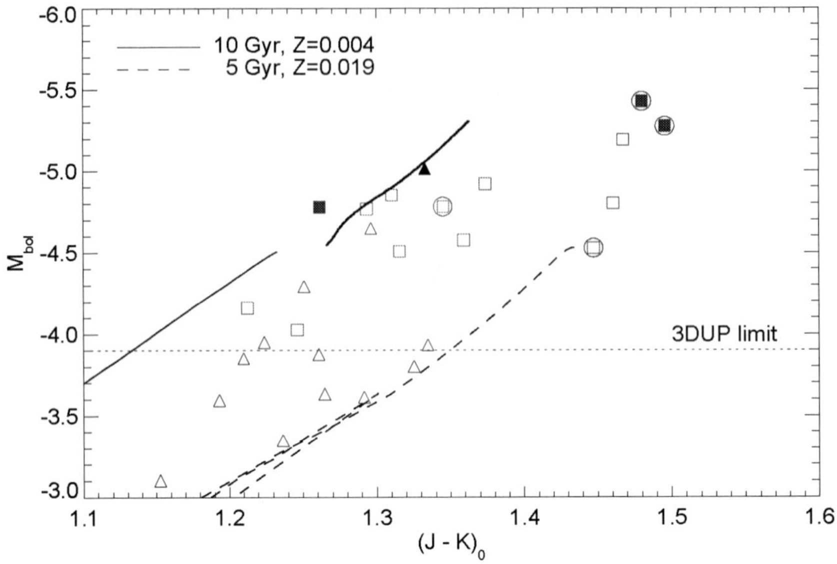

FIGURE 1. Bolometric magnitude vs. $(J-K)_0$ of the sample stars. Squares and triangles represent Mira and SR variables, respectively. The filled symbols are stars with a positive Tc detection, encircled symbols are stars with Li. The bolometric magnitude is corrected for the depth-induced scatter within the Bulge. The dotted horizontal line marks the minimum luminosity at the stage where 3DUP sets in. Isochrones from [2] are also represented in solid and dashed lines (see legend). For both isochrones, the AGB tip is well below the observed tip. The thick line is a 1.5 $M_{\odot}$ evolutionary model track from Oscar Straniero (private communication).

2. CONSTRAINTS ON THE 3DUP LUMINOSITY AND CORE MASS

The detection of Tc together with the rather accurately known luminosity derived from near-IR photometry of the sample stars and the known distance to the Galactic Bulge allows for testing predictions of AGB evolutionary models concerning the minimum luminosity at the stage when 3DUP sets in. This minimum luminosity has been estimated by [3] from models of [4] to correspond to $M_{\rm bol} = -3\overset{\rm m}{.}9$.

To compare our observed data to theoretical predictions, Fig. 1 shows a $M_{\rm bol}$ vs. $(J-K)_0$ diagram of the sample stars. The bolometric magnitude has been corrected for the depth scatter within the Bulge using the period K-magnitude relation. Since the pulsation period of the stars is more accurately known than the J and K magnitudes, we believe that it results in more realistic bolometric magnitudes. For more details on the correction and a CMD with uncorrected bolometric magnitudes we refer to the Tc-paper.

The depth scatter correction has a number of interesting effects. First, the Mira variables (squares) become on average much brighter than the SRVs (triangles, mean $M_{\rm bol}$ without correction: $\overline{M_{\rm bol}}({\rm Mira}) = -4\overset{\rm m}{.}77$, $\overline{M_{\rm bol}}({\rm SRV}) = -4\overset{\rm m}{.}19$; with correction: $\overline{M_{\rm bol}}({\rm Mira}) = -4\overset{\rm m}{.}77$, $\overline{M_{\rm bol}}({\rm SRV}) = -3\overset{\rm m}{.}79$). Also, the overlap in $M_{\rm bol}$ between SRVs and Miras is much reduced by this correction[1]. Finally, by applying the depth scatter correction, the Tc-rich stars become the brightest stars at a given $(J-K)_0$ color.

The minimum luminosity at the stage when 3DUP sets in, as estimated from evolutionary models, is indicated as dotted horizontal line in Fig. 1. All Tc-rich stars are above this limit, confirming theoretical predictions. The least luminous Tc-rich star in our sample is M626 (aka Plaut 3-626 or V3117 Sgr). After correcting for the depth scatter, its bolometric magnitude is $M_{\rm bol} = -4\overset{\rm m}{.}78$, corresponding to $L \cong 6420\,{\rm L}_\odot$. Uncertainties on the bolometric magnitude may be introduced by the intrinsic uncertainty of the near-IR photometry and the variability of the star, summing up to $\pm 0\overset{\rm m}{.}12$, uncertainties in the bolometric correction (which is probably small at the color of $(J-K)_0 = 1.26$ of M626), uncertainties in the distance to the Bulge, and uncertainties in the period K-magnitude relation used for the depth scatter correction. We estimate that the total uncertainty in the bolometric magnitude of M626 is $\pm 0\overset{\rm m}{.}2$, putting its luminosity in the range 5350 – 7720 $\rm L_\odot$. On top of this, one has to keep in mind that M626's phase in its thermal pulse cycle is unknown.

Assuming that M626 is at the moment in the quiescent phase in-between two TPs and applying a core mass – luminosity relation (CMLR) to the luminosity of M626, we can have an estimate of the core mass at the stage when 3DUP sets in. A CMLR has been first published by [7]: $L = 59250(M_{\rm C} - 0.522)$, where L is the luminosity and $M_{\rm C}$ is the mass of the H-depleted core (interior to the H-burning shell), both in solar units. Solving the equation for $M_{\rm C}$ and inserting $6420\,{\rm L}_\odot$ for the luminosity, we find a core mass of $0.630\,{\rm M}_\odot$ for the star M626. Other CMLRs have been published in the literature. The relation found by [8] reads as $L = 59250(M_{\rm C} - 0.495)$. This gives a core mass of $0.603\,{\rm M}_\odot$, just at the lower border of validity of that relation at $0.60\,{\rm M}_\odot$. [9] included a dependence on chemical composition in his CMLR: $L = 55\,320[2.3(Y-0.20)+1.0](M_{\rm C}-a)$, where $a = 0.489 + 0.23(Y-0.20) - 0.70(Z-0.02)$. Solar metallicity is a reasonable approximation to the peak metallicity found in Bulge stars ([10]; [M/H] = −0.1). We thus set $Z = 0.02$ and $Y = 0.28$, obtaining $0.605\,{\rm M}_\odot$ for the core mass of M626. [11] obtained a similar relation and solved it for the core mass:

$$M_{\rm C} = \left[\frac{L}{238\,000\mu^3 (Z_{\rm CNO})^{1/25}} + 0.1804\right]^{\frac{1}{2}} + 0.015. \quad (1)$$

μ is the mean molecular weight of the envelope with $\mu \approx 4/(5X+3-Z)$, and $Z_{\rm CNO}$ is the fractional surface abundance by mass of the CNO elements. Assuming solar composition again ($\mu \approx 0.618$, $Z_{\rm CNO} \approx 0.012$), we arrive at a core mass of $0.578\,{\rm M}_\odot$ for M626. From these considerations we see that the core mass at which 3DUP sets in probably lies in the range $0.578 - 0.630\,{\rm M}_\odot$. This range compares well with the range of core masses in full theoretical models at the stage when 3DUP sets in [4, 12].

---

[1] Note that the one SRV with Tc (filled triangle in Figure 1) has been classified as a SRV by [5] with a low quality flag, but as Mira variable by [6]. The latter is justified by the large variability amplitude and the long pulsation period of this object. We thus assume that this star is in fact a Mira variable.

Similar constraints can be found from the carbon star luminosity function (CSLF). [13] find, on basis of improved distance measurements and including mid-IR photometry, the CSLF of the Galactic Disk rising sharply at $M_{\rm bol} = -4\overset{\rm m}{.}5$. M626 is brighter than this limit of C-star formation in the Disk. It must be noted here that no C stars are known belonging to the Bulge. C star formation is probably hampered by the increased oxygen abundance of the Bulge compared to the Disk [14].

Included in Fig. 1 is an evolutionary model track kindly provided by Oscar Straniero (private communication). The track is for a 1.5 $M_\odot$ model with solar composition and a Reimers mass-loss parameter $\eta = 0.5$. The track covers only the very last interpulse evolution of this model, excluding the short-term luminosity spikes at the TPs themselves. The plotted track spans about 60 000 years, and the evolution runs from the lower left end to the upper right end of the track. The luminosity of the evolutionary model track matches that of the Tc-rich stars very closely. Note also that the track runs parallel to the sequence formed by the Mira stars. Regarding the evolution along the model track, one may speculate that M626, the least luminous Tc-rich sample star, is at the moment in the low luminosity phase just after a thermal pulse. If this is correct, the above mentioned range in core mass of this star may be underestimated and has to be shifted to somewhat higher values. However, since no information on the precise mass and age distribution of the sample stars is available, this conclusion may be premature.

We also include isochrones from [2] with two age/metallicity combinations as dashed and thin solid line in Fig. 1. The 5 Gyr isochrone covers better the observed range in $(J-K)_0$ color than the 10 Gyr isochrone. Nevertheless, even taking the luminosity variation during a thermal pulse cycle into account, both isochrones underestimate the observed AGB tip luminosity. This implies that models with large overshoot from the envelope, like the ones of [2], predict somewhat lower luminosities than what is observed.

3. LITHIUM IN LOW-MASS AGB STARS

Surprisingly, we detected stars in our sample with a strong Li I 671 nm line. From standard mixing theory, lithium is not expected to be present in low-mass AGB stars because this fragile element is burned in the convective envelope at temperatures exceeding some 2.1×10^6 K. Contrary to Tc, we were able to derive abundances of Li by fitting synthetic spectra based on COMARCS hydrostatic model atmospheres to the observed spectra around the Li line. COMARCS is a version of MARCS [15] operating with opacities from COMA [16]. Stellar parameters, particularly the effective temperature, where determined by fitting synthetic spectra to the TiO band heads in the vicinity of the 671 nm Li line. More details of the analysis and discussion can be found in ([17], Li-paper).

Table 1 summarizes some of the main characteristics of the four sample stars which were found to contain Li. For the rest of the sample stars, the spectrum is compatible with zero abundance of Li, although a Li content below the detection threshold of

TABLE 1. Characteristics of sample stars with Li detection.

Name	M_{bol}	P (days)	$\log\varepsilon(\mathrm{Li})$	$\Delta\log\varepsilon(\mathrm{Li})$	Tc?
M45	−4.53	271.02	2.0	0.5	no
M794	−4.78	303.54	1.1	0.4	no
M1147	−5.27	395.63	0.8	0.4	yes
M1347	−5.43	426.60	0.8	0.4	yes

about $\log\varepsilon(\mathrm{Li}) = 0.1$ for cool stars cannot be excluded[2]. As is obvious from Table 1, only Mira variables with long pulsation periods were found to contain Li. The error on the measured Li abundance $\Delta\log\varepsilon(\mathrm{Li})$ has been determined from varying the effective temperature of the COMARCS model atmosphere by $\pm 100\,\mathrm{K}$. Due to the ionization of Li the error is quite large. The last column of Table 1 gives information whether the respective star also shows Tc in its atmosphere.

Since Li is usually destroyed in the deep convective envelopes of evolved stars, it can be safely assumed that we do not see the initial Li content that was present when the stars were formed. Therefore, an external or internal source must have produced Li in the meantime. We identified four possible scenarios for the Li enrichment, and discuss the respective hypotheses in the following.

Internal Hot Bottom Burning. Li is naturally produced in massive AGB stars ($M > 4\,\mathrm{M}_{\odot}$) where Hot Bottom Burning (HBB) is operating. HBB denotes a situation where *no* thin layer of predominantly radiative energy transport is present between the H-burning shell and the convective envelope. Rather, the convective envelope reaches into the top of the H-burning shell with temperatures almost up to $10^8\,\mathrm{K}$. Under these conditions, Li can be produced by the so called Cameron-Fowler or $^7\mathrm{Be}$ transport mechanism [18]. As has been demonstrated by observations of massive AGB stars in the Magellanic Clouds ([19] and references therein), Li enrichment is found mainly in luminous AGB stars falling in the narrow range $-6\overset{\mathrm{m}}{.}0 > M_{\mathrm{bol}} > -7\overset{\mathrm{m}}{.}2$. Even the brightest stars with Li that we find in our sample barely reach this luminosity. Also, regarding determinations of the Bulge age [10, 1], it is highly unlikely that stars massive enough to undergo HBB are present in the outer Galactic Bulge. We thus discard the internal HBB hypothesis.

Accretion of a (Sub-)Stellar Companion. Low-mass objects like Brown Dwarfs or massive planets do not destroy their initial Li content during their slow evolution, since they do not reach high enough temperatures in their interior. The Li in our sample stars may thus have been replenished recently by the accretion of such a low-mass object into the envelopes of the observed AGB stars. This scenario and its effects on the structure and Li content of the accreting AGB star has been studied by [20]. Only a small enhancement of Li in the envelope of the AGB star, which is subsequently reduced again,

[2] The Li abundance is given on the scale $\log\varepsilon(\mathrm{Li}) = \log N(\mathrm{Li})/N(\mathrm{H}) + 12$. The solar photospheric Li abundance on this scale is $\log\varepsilon(\mathrm{Li}) = 1.1$.

is predicted by this study. The Li abundance in our most Li-rich sample star is probably incompatible with the accretion scenario.

Also, a spin-up of the envelope and enhanced mass loss as a result of such an accretion is predicted by [20]. A spin-up (short rotation period) is difficult to measure from the observed spectra because of the kinematics of the pulsating atmospheres of the sample stars, the high line density, and line blending. Apart from these difficulties, the two most Li-rich sample stars have not been detected with the IRAS satellite, contrary to a number of other sample stars with comparable pulsation period and luminosity, pointing to a non-enhanced mass loss. The two lesser Li-rich stars have been detected by IRAS, but the inferred mass loss rate of $\log(\dot{M}/(\mathrm{M}_{\odot}\,\mathrm{yr}^{-1})) \cong -6.6$ is normal for stars of this luminosity and pulsation period. We thus discard the accretion hypothesis, too.

Mass Transfer from a Former Massive Binary Companion. The Li we observe in fact may have been produced by a massive binary companion that underwent HBB and Li production when it was in the AGB phase, and Li-rich matter may have been transferred with a stellar wind to the now observed star. The massive binary companion might be a White Dwarf now. This scenario can hold only if the massive companion produced small amounts of s-process elements. If the mass transfer happened a sufficiently long time ago, we would have missed it by our search for Tc because of the short half life time of this element. However, the two Li-only sample stars do not show any enhancement of the stable s-elements either (particularly studied by means of ZrO band head strengths). S-element enhancement has been observed in the Li-rich HBB stars in the MCs by [19], but not in massive Galactic AGB stars studied by [21]. How the Bulge fits into this picture is not known.

A possible WD companion may be detectable via an excess flux in the U- or B-band. Mean B- and R-band values for the two most Li-rich sample stars were found in the literature, but they did not turn out to be conspicuously blue in $(B-R)$. We cannot exclude the massive binary companion hypothesis from accounting our observations, but regard it as rather unlikely.

Extra Mixing. A further scenario for the production of Li may be some kind of "extra mixing" below the convective envelope which subjects the radiative layer to mass circulation. Such extra mixing mechanisms are not part of standard mixing theories, but are nevertheless required to explain the low ^{12}C/^{13}C ratios (and sometimes also Li) observed in RGB stars after the first dredge-up, and isotopic ratios measured in pre-solar grains of AGB origin. The physical cause for extra mixing is not known, although various processes have been proposed. Different terms for this phenomenon are in use: "extra mixing", "cool bottom processing", "deep mixing", "extended mixing", etc. The community could not yet agree on one of the above terms because they are in differing respects inadequate to describe the processes. *We therefore propose the name "Extended Envelope Mixing" (EEM) for this phenomenon* (AGB newsletter no. 126). Under these conditions, Li can be produced by the Cameron-Fowler mechanism just like in the case of HBB, but for much lower stellar masses.

Regardless of the naming, we find that EEM best describes the observables of the sample stars which could be collected so far. In our sample, which is composed of 15 Mira and 12 SR variables, exclusively Miras are found to show Li. The statistical probability of blindly drawing four Miras out of a sample with this composition is 7.8%. This is a hint that the occurrence of Li is linked to the evolutionary state of the star. Note that in the three scenarios discussed above, Li may be detected in any star, independent of its evolutionary state and pulsation properties. Also, the fact that the occurrence of Tc and Li are not fully correlated is a clear indication that the required conditions for the onset of 3DUP and EEM are independent of each other.

Even in the absence of full theoretical models we can have an estimate of the mass circulation rate required to build up the Li content observed in the most Li-rich sample star. For doing so, we consider the ^{7}Be synthesis in the radiative layer of a model star of initially $1.5\,M_\odot$, with a Population I composition ([22, 23]; [Fe/H] = -0.3). We assume an infinite ^{3}He supply from the H-burning shell. To mimic quick transport of ^{7}Be to cool layers, where it can decay to ^{7}Li by e^- capture, the term for ^{7}Be destruction by p-capture is neglected. Integrating the thus simplified production rate of ^{7}Be over one year and over the production region (the radiative layer between H-burning shell and convective envelope) yields the total amount of Li that can be produced per year: $5.65 \times 10^{-13}\,M_\odot\,yr^{-1}$. The Li mass fraction in the most Li-rich sample star is $X(\mathrm{Li}) = 2 \times 10^{-8}$, which yields a mass circulation rate of $\dot{M} > 3 \times 10^{-5}\,M_\odot\,yr^{-1}$. Compared to values inferred from isotopic ratios of pre-solar grains this is a rather fast circulation rate. Also, in previous calculations with complete models (e.g. [24]) Li production was obtained only adopting very fast circulation rates ($\dot{M} \simeq 10^{-4}\,M_\odot\,yr^{-1}$). This result is in line with our rather rough estimate for the lower $\dot{M}$ limit. Any future full model of EEM must be capable of producing Li at the observed level, and reproduce circulation rates at the order of magnitude estimated above.

4. CONSTRAINING FLUORINE SYNTHESIS

Besides the optical spectra acquired with UVES, we could also obtain CRIRES (CRyogenic InfraRed Echelle Spectrograph) commissioning data of one of the sample stars. CRIRES is mounted to ESO's VLT/UT1 and is described in more detail in [25]. The high-resolution spectra cover ten lines of the HF molecule in the K-band. These lines allow for measuring the fluorine abundance [26]. The object observed was the Tc- and Li-rich star M1347. Contrary to genuine AGB stars in the Galactic Disk studied so far [27], the mass and luminosity of this object are constrained quite well to $M \approx 1.5M_\odot$ and $L \approx 11700L_\odot$.

Five of the observed HF lines had to be omitted from the analysis because they obviously are blended with other lines. From the five remaining lines and a few metal absorption lines we measured [F/H] = $+0.08 \pm 0.21$, and [F/Fe] = 0.25 ± 0.25. These values can be compared to model predictions for the F synthesis in low mass stars, e.g. those of [28], particularly their Figures C7 and C18. For low mass stars, only a very mild enrichment in F on the AGB is predicted. We can confirm this prediction with our measurement.

Also, the F abundance in M1347 is very similar to that in comparable Disk stars studied by [27]. The stars Y Lyn and RS Cnc have been measured by these authors to have [F/H] = +0.15 and [F/H] = +0.13, respectively. If most of the F we observe in

M1347 was indeed not produced by the star itself, it means that the nucleosynthetic source of F was operating at a very similar level in the Galactic Bulge and Disk.

ACKNOWLEDGMENTS

We acknowledge O. Straniero for providing his model track (Fig. 1), and R. H. Tipping for providing his HF line list. SU and TL acknowledge funding by the Austrian Science Fund FWF under the project P18171-N02, and BA acknowledges funding by the FWF under project P19503-N13. MTL has been supported by the Austrian Academy of Sciences (DOC-program). SP and MB acknowledge funding from MURST under contract PRIN2006-022731.

REFERENCES

1. S. Uttenthaler, J. Hron, T. Lebzelter, et al., *A&A* **463**, 251 (2007); Tc-paper
2. L. Girardi, A. Bressan, G. Bertelli, and C. Chiosi, *A&AS* **141**, 371 (2000)
3. T. Lebzelter and J. Hron, *A&A* **411**, 533 (2003)
4. O. Straniero, A. Chieffi, M. Limongi, et al., *ApJ* **478**, 332 (1997)
5. T.J.H. Wesselink, *PhD thesis, University of Nijmegen, The Netherlands*, (1987)
6. L. Plaut, *A&AS* **4**, 75 (1971)
7. B. Paczyński, *Acta Astronomica* **20**, 47 (1970)
8. P.R. Wood and D.M. Zarro, *ApJ* **247**, 247 (1981)
9. J.C. Lattanzio, *ApJ* **311**, 708 (1986)
10. M. Zoccali, A. Renzini, S. Ortolani, et al., *A&A* **399**, 931 (2003)
11. A.I.Boothroyd and I.-J. Sackmann, *ApJ* **328**, 641 (1988)
12. F. Herwig, T. Blöcker, and T. Driebe , *MemSAIt* **71**, 745 (2000)
13. R. Guandalini, M. Busso, S. Ciprini, et al., *A&A* **445**, 1069 (2006)
14. M. Zoccali, A. Lecureur, B. Barbuy, et al., *A&A* **457**, L1 (2006)
15. U.G. Jørgensen, H.R. Johnson, and Å. Nordlund, *ASPC* **26**, 540 (1992)
16. B. Aringer, *Ph.D. thesis, Univ. of Vienna, Austria*, (2000)
17. S. Uttenthaler, T. Lebzelter, S. Palmerini, et al., *A&A* **471**, L41 (2007); Li-paper
18. A.G.W. Cameron and W.A. Fowler, *ApJ* **164**, 111 (1971)
19. V.V. Smith, B. Plez, D.L. Lambert, and D.A. Lubowich, *ApJ* **441**, 735 (1995)
20. L. Siess and M. Livio, *MNRAS* **304**, 925 (1999)
21. D.A. García-Hernández, P. García-Lario, B. Plez, et al., *A&A* **462**, 711 (2007)
22. M. Busso, R. Gallino, and G.J. Wasserburg, *PASA* **20**, 356 (2003)
23. O. Straniero, I. Domínguez, S. Cristallo, and R. Gallino, *PASA* **20**, 389 (2003)
24. A.I. Boothroyd and I.-J. Sackmann, *ApJ* **510**, 232 (1999)
25. H.U. Käufl, P. Ballester, P. Biereichel, et al., *SPIE* **5492**, 1218 (2004)
26. S.Uttenthaler, B. Aringer, T. Lebzelter, et al., *ApJ submitted*, (2007)
27. A. Jorissen, V.V. Smith, and D.L. Lambert, *A&A* **261**, 164 (1992)
28. A.I. Karakas, *PhD thesis, Monash University, Australia*, (2003)

SESSION IX:

CIRCUMSTELLAR ENVELOPES AND MASS LOSS

The Connection Between Mass Loss and Nucleosynthesis

Jacco Th. van Loon

Lennard-Jones Laboratories, Keele University, Staffordshire ST5 5BG, United Kingdom

Abstract. I discuss the relationship between mass loss and nucleosynthesis on the Asymptotic Giant Branch (AGB). Because of thermal pulses and possibly other mixing processes, products of nucleosynthesis can be brought to the surface of AGB stars, increasingly so as the star becomes more luminous, cooler, and unstable against pulsation of its tenuous mantle. As a result, mass loss is at its most extreme when dredge-up is too. As the high rate of mass loss truncates AGB evolution, it determines the enrichment of interstellar space with the AGB nucleosynthesis products. The changing composition of the stellar atmosphere also affects the mass-loss process, most obviously in the formation of dust grains — which play an important rôle in driving the wind of AGB stars.

Keywords: Molecular and Chemical Processes and Interactions, Dust Processes, Stellar Structure, Interiors, Evolution and Nucleosynthesis, Stellar Atmospheres, Circumstellar Shells and Expanding Envelopes, Mass Loss and Stellar Winds, Pulsations, Abundances and Chemical Composition, Giant Stars, Carbon Stars, S Stars and Related Types, Long-period Variables (Miras and Semiregulars)
PACS: 95.30.Ft, 95.30.Wi, 97.10.Cv, 97.10.Ex, 97.10.Fy, 97.10.Me, 97.10.Sj, 97.10.Tk, 97.20.Li, 97.30.Hk, 97.30.Jm

1. INTRODUCTION

Asymptotic Giant Branch (AGB) stars span a range in masses from $< 1\ M_\odot$ to $\sim 8\ M_\odot$, which makes them important diagnostics of the star formation history and important players in galactic chemical enrichment from ages as young as < 100 Myr up to (nearly) a Hubble time. They shed much, often most, of their initial mass in the form of dusty winds before leaving a white dwarf behind.

In this review I will first ask the question why we care about mass loss, with a bias towards those amongst us who are interested in nucleosynthesis. I then briefly discuss what determines the mass-loss rate, and comment on ways in which nucleosynthesis may, or may not, alter the conditions for the mass-loss mechanism.

2. WHY DO WE CARE ABOUT MASS LOSS?

Besides its importance for galactic recycling processes and the build-up of chemical richesse, mass loss greatly affects the star itself. Although mass loss may be seen as an outer boundary condition, it does affect the stellar interior.

As a result of mass loss the convective mantle reduces in mass as well as density. This enhances the effects from 3rd dredge-up, as a given mass of dredged-up product of nucleosynthesis is going to lead to a larger enrichment of the mantle (and photosphere) if it is less diluted by pre-existing material. Massive AGB stars (roughly $M_{\rm init} > 4 M_\odot$)

CP1001, *9th Torino Workshop on Evolution and Nucleosynthesis in AGB Stars and the 2nd Perugia Workshop on Nuclear Astrophysics*, edited by R. Guandalini, S. Palmerini, and M. Busso experience Hot Bottom Burning (HBB), which converts carbon into nitrogen (and oxygen), which is the reason why they do not normally become carbon stars. However, if mass loss depletes the mantle to near extinction, HBB ceases and a final thermal pulse will almost certainly turn the star into a carbon star.

Once the mantle mass is depleted to the extent that it becomes transparent, the star leaves the AGB and gradually dies. The higher the rate of mass loss on the AGB, the earlier this happens and the fewer thermal pulses the AGB stars will have experienced. This affects the photospheric abundances observed near the tip of the AGB, as well as the chemical and isotopic yields from AGB stars. Mass loss will also truncate the growth of the core, and thus affect the mass distribution of white dwarfs.

Finally, as the convective envelope *expands* as a reaction to its diminished density, the surface gravity and temperature drop, which would appear to facilitate further mass loss. Stars for which this becomes important are generally seen to pulsate radially in the fundamental mode, and their pulsation period increases with increasing radius. If the radius increases to the extent that the pulsation period approaches the thermal timescale then the pulsating mantle will continuously adapt its thermal structure as it expands, and this might detach the mantle to form a planetary nebula (PN).

The observed enrichment of the interstellar medium and subsequent generations of stars which is due to AGB stars depends on the AGB mass loss. This is true for elements as well as isotopic ratios; [1], for example, show that the [2] mass-loss formalism leads to a 75% reduction in the yields of elements like silicon and aluminium compared to using the Reimers' Law [3].

The dust production by AGB stars is also affected by the mass-loss history, but in a slightly different way [4]. We shall later see how the dust depends on the detailed chemical composition of the star, but it is already obvious that carbon stars would form very different dust from other, oxygen-rich AGB stars. Depending on how the mass loss varies in time, this may be accompanied by more or less dust mass, and more or less of the dust produced whilst the star is a carbon star (if it ever becomes one). This may all depend on initial metallicity; [4] predict that metal-poor AGB stars produce predominantly carbonaceous dust also if they are massive.

The Initial-Final Mass (IFM) relation depends on how quickly the mass loss truncates the growth of the core on the AGB. Measurements of the IFM relation in galactic open clusters [5] and central stars of PNe in the metal-poor Magellanic Clouds [6] suggest that it varies very little with metallicity. As a consequence, the total amount of mass shed on the AGB is quite well known, e.g., $\Delta M = 4.1 \pm 0.1$ $M_\odot$ for $M_{\rm init} = 5$ $M_\odot$. This is because mass loss ($\dot{M} \gg 10^{-7}$ $M_\odot$ yr^{-1}) outpaces nuclear burning for long enough ($\sim$0.1–1 Myr), causing negligible core growth whilst the mantle is shed. This is true for the solar neighborhood, with $Z_{\rm init} = Z_\odot$, but also in the LMC, $Z_{\rm init} = 0.3$–0.5 $Z_\odot$, and SMC, $Z_{\rm init} = 0.1$–0.2 $Z_\odot$ [7, 8] and other metal-poor dwarf galaxies [9, 10, 11], and even if mass-loss rates were reduced by an order of magnitude[1].

For low-mass AGB stars the preceding first ascent red giant branch (RGB) phase also involves substantial mass loss. This affects the mantle mass with which the star ascends the AGB, and thus its AGB evolution and mass loss. It has been suggested recently

[1] Note that this is *not* the case for red supergiants, which therefore explode as core-collapse supernovae.

[12, 13] that super-solar metal-rich RGB stars may lose mass at a higher rate, truncating their RGB evolution, evading the helium flash and producing low-mass white dwarfs. These stars would never become AGB stars. Note that only a doubling of the mass lost on the RGB is required for this to happen. RGB mass loss seems to be fine-tuned within a narrow range though, as even in very metal-poor globular clusters it is inferred to explain the blue horizontal branches whilst still having to allow for the observed AGB stars and a PN in M 15, $Z = 0.005\ Z_\odot$ (cf. [14]).

3. WHAT DETERMINES THE MASS-LOSS RATE?

Several mechanisms that could, in principle, drive mass loss do not work in AGB stars. The main and efficient driver of hot star winds, radiation pressure in atomic and ionic transitions does not work in AGB stars because they are not hot enough to produce enough opacity and not luminous enough to provide enough pressure. AGB stars do not reach the Eddington luminosity as defined in terms of the radiation pressure in their optically thick mantles, nor do they rotate fast enough to initiate or significantly facilitate mass loss. Radiation pressure in molecular bands is not thought to be sufficient to support a massive outflow either: because the bands are broad and the wind will be slow, the wind never accelerates out of its own shadow, and therefore only utilizes the minor fraction of the luminosity that is emitted within the molecular bands.

What does work, at least in some cases, is the two-stage mechanism of pulsation-induced levitation, and radiation pressure on dust forming in this cool, dense molecular atmosphere. The pulsation is the result of oscillations in the partial ionization zône of hydrogen within the mantle, as the ionization stage determines the opacity in the mantle and thus the balance between gravity and the outward-directed radiation pressure, the so-called κ-mechanism. The pulsation is stronger at lower temperature and higher luminosity. Dust forms preferentially in the dense shocks during part of this pulsation cycle (which has a timescale of 10^{2-3} days), and *contributes* to driving a wind via radiation pressure and collisional coupling with the gas [15, 16].

The dust-driven wind regimes in the Hertzsprung-Russell diagram are set by the Eddington luminosity above which the radiation pressure is sufficient to support such a dust-driven wind [4]. This depends on the dust optical depth compared to the total mass in the wind, i.e. the opacity of the dust species, dust:gas mass ratio and total mass-loss rate. Dust only forms below a certain temperature and provided the density is high enough. On the RGB, a dust-driven wind is only possible if carbonaceous dust is formed as it has a relatively high opacity, but no carbon stars are found on the RGB (although carbon stars do exist in old globular clusters, [14]). Silicate dust can be used to drive a wind only for AGB stars significantly more luminous than the tip of the RGB [4]. This also shows that the exact evolution of a star in terms of its luminosity and especially temperature is critical for its ability to form dust and use it to drive a wind.

What happens if the pulsation and/or dust formation fail? Because these tend to be warmer, more compact stars, which are known to be chromospherically active, it is believed that chromospheric processes are likely at the origin of the driving mechanism. Instead of radiation, Alfvén (electromagnetic) or acoustic (pressure) waves are invoked. Mass-loss rates are very difficult to measure for these stars because of the sensitivity to the excitation/ionization balance and deviations from local thermodynamic equilibrium. A semi-empirical relation has been suggested by [17], which would lead to higher mass-loss rates of metal-poor stars, because they are warmer.

Observations of RGB stars do suggest that non-dusty winds may be at least as efficient as dusty winds. Modelling of optical absorption line profiles in globular cluster red giants by [18] yields mass-loss rates in excess of Reimers' Law and in agreement with emission-wing diagnostics [19] and dust-derived mass-loss rates where available [20]. Indeed, [21] show that the combined energy involved in radiation and wind is similar for chromospherically and dust-driven winds, operating in different temperature regimes but with a smooth transition zône (cf. [22]).

The luminosity and temperature are obvious measures of the radiation pressure and ability to form dust. Because evolutionary tracks do not cross, the mass is not needed as an independent parameter, $M = M(L,T)$. Both observations and hydrodynamical computations support a parametrization of the mass-loss rate of the form $\dot{M}(L,T)$, with a steep dependence on temperature but roughly proportional to luminosity [7, 23, 16] — where it must be realized that the temperature varies less than a factor two.

The mass-loss rate has also been parameterized as a function of the pulsation period, P [2]. This works reasonably well at least for the sample for which it was derived, partly because $P = P(L,T)$ as it is a measure of the stellar radius, R, and $L = 4\pi\sigma R^2T^4$. The period sets the timescale for dust to form and grow, the *amplitude* of pulsation may be at least as relevant [22].

[21] suggest that irrespective of the driving mechanism, the mass-loss rate may be parameterized as a function of the surface gravity, g. Again, it is not obvious whether the gravity, or escape velocity, is the parameter of fundamental importance, as $g = g(L,T)$ because both the radius and mass are functions of luminosity and temperature.

Complications arise if the AGB star is in a binary system; the mass-loss rate may be enhanced equatorially, a circumbinary disc may trap ancient material, a fast(er) bipolar wind may develop, and mass transfer may enrich the surface of the companion star. In novae, mass loss from the donor indirectly contributes to explosive nucleosynthesis, and a merger of the two stars may induce an eruption.

4. HOW DOES NUCLEOSYNTHESIS ALTER THE CONDITIONS FOR THE MASS-LOSS MECHANISM?

Nucleosynthesis and thermal pulses affect the luminosity and temperature and thus the mass-loss rate. The evolutionary track in a Hertzsprung-Russell diagram depends on a star's initial metallicity, Z, helium abundance, and a changing carbon:oxygen ratio. During a thermal pulse the luminosity and temperature change in a complex manner, and depending on the $\dot{M}(L,T)$ formalism the star is normally expected to experience a brief episode of more intense mass loss. [2] predict less mass loss during the first thermal pulses, but a rapid truncation of AGB evolution, compared to Reimers' Law which allows more thermal pulses and a more balanced mass loss over the AGB evolution. If a final thermal pulse occurs in a massive AGB star after HBB ceased [24, 25], a massive carbon star is formed and fresh dust will be carbonaceous. It is unclear whether this would affect the mass-loss rate. A final thermal pulse occurring whilst on the post-AGB

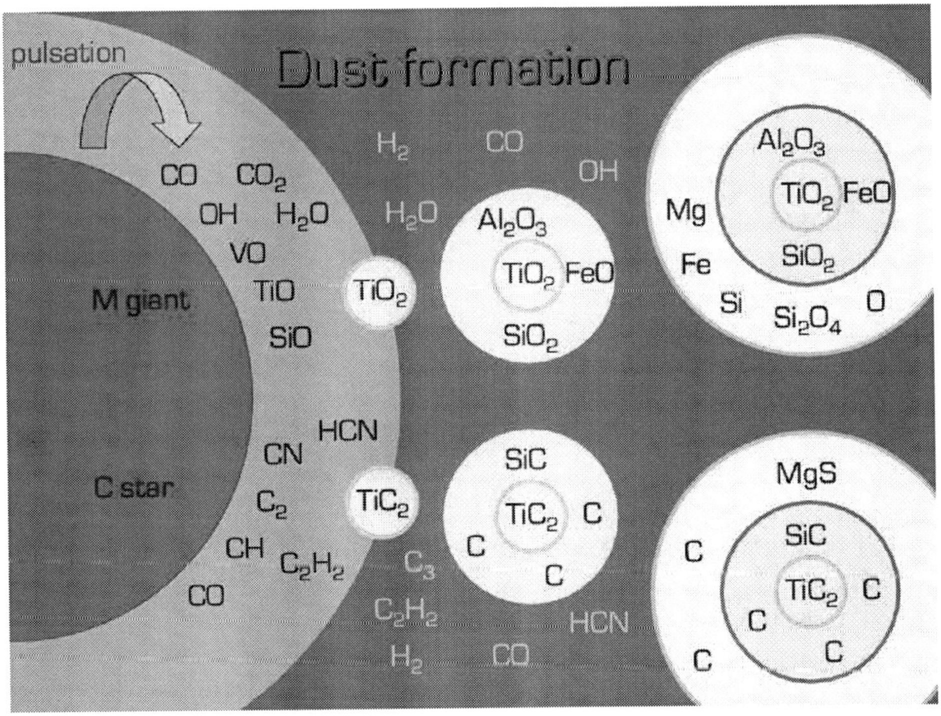

FIGURE 1. Schematic overview of the dust formation process in oxygen- (top) and carbon-dominated (bottom) circumstellar environments. Seeds involving trace chemical elements such as titanium act as condensation nuclei. First coated with highly refractory molecular species (e.g., alumina or silicon carbide), these in turn are coated with more volatile species (e.g., silicates or magnesium sulfide). The fraction of encapsulated "proper" metals such as iron may vary, affecting the transparency of the grains. Little is known about the initial nucleation and its relation to the composition of the molecular atmosphere.

track may sent the star back onto the AGB, rejuvenating its mass loss. Clearly, thermal pulses could have a great impact if they occur in a symbiotic system.

Dust formation is a complex process too. It takes place in an either oxygen-dominated or carbon-dominated environment depending what element is left after the other is exhausted in the formation of the carbon-monoxide molecule (CO) in the cool, dense lower layers of the stellar atmosphere (Fig. 1). Molecules do not directly bind together to form solid particles, but need a nucleation seed. This is confirmed by meteoritic evidence (cf. Hoppe, these proceedings) to be based on titanium, and possibly zirconium and silicon. Neither of these are produced inside AGB stars, so their abundance is set by the initial metallicity, $Z_{\rm init}$. The grain growth follows a sequence of decreasing condensation temperature, with aluminium-oxides and silicon-carbide (SiC) condensing before silicates and amorphous carbon (in oxygen- and carbon-rich environments, respectively). Apart from composition and size, the grain shape and degree of crystalline structure further affect the optical properties and thus the ability of the grains to drive a wind.

None of the dust species involve hydrogen or helium, which leaves it up to a minor

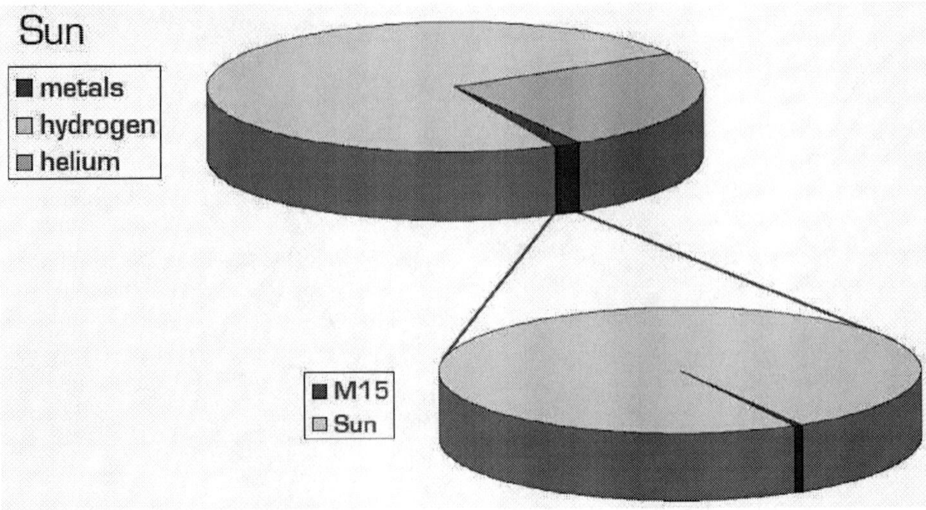

FIGURE 2. Dust is formed out of metals, which constitute only a minor fraction of any star. In metal-poor globular clusters, this fraction is truly minute yet dust is observed in at least one such cluster, M 15.

mass fraction of the wind to provide the agents for the interception of the momentum from the radiation field. Even in the Sun it is only ~1% (not all metals are used for dust formation, and CO takes away a large fraction of it too), but in very metal-poor environments such as in M 15, with $Z = 0.005\ Z_{\odot}$, dust is seen [26]. To appreciate how little mass in condensable material this is, just look at Fig. 2!

Oxygen-rich AGB stars pose a challenge: [27] reminds us of earlier findings that silicates are transparent; iron seems to be required to provide sufficient opacity where the stellar radiation peaks. [28] propose an alternative solution, if some carbonaceous dust could still be formed. Although most of the dust mass would be oxygenous, most of the opacity in the near-infrared would be due to carbonaceous dust.

Oxygenous minerals present a rich diversity, due to various admixtures of aluminium, silicon, iron, magnesium, even calcium. Mixing of the products of CNO processing could affect the amount of oxygen available, but it is generally the other elements it binds with, that are the critical component (Fig. 3). Changes in the aluminium abundance could affect the initial coating of the nucleation seeds, but it is unclear whether the later coating by silicates is affected. Hence, titanium is critical as seed, silicon for grain growth, and iron for opacity. All these scale directly with initial metallicity. This is consistent with the simple and predicted behaviour of dust-driven winds of massive AGB stars: $\dot{M}$ is insensitive to $Z_{\rm init}$, but the dust:gas ratio $\psi \propto Z_{\rm init}$ [29, 8, 30].

Metal-poor carbon stars exhibit strong molecular absorption [31, 32, 33, 34], but there is no convincing evidence for enhanced dust formation [35, 36, 37, 38]. These works show that, in sufficiently cold envelopes, magnesium-sulfide (MgS) forms in a similar ratio to the amorphous carbon which constitutes the bulk of carbonaceous dust. This suggests that if MgS is depleted in metal-poor winds because of lower

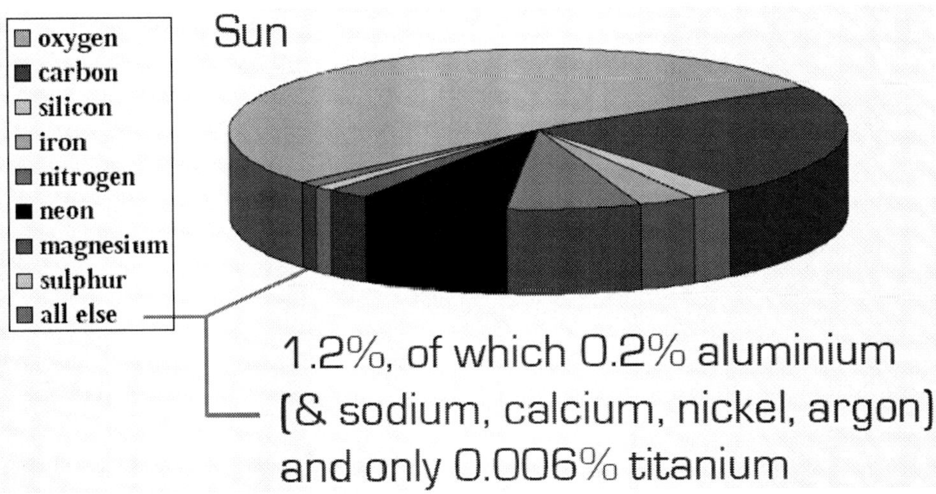

FIGURE 3. Approximate fractions of the metals in the Sun, of various elements. Oxygen and/or carbon, which form the basis of most of the circumstellar dust produced in AGB winds, are always abundant. Important constituents of certain dust species, iron, silicon, magnesium and aluminium are much less abundant and thus limit grain growth. As a nucleation seed, titanium is a key element but extremely rare.

abundances of magnesium and sulphur, amorphous carbon must be depleted too. SiC is more complicated to interpret as it may be coated by amorphous carbon.

In carbon star winds too, the nucleation seeds depend on titanium and similar elements. The strong molecular bands especially of the slightly larger molecules such as acetylene and C_3 suggest that the density of molecules may not be larger, but they may be found throughout a larger part of the wind [34]. The grain growth may thus not be any different. Indeed, the observed spectral energy distributions are consistent with $\psi \propto Z_{\rm init}$ and $\dot{M}$ insensitive to $Z_{\rm init}$ [29, 8], and this is also supported by considerations of the molecular mass-loss rate [33], the availability of silicon in gas phase and SiC [38], and the lower dust production seen in magellanic PNe [39].

S stars, which have a carbon:oxygen ratio approaching unity, are a relatively unimportant contributor to chemical enrichment and unimportant phase for the star itself, because the mass-loss rate is not particularly high and the phase lasts very few thermal pulse cycles. They may teach us something about dust formation and growth, though, when oxygen or carbon have become critical elements in the chain of events.

ACKNOWLEDGMENTS

I would like to thank Maurizio Busso for inviting me to the workshop to talk about AGB mass loss, all the participants for a very interesting and joyous time together, and Joana for putting up admirably with my busy travel diary. This talk was presented on 26 October 2007, which was also the "Wear it Pink" awareness day of the british Breast Cancer Campaign, to which this talk is dedicated. Oh yes, and the food was very good!

REFERENCES

1. A.I. Karakas, M.A. Lugaro, M. Wiescher, J. Görres, and C. Ugalde, *ApJ* **643**, 471 (2006)
2. E. Vassiliadis and P.R. Wood, *ApJ* **413**, 641 (1993)
3. D. Reimers, *Societé Royale des Sciences de Liège Memoires* **8**, 369 (1975)
4. A.S. Ferrarotti and H.-P. Gail, *A&A* **447**, 553 (2006)
5. K.A. Williams, *in:* 15th *European Workshop on White Dwarfs, eds. R. Napiwotzki & M. Burleigh, ASPC* **372**, 85 (2007)
6. E. Villaver, L. Stanghellini, and R.A. Shaw, *ApJ* **656**, 831 (2007)
7. J.Th. van Loon, M.A.T. Groenewegen, A. de Koter, N.R. Trams, L.B.F.M. Waters, A.A. Zijlstra, P.A. Whitelock, and C. Loup, *A&A* **351**, 559 (1999)
8. J.Th. van Loon, *in: Stellar Evolution at Low Metallicity, eds. H. J. G. L. M. Lamers, N. Langer, T. Nugis, and K. Annuk, ASPC* **353**, 211 (2006)
9. D.C. Jackson, E.D. Skillman, R.D. Gehrz, E. Polomski, and C.E. Woodward, *ApJ* **656**, 818 (2007)
10. D.C. Jackson, E.D. Skillman, R.D. Gehrz, E. Polomski, and C.E. Woodward, *ApJ* **667**, 891 (2007)
11. E. Lagadec, A.A. Zijlstra, M. Matsuura, J.W. Menzies, J.Th. van Loon, and P.A. Whitelock, *MNRAS, in press*, (2007)
12. B.M.S. Hansen, *ApJ* **635**, 522 (2005)
13. J.S. Kalirai, P. Bergeron, B.M.S. Hansen, D.D. Kelson, D.B. Reitzel, R.M. Rich, and H.B. Richer, *ApJ* **671**, 748 (2007)
14. J.Th. van Loon, F. van Leeuwen, B. Smalley, A.W. Smith, N.A. Lyons, I. McDonald, and M.L. Boyer, *MNRAS* **382**, 1353 (2007)
15. G.H. Bowen and L.A. Willson, *ApJ* **375**, L53 (1991)
16. A. Wachter, K.-P. Schröder, J.M. Winters, T.U. Arndt, and E. Sedlmayr, *A&A* **384**, 452 (2002)
17. K.-P. Schröder and M. Cuntz, *ApJ* **630**, L73 (2005)
18. I. McDonald and J.Th. van Loon, *A&A* **476**, 1261 (2007)
19. J.G. Cohen, *ApJ* **203**, L127 (1976)
20. L. Origlia, F.R. Ferraro, F. Fusi Pecci, and R.T. Rood, *ApJ* **571**, 458 (2002)
21. P.G. Judge and R.E. Stencel, *ApJ* **371**, 357 (1991)
22. J.Th. van Loon, *in: XXI Century Challenges for Stellar Evolution, eds. S. Cassisi & M. Salaris, MemSAIt* **79/2** (2008)
23. J.Th. van Loon, M.-R.L. Cioni, A.A. Zijlstra, and C. Loup, *A&A* **438**, 273 (2005)
24. J.Th. van Loon, et al., *A&A* **329**, 169 (1998)
25. C.A. Frost, R.C. Cannon, J.C. Lattanzio, P.R. Wood, and M. Forestini, *A&A* **332**, L17 (1998)
26. M.L. Boyer, C.E. Woodward, J.Th. van Loon, K.D. Gordon, A. Evans, R.D. Gehrz, L.A. Helton, and E.F. Polomski, *AJ* **132**, 1415 (2006)
27. P. Woitke, *A&A* **460**, L9 (2006)
28. S. Höfner and A. Andersen, *A&A* **465**, L39 (2007)
29. J.Th. van Loon, *A&A* **354**, 125 (2000)
30. J.R. Marshall, J.Th. van Loon, M. Matsuura, P.R. Wood, A.A. Zijlstra, and P.A. Whitelock, *MNRAS* **355**, 1348 (2004)
31. J.Th. van Loon, A.A. Zijlstra, and M.A.T. Groenewegen, *A&A* **346**, 805 (1999)
32. M. Matsuura, et al., *A&A* **434**, 691 (2005)
33. M. Matsuura, et al., *MNRAS* **371**, 415 (2006)
34. J.Th. van Loon, J.R. Marshall, M. Cohen, M. Matsuura, P.R. Wood, I. Yamamura, and A.A. Zijlstra, *A&A* **447**, 971 (2006)
35. G.C. Sloan, K.E. Kraemer, M. Matsuura, P.R. Wood, S.D. Price, and M. P. Egan, *ApJ* **645**, 1118 (2006)
36. A.A. Zijlstra, et al., *MNRAS* **370**, 1961 (2006)
37. E. Lagadec, et al., *MNRAS* **376**, 1270 (2007)
38. M. Matsuura, et al., *MNRAS* **382**, 1889 (2007)
39. L. Stanghellini, P. García-Lario, D.A. García-Hernández, J.V. Perea-Calderón, J.E. Davies, A. Manchado, E. Villaver, and R.A. Shaw, *ApJ, in press*, (2007)

Spitzer/IRAC Observations of AGB Stars

Massimo Marengo*, Megan Reiter† and Giovanni G. Fazio*

*Harvard-Smithsonian Center for Astrophysics, Cambridge, MA
†University of California, Berkeley, CA

Abstract. We present here the first observation of galactic AGB stars with the InfraRed Array Camera (IRAC) onboard the *Spitzer* Space Telescope. Our sample consists of 48 AGB stars of different chemical signature, mass loss rate and variability class. For each star we have measured IRAC photometry and colors. Preliminary results shows that IRAC colors are sensitive to spectroscopic features associated to molecules and dust in the AGB wind. Period is only loosely correlated to the brightness of the stars in the IRAC bands. We do find, however, a tight period-color relation for sources classified as "semiregular" variables. This may be interpreted as the lack of warm dust in the wind of the sources in this class, as opposed to Mira variables that show higher infrared excess in all IRAC bands.

Keywords: Infrared: Stars; Stars: AGB and Post-AGB, Carbon, Mass Loss, Variables: Other
PACS: 95.85.Hp; 97.10.Me; 97.30.Hk; 97.30.Jm

1. INTRODUCTION

The *Spitzer* Space Telescope [1], and in particular its InfraRed Array Camera (IRAC, [2]), is an ideal facility to study the distribution of AGB stars in our own and other galaxies, due to its proficiency in surveying vast areas of the sky in a short time and its ability to detect sources with infrared excess. The IRAC colors of AGB stars, however, are not known very well. The Infrared Space Observatory (ISO) Short Wavelength Spectrometer (SWS, [3]) observed a sample of AGB stars, finding numerous molecular absorption features in the spectral region covered by the IRAC photometric system [4]. Just to mention a few, H_2O, SiO, CO_2, CO and silicate dust features have been detected in stars with atmospheric C/O ratio < 1, while C_2H_2, HCN, CS, C_3 and carbonaceous dust (SiC and amorphous carbon) have been found in carbon stars, having C/O > 1. The presence and strength of these spectral features depends on the chemistry of the stellar atmosphere [5, 6] and the mass loss rate, and can change with time [7] because AGB stars are Long Period Variables generically classified as Mira, semiregular or irregular pulsators.

As a result of dust and molecular features, the IRAC colors of AGB stars can be quite different from the "reddened photospheres" that one would expect for mass-losing giants of late spectral type. Simulations we made, based on the available ISO SWS spectra [8], show that AGB stars with different chemical signature can in some case be differentiated on the basis of the IRAC colors alone. IRAC colors also appear to be sensitive to the mass loss rate of the sources, which has also been noted by [9]. Given the importance of reliably identifying AGB stars in already available IRAC data from galactic and extragalactic surveys, the need to study in detail the color variations of AGB stars is clear.

CP1001, *9th Torino Workshop on Evolution and Nucleosynthesis in AGB Stars and the 2nd Perugia Workshop on Nuclear Astrophysics*, edited by R. Guandalini, S. Palmerini, and M. Busso

With this in mind we proposed the observations of a representative sample of well characterized AGB stars as part of the Cycle-3 IRAC Guaranteed Time Observations. The goal of the program was to obtain high-precision IRAC absolute magnitudes and colors. The sample was chosen to explore the variations of IRAC photometry with time, variability class, chemical type and mass loss rate. This will enable longer term goals such as identifying AGB stars in other IRAC datasets, allowing more detailed studies of the AGB phase as part of galactic population synthesis models and studies of the galactic chemical evolution of the diffuse matter in the interstellar medium. Reliable prescriptions for discriminating AGB stars from other classes of red sources will also help other projects whose samples are contaminated by background AGB sources. Given that a fraction of our targets ($\sim 60\%$) have been observed with ISO, and their SWS spectra are publicly available, our observations will also provide an independent dataset for the cross calibration between IRAC and the ISO SWS. This is particularly important in the case of "red" point sources as mass losing AGB stars, since IRAC has been absolutely calibrated by using a sample of "bluer" stars of spectral type A [10]. The results of this program will be used to improve the absolute calibration of IRAC for red stellar sources.

The program will be completed in the first months of 2008, and we are presenting here the first results derived from $\sim 74\%$ of the total scheduled observations. Once the whole sample of stars will be observed, the catalog will be made available to the community, with the goal of aiding the identification of AGB stars in already available and future IRAC survey.

2. TARGET SELECTION

Our target list was selected from a number of AGB star catalogs available in the literature ([11, 12, 13, 14, 15] and references therein) with the intention of representing all main categories of AGB stars. An important requirement for the selection was the availability of reliable distance estimates, to allow for the measurement of absolute magnitudes. The distances for our target stars were derived either from astrometric [16] or interferometric [17, 18] observations or from models of radio emission and bolometric luminosity [14]. The distance of our selected targets varies from 0.17 to 4.1 kpc.

The sample is divided in 4 categories: 22 O-rich AGB stars (MIII spectral type), 7 intrinsic S stars, 19 carbon stars and 4 M-type cool supergiants. The supergiants have been added as a comparison sample of red mass-losing stars with O-rich chemistry outside the AGB. In each category we have a similar number of Mira, semiregular and irregular variables, with period ranging from 50 to 822 days. The mass loss rates of our target stars (estimated from radiative transfer modeling of their circumstellar emission, see [11] and references therein) range from 10^{-8} to 10^{-4} $M_{\odot}$/yr in each category.

Given the variability of the sources, we have requested two epochs (six months apart, as constrained by the *Spitzer* visibility windows) for each target, in order to check for variations in the IRAC photometry on timescales of several months. As of November 2007, 48 sources have been observed in at least one epoch (with 4 carbon stars missing), with 29 targets (56% of the total) observed in both epochs.

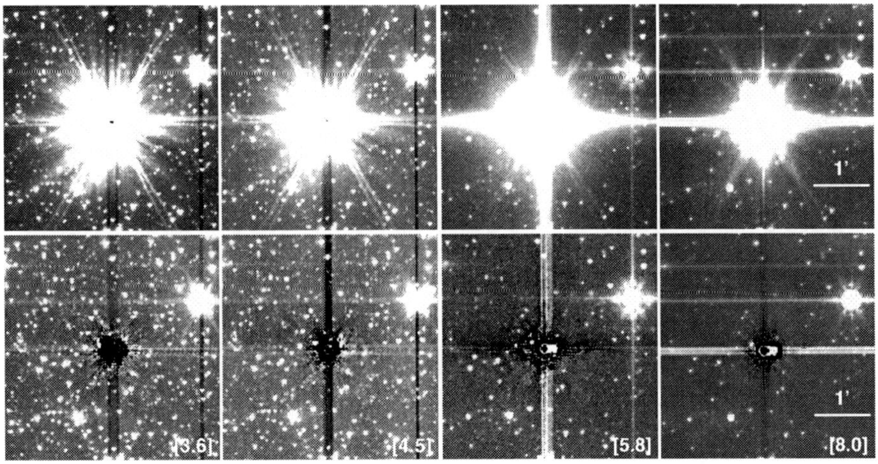

FIGURE 1. Images of the M-type AGB star CZ Ser in each of the IRAC bands before (top panels) and after (bottom panels) PSF subtraction.

3. DATA PROCESSING

The processing of this dataset presented peculiar challenges, due to the brightness of the sources. Even though we selected our targets to have a K magnitude larger than 3, to avoid excessive saturation of the IRAC images, all our stars still saturate the IRAC arrays. The choice of AGB stars in this brightness range was motivated by the necessity of having nearby targets, the ones for which a reliable estimate of the distance is available. Each source was observed with 5 dithered 2 sec full frame exposures, to limit excessive saturation while avoiding sub-array exposure that, while in many cases still saturated, have less field of view for PSF fitting photometry.

Starting from the Basically Calibrated Data (BCD) provided by the *Spitzer* Science Center, we have produced a mosaic image for each source, in each band and for each one of the two epochs, separately, using our own post-BCD software *IRACproc* [19].

Due to saturation, aperture photometry was not feasible, so we used a Point Spread Function (PSF) fitting technique developed specifically to treat IRAC images with heavy saturation. As described in [8], we have generated a high dynamic range image of the IRAC PSF in each band, by combining the individual images of a set of stars with different brightness (Sirius, Vega, Fomalhaut, ε Eridani, ε Indi and the IRAC calibrator BD+681022). This PSF is rescaled to the actual image of Vega, providing an absolute photometric reference. By fitting the unsaturated parts (diffraction spikes and PSF "tails") of each target star image, we have measured the flux ratio between the stars in our sample and Vega, which we have then converted into Vega magnitudes for each source. The typical uncertainty of this procedure is of the order of 1–3%, better than the typical absolute aperture photometry of unsaturated stars with IRAC.

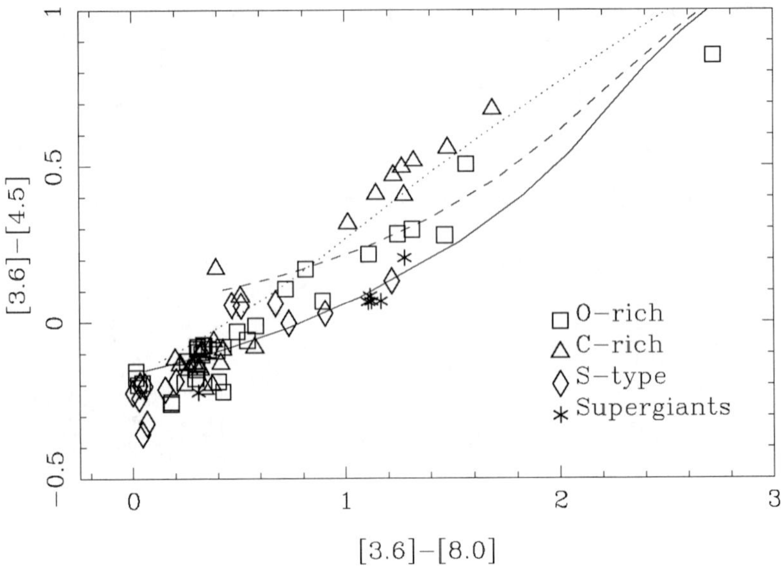

FIGURE 2. IRAC [3.6]-[8.0] vs. [3.6]-[4.5] color diagram of all the available observations to-date. The tracks are models from [9]. Dotted line is for carbon stars with carbonaceous circumstellar dust of increasing optical depth. Solid and dashed line are for silicate envelopes around stars of M0 III and M10 III spectral type respectively.

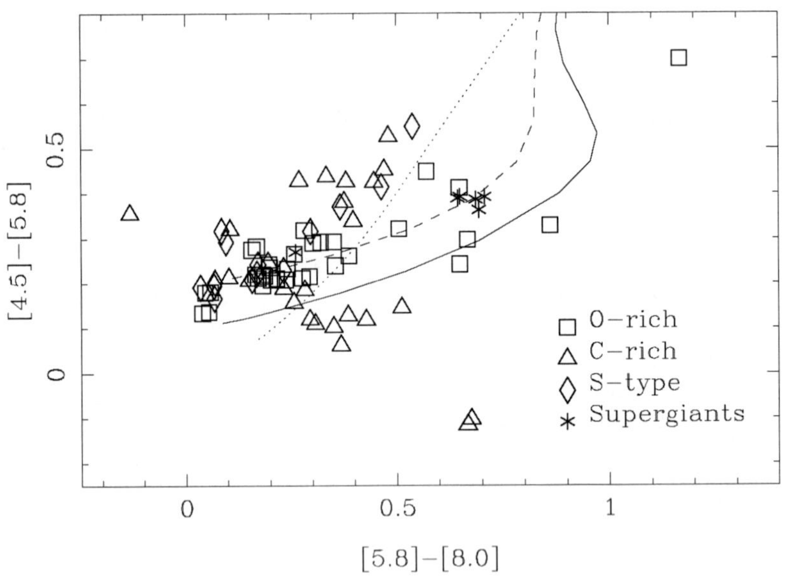

FIGURE 3. Same as Figure 2, but with colors including the IRAC 5.8 μm band, causing dispersion of "naked photospheres" carbon stars from the sequence of carbon stars with higher mass loss.

4. IRAC COLORS OF AGB STARS

Figure 2 and 3 show IRAC colors for all the available observations as of November 2007. The sources for which both epochs are already available have been represented by two separate points in the diagrams.

As shown in [8] using synthetic IRAC colors derived from ISO SWS spectra, and by [9] using radiative transfer modeling of dusty circumstellar envelopes, the IRAC colors appear to be sensitive to the optical thickness of the dusty shells surrounding the sources. The [3.6]-[8.0] color, in particular, provides the largest spread in function of the optical thickness of the envelope, due to the thermal radiation from the circumstellar dust, source of the infrared excess which "flattens" the spectrum in the IRAC wavelength range. The [3.6]-[4.5] color shows a similar trend, and the combination of these two colors describes a sequence of increasing excess closely following the model tracks derived by [9]. The diagrams show a separation of carbon stars from stars with silicate envelopes (M and S AGB stars, and red supergiants) for sources with [3.6]-[8.0] color in excess to 1 mag. Inspection of ISO SWS spectra of individual AGB stars with different chemistry shows that this is caused by two simultaneous factors: sources with higher excess present a growing 10 μm silicate feature that is partially contributing to the 8.0 μm IRAC band, while at the same time the 3.6 μm flux is depressed by deeper H_2O features. Carbon stars, lacking any prominent dust feature in the IRAC wavelength range, for the same amount of [3.6]-[4.5] color have a smaller [3.6]-[8.0] excess, causing them to be aligned on a sequence above the sources with O-rich chemistry. The infrared excess, for sources with [3.6]-[8.0] $>$ 1, correlates well with the mass loss rate of the individual stars, albeit with a large scatter which can be attributed to the large uncertainties in the mass loss rate estimates.

Figure 3 shows a trend similar to Figure 2, with a twist: carbon stars with very small mass loss rate align on a separate sequence with bluer [4.5]-[5.8] and redder [5.8]-[8.0] colors. This was already shown in [8], and attributed to a broad absorption feature depressing the 5.8 μm flux. Based on [4] we give a tentative identification of the feature as being C_3 developing in the atmospheres of these dust poor carbon stars. The feature disappears in stars with thicker circumstellar envelopes, either because it is filled by the continuum dust emission, or because the molecule responsible is depleted. As noted in [8], this feature appears to be transient, as some sources observed with ISO in multiple epochs do not show it in all spectra. This may be an indication that the feature variability can be related to changes in the abundances of the molecule in the stellar atmosphere as the star pulsate, rather than changes in the dust content of the circumstellar envelope (which is unlikely to show large scale variations on the short time-scales of the ISO repeated observations).

5. IRAC COLORS AND VARIABILITY

Figure 4 shows the 8.0 μm absolute magnitude of all sources plotted against their Mira (circles) or semiregular (triangles) period. Again, sources already observed in both epochs have separate points, one above the other (same period, different brightness). Note that all semiregular variables (characterized by small amplitude) show very little

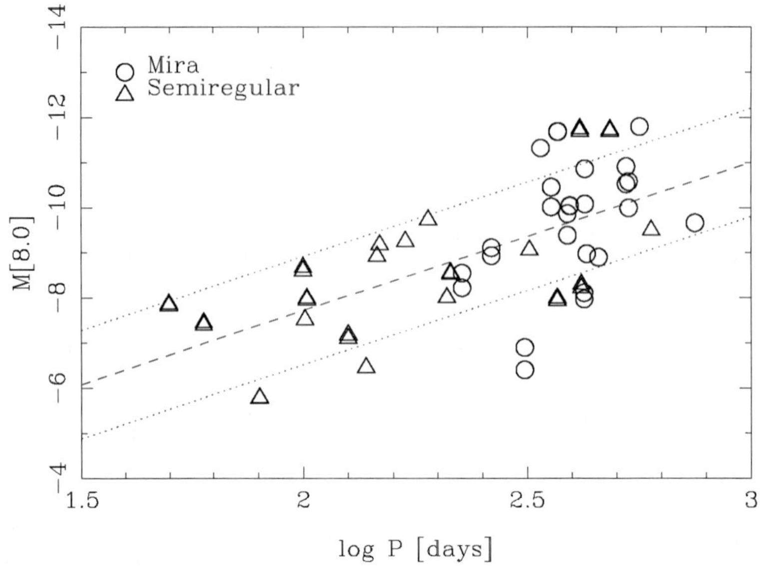

FIGURE 4. Period - 8.0 μm absolute magnitude relation for all the observations available to-date. Circles are sources classified as Mira variables in the GCVS [21]. Triangles are sources classified as semiregular. The solid line is the best linear fit for all data, with *RMS* dispersion of $\sim$ 1.2 mag indicated by the two dotted lines

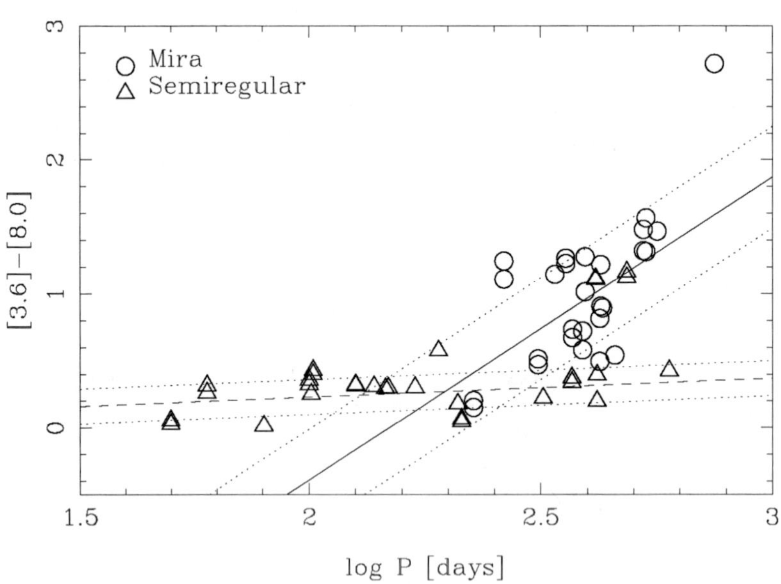

FIGURE 5. Period - [3.6]-[8.0] color relation for all the observations available to-date. Symbols are as in Figure 4. The dashed line is the best fit for the color - period relation of semiregular stars, while the best fit for Miras is indicated by a solid line. The dotted lines show the dispersion in the best fit relations.

change between the two IRAC epochs. Only Miras (larger amplitude and longer periods) show variations that can be as high as ~ 1 mag. The plot shows that there is trend between the period and the luminosity at 8.0 μm (dashed line), with longer period sources (generally Mira variables) having higher brightness. The dispersion of ~ 1.2 mag may be a consequence of the varying amount of mass loss of the different stars, not necessarily correlated to their period, responsible for different degrees of infrared excess. Diagrams made with other IRAC bands show a similar trend.

The IRAC period - color relation of all the observed sources is instead shown in Figure 5. As color we have chosen the [3.6]-[8.0] combination, which is the most sensitive to dust thermal emission from the circumstellar envelope. Again, Mira variables are plotted with a different symbol than semiregulars. We have derived separate linear fits for the period - color relation of the two classes. Mira and semiregulars show a very different trend: the former have a very steep relation (solid line) with a large dispersion (~ 0.38 mag), while the latter follow a very tight ($RMS \sim 0.13$ mag) trend (dashed line).

6. DISCUSSION AND CONCLUSIONS

The IRAC photometry of the observed sources is in general agreement with models of AGB photospheres surrounded by dusty circumstellar envelopes of different optical depths, as modeled by [9]. As expected, the main trend shown by IRAC colors is an increasing infrared excess for larger optical thickness of the dust shell, and separation between sources with silicate or carbonaceous dust. With enough photometric accuracy, however, the data show deviations from the models that can be attributed to molecular spectral features in the stellar atmosphere or wind. These features show time variability when observed at different phases of the star.

The IRAC absolute magnitudes and colors are also correlated with the Mira or semiregular period of the stars. In general longer period variables are brighter, but there is a rather large spread in this relation, probably due to the variations in mass loss rate, and thus infrared excess, generated by the dusty envelope.

While the overall mass loss rate distribution of Mira and semiregular variables in our sample is similar, the period - color relation shows different trends. This may be interpreted as lack of hot dust around semiregular variables which translates in a smaller infrared excess. In [22] a similar result was suggested by the shape of the silicate feature in a large sample of O-rich AGB stars, and was interpreted as a sign of time-variability in the silicate dust production in semiregulars. The new IRAC data supports this interpretation and extends it to carbon stars (making more unlikely that this effect is due to differences in the dust composition between semiregulars and Miras).

The GCVS [21] Mira/semiregular classification is somewhat outdated, and a more accurate description of the AGB variability should be done in terms of pulsational modes and amplitudes. In these terms we can rephrase the above conclusions by noticing that semiregular variables are the ones with smaller amplitudes, while Miras have in general longer periods and amplitudes. This may imply that the stars with larger amplitudes may have more efficient dust production rates, resulting in larger quantities of hot dust in proximity of the stellar photosphere, which are responsible for the observed larger IRAC infrared excess.

The observing program is scheduled for completion for the first months of 2008. Once the data for all stars, both epochs, will be available, the complete dataset will be published as template photometry of galactic AGB stars, for general use of the community.

ACKNOWLEDGMENTS

This work is based on observations made with the *Spitzer* Space telescope, which is operated by the Jet Propulsion Laboratory, California Institute of Technology, under a contract with NASA. Support for this work was provided by NASA through an award issued by JPL Caltech. This work is also supported by the National Science Foundation's Research Experience for Undergraduates program.

REFERENCES

1. M.W. Werner, et al., *ApJS* **154**, 1(2004)
2. G.G. Fazio, et al., , *ApJS* **154**, 10(2004)
3. E.A. Valentjin, et al., *A&A* **315**, 60(1996)
4. L.B.F.M. Waters, et al., in proc. *"The Universe as seen by ISO"*, (ESA SP-427), p. 219 (1998)
5. G.C. Sloan and S.D. Price, *ApJS* **119**, 141 (1998)
6. G.C. Sloan, I.R. Little-Marenin, and S.D. Price, *AJ* **115**, 809 (1998)
7. T. Onaka, T. de Jong, and I. Yamamura, *A&A* **388**, 573 (2002)
8. M. Marengo, et al., in proc. *"Why Galaxies Care about AGB Stars"*, Vienna 7-11 August 2006, arXiv:astro-ph/0611346 (2006)
9. M.A.T. Groenewegen, *A&A* **448**, 181 (2006)
10. W.T. Reach, *PASP* **117**, 978 (2005)
11. R. Guandalini, et al., *A&A* **445**, 1069 (2006)
12. A.M. Heras and S. Hony, *A&A* **439**, 171 (2005)
13. S.J. Adelman and D.W. Maher, *IBVS* **4591**, 1 (1998)
14. C. Loup, et al., *A&AS* **99**, 291 (1993)
15. F. Kerschbaum and J. Hron, *A&A* **308**, 486 (1996)
16. M.A.C. Perriman, et al., *A&A* **323**, L49 (1997)
17. Millan-Gabet, et al., *ApJ* **620**, 961 (2005)
18. G.T. van Belle, R.R. Thompson, and M.J. Creech-Eakman, *AJ* **124**, 1706 (2002)
19. M. Schuster, M. Marengo, and B. Patten, SPIE Meeting, Orlando, #6720-65 (2006)
20. M. Marengo, S.T. Megeath, G.G. Fazio, K.R. Stapelfeldt, M.W. Werner, and D.E. Backman, *ApJ* **647**, 1437 (2006)
21. P.N. Kholopov, et al., *General Catalog of Variable Stars*, 4th edn. 1985 Nauka, Moscow (1988)
22. M. Marengo, Z. Ivezic, and G.R. Knapp, *MNRAS* **324**, 1117 (2001)

MS, S and C Stars in the Infrared. Luminosities and Mass Loss Rates.

R. Guandalini

Department of Physics, University of Perugia, Via A. Pascoli, 06123 Perugia, Italy

Abstract. In this note I present an outline of infrared (IR) photometric AGB properties, based on two samples of Galactic Long Period Variables (C- and S-type respectively). I show the various selection criteria used during the choice of the sources and describe the motivations of observing them at near- and mid-IR wavelengths. I discuss the problems encountered in estimating their luminosity and distance and motivate the methods I choose for this purpose. Properties of the luminosity functions and of the Hertzsprung-Russell (HR) diagrams obtained from the analysis are discussed. Finally, the choices made for estimating of the mass loss rates are described and preliminary results concerning them are shown.

Keywords: Stars: AGB; Stars: Miras; Stars: Evolution; Stars: Luminosity; Stars: Mass Loss; Infrared
PACS: 97.20.Li; 97.30.Jm; 97.10.Cv; 97.10.Ri; 97.10.Me; 95.85.Hp

1. INTRODUCTION

The Asymptotic Giant Branch (hereafter AGB) phase lies at the end of the active life for Low- and Intermediate-Mass Stars (mass between $0.8M_{\odot}$ and $8M_{\odot}$). In this stage their energy is released by two shells (burning H and He, respectively) that ignite alternatively. For more details about nucleosynthesis and the physical properties inside AGB stars see e.g. [1, 2, 3] and references therein. Unfortunately, quantitative knowledge of crucial chemical and physical parameters of AGB sources is still poor; among the uncertain issues we emphasize particularly the mass loss rate and the estimate of stellar luminosity, which are hampered by the difficulties in measuring the distances for these dust-enshrouded objects.

We are currently performing an analysis of Galactic AGB stars of different chemical composition, looking for correlations between their basic parameters (bolometric luminosity, mass loss rate, photometric colors ...). Our main aim is that of establishing observationally-based criteria permitting a more quantitative determination of mass loss rates and luminosities for the various types of AGB stars thus providing general rules that could help in the improvement of stellar codes.

Extensive samples of Galactic AGB stars (M, S-type, C rich) have been collected. They must be large enough so that conclusions on them have a good statistical significance; they also need to have detailed and accurate Spectral Energy Distributions (SEDs) at near- and mid-IR wavelengths and, whenever possible, reliable measurements of mass loss rates and distances. Results of this research have been already published for carbon-rich stars ([4], hereafter Paper I), and are in press for S-type stars ([5], hereafter Paper II).

CP1001, *9th Torino Workshop on Evolution and Nucleosynthesis in AGB Stars and the 2nd Perugia Workshop on Nuclear Astrophysics*,
edited by R. Guandalini, S. Palmerini, and M. Busso

In the following section (Section 2) we present the two samples analyzed in Paper I and Paper II, explaining in detail through which criteria they have been divided into the different sub-samples. In Section 3 we show the importance of performing observations of AGB stars at mid-IR wavelengths (at least up to 20 μm). The methods adopted to estimate the luminosity and/or distance are described in Section 4. In Section 5 we discuss the results on luminosity functions and HR diagrams (in particular for S-type stars). In Section 6 we comment briefly on our choices to estimate mass loss rates and show the first results for S-type stars from our research. Finally, in Section 7 preliminary conclusions are derived.

2. SAMPLES OF AGB STARS

Two samples of AGB stars have been created to perform this analysis: one of carbon-rich sources (as discussed in Paper I) and the second of O-rich but s-element enhanced sources (S-type stars, as described in Paper II). Both samples have photometric observations available at mid-IR wavelengths (at least in the 8 μm region). They are subsequently organized in sub-samples according to the quality and extension of the database available.

2.1. Carbon-rich Stars

AGB sources that have been considered in this sample are chemically carbon-rich. Moreover, they have at least one published estimate of the mass loss rate, done according to the criteria we consider most suitable for the purpose (see Section 6 for details).

We have divided the carbon stars into three sub-samples according to the available knowledge of their variability and distance:

- **Sub-sample 1** (65 sources) is made out of our "best" sources. They are characterized by reliable estimates of the distance, obtained through recent updates of original Hipparcos measurements [6, 7]. Moreover these sources are all of a known variability type.
- **Sub-sample 2** (50 sources). Here the estimates of the distances derive from various methods (mainly from the application of the period-luminosity relations valid for C-rich stars). All the stars of this sub-sample are again of known variability types.
- **Sub-sample 3** (150 sources). These sources are the less studied of the whole sample. Their distances are computed through various methods, as found in the literature, the most commonly used being the one from [8]. They are all of "unknown" variability types.

For a more extensive discussion of this sample of Galactic C-rich AGB stars the reader is referred to Paper I.

2.2. S-type Stars

In order to prepare the sample under analysis for the S-type Galactic stars, we started from the extended lists by [9, 10], containing O-rich evolved red giants with known or suspected "S star-like" chemical peculiarities. We selected a total of $\simeq$600 sources for which measurements in the near-IR (from 2MASS) and in the mid-IR (from ISO, MSX or IRAS–LRS) are available. The chosen sources fall again into different categories, depending on the information we could collect on their IR colors, distance, variability type and period.

The S-type stars have therefore been organized in four sub-samples:

- **Sub-sample *A*** (21 sources): stars with ISO-SWS observations at mid-IR wavelengths up to 45 μm. They are fundamental for the calculation of bolometric corrections.
- **Sub-sample *B*** (26 sources). These sources have mid-IR measurements from MSX and/or IRAS–LRS (not from ISO–SWS). Their distances are estimated astrometrically from Hipparcos measurements.
- **Sub-sample *C*** (40 sources). This class contains only stars of Mira-type variability with reliable studies of their light-curves and therefore, good estimates of their period. Photometric observations in the mid-IR have been performed by MSX or IRAS–LRS. Distances are estimated through period-luminosity relations valid for O-rich stars (see Section 4).
- **Sub-sample *D*** ($\simeq$500 sources). This class includes all the other sources. They have "good" photometric observations at mid-IR wavelengths (at least 8.8 μm) but we could not find estimates of the distance.

The four sub-samples are divided in further sub-categories according to:

1. the spectral type (usually S, MS, (M)(S)[1], SC, plus unknown sources and anomalous cases);
2. the variability type (Mira, Semiregular, Irregular; unknown sources exist also in this case).

A detailed analysis of this large sample of S-type stars is presented in Paper II.

3. THE IMPORTANCE OF IR FOR AGB STARS

Evolution during the AGB phase has many "much-debated" topics: their study could be greatly helped by photometric observations at IR wavelengths.

In particular, an extended wavelength coverage is fundamental in the exam of the SEDs of AGB stars. Indeed, Fig. 1 shows that evolved AGB stars (Mira variables and post-AGBs) emit large part of their flux at mid-IR wavelengths. This fact is particularly evident for C-rich sources, to the point that mid-IR observations are necessary to

[1] (M)(S) indicates an uncertain classification between the two classes above

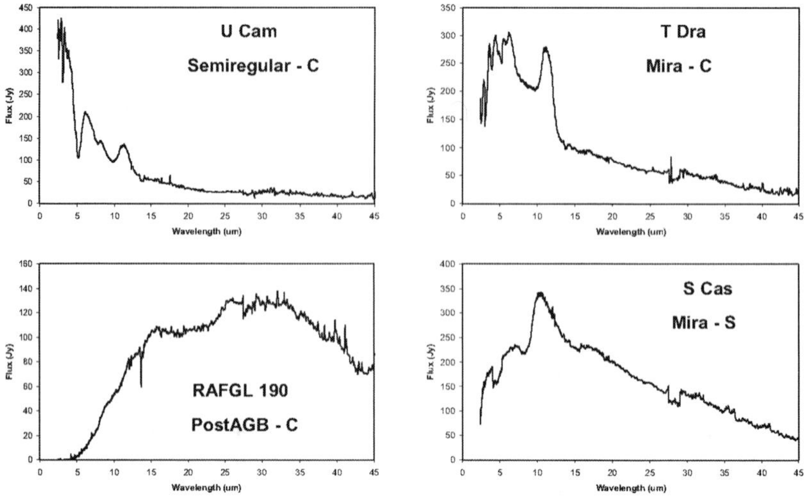

FIGURE 1. ISO-SWS spectra for a C-rich Semiregular variable (U Cam, top left), a C-rich Mira variable (T Dra, top right), a C-rich post-AGB star (RAFGL 190, bottom left) and S-type Mira variable (S Cas, bottom right).

obtain good estimates of bolometric luminosity, that otherwise would be significantly underestimated.

AGB stars are characterized by strong variability at optical wavelengths. Variability of the flux at IR wavelengths is less relevant if compared with the optical region, because in IR the contribution from the dynamical photosphere is less relevant. However, IR variability is not negligible. Indeed, the few Mira sources that have been observed several times by ISO with the SWS spectrometer show remarkable fluctuations in the mid-IR region. By contrast, SR variables do not show any appreciable variation of the flux (see [11] and its figures for more details). There is not yet an agreement on the origin and properties of this phenomenon: it needs to be examined in more detail in the mid-IR to understand its nature.

4. LUMINOSITY AND DISTANCE

The SEDs from ISO-SWS available for the two samples of AGBs plus the near-IR photometric points (from 2MASS) can be integrated in order to infer a reliable (apparent) bolometric magnitude, making use of the relations of fundamental photometry (Glass, [12]). If we consider the S-type sample, the sources of the sub-sample *A* give us the tools for computing the bolometric corrections, which turn out to be tightly correlated with near-to-mid infrared colors and can be used for deriving accurate bolometric magnitudes for the stars of sub-sample *B*, where the ISO-SWS measurements are not available (see Paper II for details). The same approach was adopted in Paper I in analyzing C star luminosities.

With the procedure described above, we can compute the bolometric magnitude whenever a source has observations in at least one of the relevant IR colors.

S-type stars that belong to the sub-sample C offer the possibility of an independent way for estimating luminosities. In fact, their Mira variability, of known period, allows for the application of period-luminosity relations, yielding directly the absolute magnitudes.

Concerning period-luminosity relations, for Galactic C-rich Miras they have been computed by Whitelock et al. (see [13] and references therein) and summarized in the relation:

$$M_{bol} = -2.54log(P) + 1.87 \tag{1}$$

where P is the variability period of the Mira and M_{bol} is its absolute luminosity. Instead, searching a similar period-luminosity relation for the O-rich Miras (and also S-type stars) is definitely more difficult. In the literature there is no affordable, unique estimate of this relation. We choose (mainly for the sake of homogeneity) to follow the indications from Feast and Whitelock (see [14] and references therein for a general review on this topic). Obviously the P-L relation for O-rich Miras (and S-type stars) will be different from the one obtained for C-rich Miras. Our choice fell on two different P-L relations obtained using period correlations either to the absolute bolometric luminosity ([15], Eq. 2) or to the absolute K-band luminosity ([14], Eq. 3):

$$M_{bol} = -3.00log(P) + 2.80 \tag{2}$$

$$M_K = -3.47log(P) + 1.00 \tag{3}$$

We made use of both methods, because they have both advantages and disadvantages. In principle Eq. 2 (from [15]) would be probably the best relation available, as it involves M_{bol} directly. However, its last version is quite "old" (1994, [15]). Instead Eq. 3 (from [14]) has been more recently updated and extensively examined [14] but gives us "only" an estimate of the absolute magnitude in the K–band. We adopted both relations to calculate two independent estimates of distance:

- First case – using Eq. 2. Here we find an estimate of the apparent bolometric magnitude through the bolometric corrections obtained in Paper II, then we compare it with the absolute bolometric magnitude from Eq. 2 to estimate the distance of the S Mira.
- Second case – using Eq. 3. Here we apply a similar procedure comparing the apparent K–magnitude from 2MASS with the absolute K–magnitude from Eq. 3 to obtain another estimate of the S Mira's distance.

Finally, we make an average between the two distance estimates and with the final result we calculate the absolute magnitude of the star applying the "final" distance to the apparent bolometric magnitude obtained form our bolometric corrections. See Paper II for a more complete discussion of this topic.

We are trying to improve our knowledge on the variability properties of Mira and Semiregular AGB stars. One of the analyses that we are performing is centered on the re-observation of Maffei's stars with Small–IRAIT, in order to confirm the periods

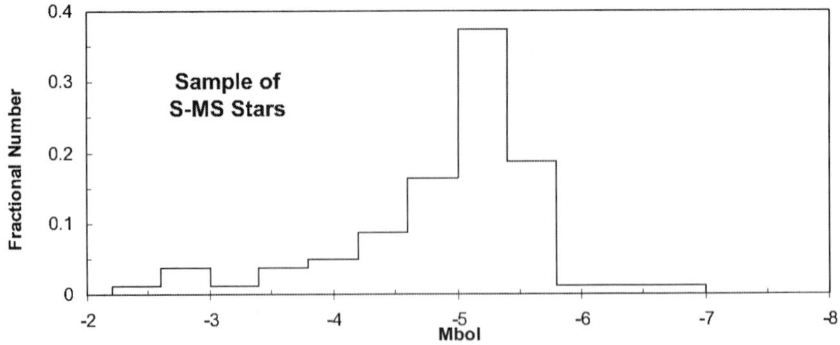

FIGURE 2. The histogram showing the fractional number of S-type stars per magnitude interval.

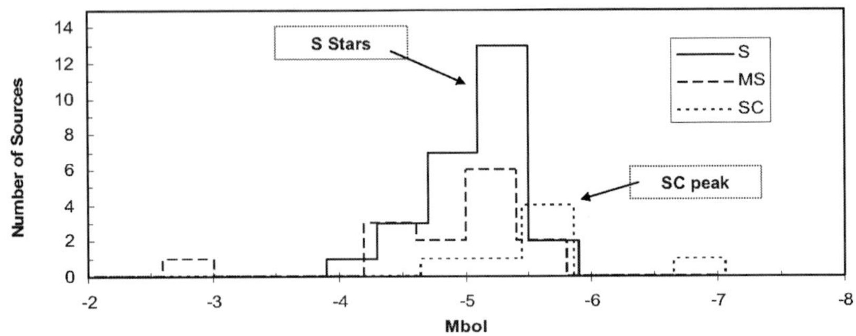

FIGURE 3. Luminosity Functions from the same sample already shown in Fig. 2. The original data have been divided into 3 sub-samples, according to the source spectral type (S–MS–SC).

and light–curves obtained by Maffei and Tosti (the catalogue is available online at *http://astro.fisica.unipg.it/atlasmaffei/main.htm*). See the contribution from Briguglio in this conference for more details [16].

5. LUMINOSITY FUNCTIONS AND HR DIAGRAMS

Here we present few results obtained using the bolometric magnitudes computed through the methods presented in the previous section. We do not discuss in detail the luminosity function for Carbon stars because it has been already presented in Paper I. Here we remark only that, according to our data, Galactic C-rich AGB stars show luminosities in the range $-4.5/-6$. This is in agreement with the theoretical predictions from the FRANEC model (see [17] and the contribution from Cristallo, this conference). We don't find any problem of underluminous C-rich stars, contrary to suggestions in the past ([18], see also Paper I)

In Fig. 2 the luminosity function for the S-type stars from sub-samples *A–B–C* is shown. The spread in the luminosities for these 80 sources is again in agreement with

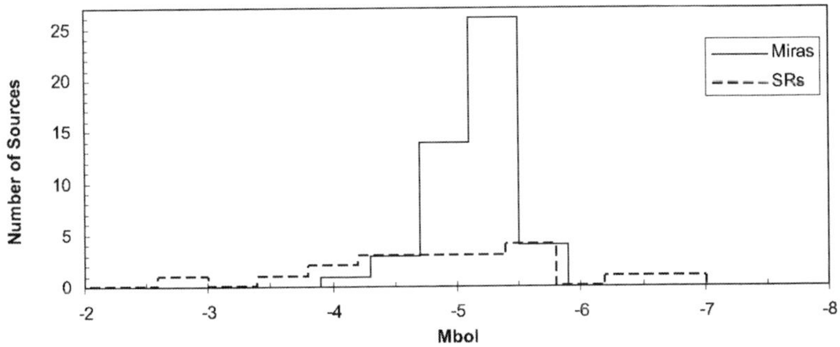

FIGURE 4. Luminosity Functions from the same sample already used for Fig. 2. We present two sub-samples formed according to the variability type (Mira–Semiregular).

the FRANEC models. The luminosities are in the range $-4.2/-5.7$ while the main peak is between -4.9 and -5.2: Galactic S stars seem to be on average only slightly less luminous of the C-rich ones. At this stage we cannot say whether this apparent difference is statistically significant: in fact, the two Luminosity Functions are overlapping. It is clear that attributing the spectral distinction among the classes MS–S–SC–C to a simple evolutionary sequence, is not sufficient.

We divide the sample used for the luminosity function of Fig. 2 in three sub-groups according to the different spectral types (S–MS–SC). In this way we obtain three luminosity functions that are shown in Fig. 3. The S stars present a peak in their distribution in the range $-4.7/-5.5$, the SC stars occupy the high luminosity tail of the distribution with a peak at magnitudes around -5.6. Instead, the MS sample is overlapping both the S and SC sub-groups without a well-defined peak.

Figure 4 shows other luminosity functions. However in this case the sample of S-type stars has been divided according to the variability type of the sources (Mira or Semiregular). The Mira's Luminosity Function presents a well defined peak and cover mainly the range between -4.8 and -5.5. On the other hand, the distribution of Semiregulars is more or less uniform over all the range of luminosities.

Finally, we present in Fig. 5 a preliminary result on the S-star HR diagram, adopting a near-to-mid-IR color as the abscissa. We refer to Paper II for a more complete discussion. Here we emphasize a few main features after dividing the points according to the spectral classification and the variability types. We note that semiregular S stars are essentially unreddened, while the Miras do show some reddening. However, if we compare S stars with the Carbon-rich sample (see Paper I), C stars are in general more deeply enshrouded by dust.

6. MASS LOSS

As a last topic of this contribution we present a rough analysis of mass loss rates from AGB stars including a preliminary result for the S-type stars. For C-stars see Paper I.

Longstanding efforts have been devoted to describe the mass loss mechanisms, either

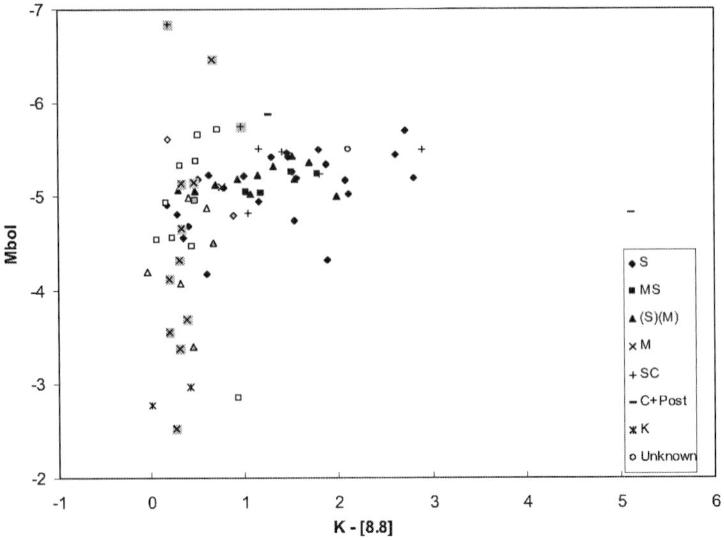

FIGURE 5. The HR diagram for the sample of S stars adopting the near-to-mid-IR colour K–[8.8] as abscissa. The sources are divided into several sub-groups according to their spectral type, as shown in the legend inside the plot. For each sub-group, filled points are Mira variables, while blank points are Semiregular or Irregular ones.

with phenomenological models or with sophisticated hydrodynamical approaches [19, 20, 21, 22]. Despite this, our quantitative knowledge of AGB winds is still poor. Our choice has been that of using mass loss rates available form the literature. We considered only those computed through observations at radio-wavelengths of the CO lines: this type of approach is known to give an estimate of the "average" rate of the AGB source's stellar wind, integrated over a long time. We upgraded the estimates of the mass loss rate adopting updated measurements of the distance or, in the case of the S-type sample (see Paper II) giving directly our estimate of the distance through the use of the Period-Luminosity relations (see also Section 4).

The approaches to obtain mass loss estimates through the observations of CO lines that are available in the literature are several. In particular we adopt:

1. *First Choice* (our best choice): The analysis performed by the Stockolm's group: we refer to Schoier and Olofsson [23] for C-rich stars, Olofsson et al. [24] for M stars and Ramstedt et al. [25] for S-type stars.
2. *Second Choice*: the method used by Loup et al. [26] and Winters et al. [21]. In this case the mass loss rate and the size of the CO envelope are both estimated in a bi-parametric system.
3. *Third Choice*: the method used by Groenewegen et al. ([27] for S stars, [8] for C-rich stars). Here a constant size for the CO envelope is adopted for all the sources. However, this method is more reliable for stars with an optically thick CO emission. Due to this fact this method can sometime underestimate the wind efficiency by

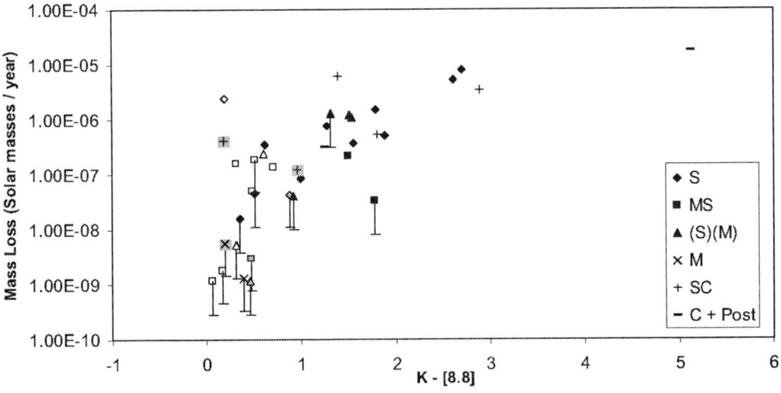

FIGURE 6. Relation linking the mass loss rates of S-type stars to the near-to-mid-IR colour K–[8.8]. The sources are divided into several sub-groups according to their spectral types as shown in the legend inside the plot. For every sub-group, filled points are Mira variables, while blank points are Semiregular or Irregular ones. The data with the downward-pointing error bars are upper limits.

factors 3–4 (as said by [23]).

The analysis performed on the carbon-rich sample is shown in Paper I.

Figure 6 displays a preliminary result obtained plotting the mass loss rates against a near-to-mid-IR color (the same one already used as abscissa of Fig. 5) in the case of the sample of S-type stars. The sources show a behaviour similar to the one already analyzed in the C-rich sample (see Paper I); here we remark a few main properties: 1) Miras have stronger mass loss rates and are redder if compared with Semiregulars and Irregulars; 2) a relation of the same type as found for C stars exists for S-stars suggesting the association of strong mass loss with strong IR excess for AGB stars of any type, as expected is the wind is driven by radiation pressure on dust grains; 3) this relation, if real, is however very scattered for both samples; 4) van Loon et al. [28] have suggested that the mass loss rate should be expressed as a function of both the bolometric luminosity and the stellar effective temperature (photometric color); 5) we remind that mass loss processes in AGB stars are surely not "simple": the large variations of the mass loss rates, due to different causes, that occur during the AGB phase certainly contribute largely to the scattering that we see in Fig. 6. Finally, we remark that there is no clear distinction in the mass loss rates of S-type and C-rich stars, but carbon stars become much redder by a large factor (see Paper I and Paper II).

7. CONCLUSIONS

We have analysed two sample of Galactic AGB stars: the first is made of carbon rich sources and the second one of S-type stars. We looked for methods to obtain reliable estimates of luminosities and distances (bolometric corrections and Period–luminosity relations). The problem of the large IR excess typical of evolved AGBs has been shown together with its implications in the estimates of the stellar luminosity.

C- and S-stars Luminosity Functions are in the ranges -4.5/-6 and -4.2/-5.7 respectively (in agreement with expectations from the FRANEC model). The Luminosity Functions of the two samples are overlapping (with S stars being slightly less luminous). It is clear that attributing the spectral distinction among the classes MS–S–SC–C to a simple evolutionary sequence, is not sufficient.

SC stars occupy the high luminosity tail of the S-type stars distribution. It is highly uncertain if this is due to evolutionary properties. Semiregular S stars are essentially unreddened and S stars in general are less reddened than C stars.

There is no clear distinction between S and C classes in what concerns the relation of mass loss rates to near-to-mid-IR colors, but C stars become much redder. High-reddened C stars (evolved Miras and post-AGBs) possibly lose more mass than S stars but the scatter and the uncertainties of the rates are large.

ACKNOWLEDGMENTS

R.G. acknowledges support in Italy by MIUR under contract PRIN2006-022731 and the University of Perugia for a postdoctoral fellowship.

REFERENCES

1. M. Busso, R. Gallino, and G.J. Wasserburg, *ARA&A* **37**, 239 (1999)
2. K.M. Nollett, M. Busso, and G.J. Wasserburg, *ApJ* **582**, 1036 (2003)
3. M. Busso, G.J. Wasserburg, K.M. Nollett, and A. Calandra *ApJ* **671**, 802 (2007)
4. R. Guandalini, M. Busso, S. Ciprini, G. Silvestro, and P. Persi *A&A* **445**, 1069 (2006) [Paper I]
5. R. Guandalini and M. Busso, *A&A* in press (2008) [Paper II]
6. J. Bergeat, A. Knapik, and B. Rutily, *A&A* **390**, 967 (2002)
7. J. Bergeat and L. Chevallier, *A&A* **429**, 235 (2006)
8. M.A.T. Groenewegen, M. Sevenster, H.W.W. Spoon, and I. Pérez, *A&A* **390**, 511 (2002)
9. C.B. Stephenson, "A General Catalogue of Galactic S-Stars - ED.2", Publ. of the Warner & Swasey Observatory **3**, 1 (1984)
10. C.B. Stephenson, *AJ* **100**, 569 (1990)
11. M. Busso, R. Guandalini, P. Persi, L. Corcione, and M. Ferrari-Toniolo, *AJ* **133**, 2310 (2007)
12. I.S. Glass, "Handbook of Infrared Astronomy", Cambridge University Press (1999)
13. P.A. Whitelock, M.W. Feast, F. Marang, and M.A.T. Groenewegen, *MNRAS* **369**, 751 (2006)
14. M. Feast, *ASP Conference Proceedings* **310**, 304 (2004)
15. P. Whitelock, et al., *MNRAS* **267**, 711 (1994)
16. R. Briguglio, this conference (2008)
17. O. Straniero, I. Domínguez, S. Cristallo, and R. Gallino, *PASA* **20**, 389 (2003)
18. E. Costa and J.A. Frogel, *AJ* **112**, 2607 (1996)
19. E.E. Salpeter, *ApJ* **193**, 585 (1974)
20. G.R. Knapp and M. Morris, *ApJ* **292**, 640 (1985)
21. J.M. Winters, T. Le Bertre, K.S. Jeong, L. -Å. Nyman and, N. Epchtein, *A&A* **409**, 715 (2003)
22. A. Wachter, K.-P. Schröder, J.M. Winters, T.U. Arndt, and E. Sedlmayr, *A&A* **384**, 452 (2002)
23. F.L. Schöier and H. Olofsson, *A&A* **368**, 969 (2001)
24. H. Olofsson, D. González Delgado, F. Kerschbaum, and F.L. Schöier, *A&A* **391**, 1053 (2002)
25. S. Ramstedt, F.L. Schöier, H. Olofsson, and A.A. Lundgren, *A&A* **454**, L103 (2006)
26. C. Loup, T. Forveille, A. Omont, and J.F. Paul, *A&AS* **99**, 291 (1993)
27. M.A.T. Groenewegen and T. de Jong, *A&A* **337**, 797 (1998)
28. J.Th. van Loon, M.-R.L. Cioni, A.A. Zijlstra, and C. Loup, *A&A* **438**, 273 (2005)

Circumbinary Discs Around Post-AGB Binary Stars: a Common Phenomenon

Hans Van Winckel*, Pieter Deroo*,†, Clio Gielen*, Maarten Reyniers*,**, Els van Aarle* and Edgardo Vidal‡

*Instituut voor Sterrenkunde, Katholieke Universiteit Leuven, Celestijnenlaan 200D, 3001 Leuven, Belgium
†JPL, California Institute of Technology, MS 183-900, Pasadena, CA 91109, USA
**KMI, Departement Waarnemingen, Ringlaan 3, 1180 Brussel, Belgium
‡Sterrenkundig Observatorium, Universiteit Gent, Krijgslaan 281 S9, B-9000 Gent, Belgium

Abstract. In this contribution we report on the results of our studies on the diverse impact of the circumstellar material on the evolution of binary post-AGB stars. Binary post-AGB stars are common and thanks to our radial velocity monitoring program we could identify the binary observables (period, eccentricity, mass function) of a significant fraction of the whole Galactic sample. We argue that the circumstellar material around all the studied objects is dominated by a circumbinary disc. Using the newly available interferometric capabilities at ESO, we were able to resolve the discs and study their structure in detail. We argue that the discs are stable and in hydrostatic equilibrium. The impact of the disc on the evolution of the system can hardly be overestimated and the formation and evolution of the disc has a fundamental impact on the evolution of the system. We touch upon the poorly understood connection between the binary post-AGB stars and the other groups of evolved binaries.

Keywords: Stars: AGB, Post-AGB - Stars: Evolution - Stars: Binaries - Stars: Circumstellar Matter
PACS: 95.75.Kk - 95.85.Hp - 97.10.Fy - 97.10.Tk - 97.80.Fk

1. INTRODUCTION

In the canonical picture of stellar evolution theory, the post-AGB stars are short-lived descendants of Asymptotic Giant Branch (AGB) stars. Post-AGB stars are surrounded by a physically-detached circumstellar dust shell, which is a relic of the strongly enhanced superwind mass-loss phase at the end of the AGB. The central stars of post-AGB stars are not yet hot enough to ionize this circumstellar environment and the resolved reflection nebulae found around some post-AGB stars are dubbed proto-planetary nebulae (PPNe).

The evolutionary time-scales of post-AGB stars (e.g.[1]) are short and hence not very many are known. A compilation of candidates can be found in `http://www.ncac.torun.pl/postagb` from [2]. Since the thermal infrared emission of the circumstellar dust is a defining characteristic of post-AGB stars, it was no surprise that the all sky survey of IRAS was instrumental in the sample selection of Galactic post-AGB star candidates.

Despite being rare, the Post-AGB stars are very useful probes to many interesting and very poorly understood processes: their photospheric content is the end-product of the chemical enrichment induced by the thermal pulses of AGB stars and as such post AGB

CP1001, *9th Torino Workshop on Evolution and Nucleosynthesis in AGB Stars and the 2nd Perugia Workshop on Nuclear Astrophysics*, edited by R. Guandalini, S. Palmerini, and M. Busso

stars are ideal tracers for the 3rd dredge-up nucleosynthesis (e.g. see also contribution by Reyniers et al., these proceedings). Moreover, the onset of the creation process of the panoply of structures and shapes observed in planetary nebulae occurs during the short post-AGB evolution. The physical nature of all the processes involved are still very badly understood.

Since the post-AGB evolution is so short, studies of individual objects prevail. The main problem in the interpretation of the many detailed studies of individual objects is, that it is often quite unclear how representative detailed characteristics are in the complete picture of stellar evolution. Anyones pet objects are not necessarily very representative examples of a 'typical' post-AGB evolution. Recent reviews on post-AGB research can be found in [3, 4, 5].

Stellar evolution theory predicts that the end of the superwind phase coincides with the end of a dusty mass loss of the central star. The dust envelope is expected to detach and cool fast while expanding. The full Spectral Energy Distribution (SED) of even a very young post-AGB star, should therefore not show thermal dust emission in the near-infrared. This is not always observed, however, and in this contribution we focus on the specific, but quite numerous, sample of post-AGB stars displaying a near-IR excess due to thermal emission of hot dust. These objects turned out to be binaries and our study aims at contributing to a better understanding of the final evolution of binary stars.

In this contribution we start with introducing the sample as well as the results of our radial velocity monitoring. The results of our interferometric survey are discussed in the next sections. We end with a short discussion and an outlook on the prospects ahead.

2. SAMPLE OF BINARY POST-AGB STARS

The first binary post-AGB stars were serendipitously discovered and inspired by the SED characteristics of objects like HR4049 and HD44179 (e.d. [6]), we initiated a systematic search for new binary candidates.

The discriminating SED characteristic of this, initially very limited sample, turned out to be exposed best in the near-IR: all objects display a broad IR excess, which often starts already at H or K, due to thermal emission of dust. The colour temperature of the hot dust component is near sublimation temperature, showing that there must be a component of the circumstellar material which is very close to the central star.

We started a systematic search for binary post-AGB stars based on sample selection criteria which included the observational evidence for the presence of this hot dust component. Initially we selected 51 objects [7] but this increased steadily since then. Another example is given in Fig. 1.

We initiated a very extensive multi-wavelength study of those systems, including radial velocity monitoring, high spectral resolution optical studies, infrared spectral dust studies, sub-mm bolometric observations and high spatial resolution interferometric experiments with the VLTI. The analysis of infrared spectra is given by Gielen, these proceedings.

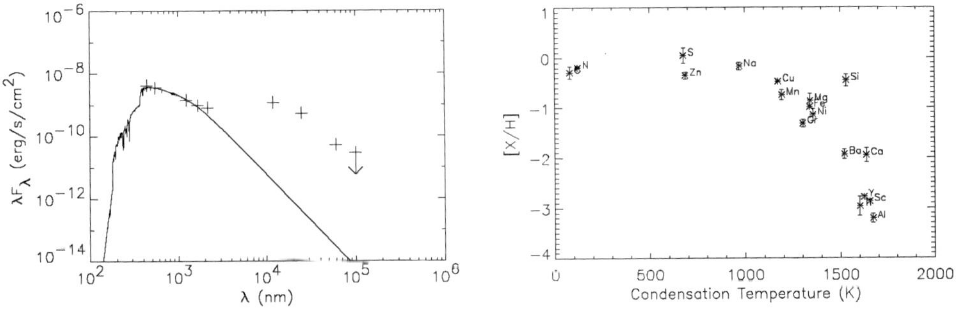

FIGURE 1. Left: the broad-band SED of HP Lyr. Note the near-IR excess which is a characteristic feature of the objects under study. Right: the photospheric chemical composition of HP Lyr. The abundances with respect to the Solar values are given with respect to the dust condensation temperature of the element according. The abundances are from [8] while the dust condensation temperature is from [9].

3. BINARY DETECTION AND RESULT

To test our suspicion that the infrared selected stars are binaries, we have set-up an extensive radial velocity monitoring campaign. Thanks to the Flemish-Swiss collaboration in exploiting the twin telescopes Mercator-Euler, we have regular access to the CORALIE radial velocity spectrograph mounted on the Swiss 1.2m Euler telescope. Moreover, in the near-future we will be able to extend our monitoring efforts to fainter objects in the Northern hemisphere, when our high resolution optical spectrograph HERMES will be commissioned in the fall of 2008 [10].

We obtain spectra of limited signal-to-noise but using optimised cross-correlation masks, we are able to derive accurate radial velocity measurements in a limited amount of telescope time. The line masks are tuned for every individual star on the basis of the high signal-to-noise spectra obtained for our photospheric chemical studies [11, 12, 13].

The major difficulty is that most objects are on, or near, the population II Cepheid instability strip in the HR-diagram. This means that they display significant photospheric motions due to their atmospheric pulsations. The objects with the largest pulsational amplitudes turned out to be the dusty RV Tauri stars. The binary rate of the low amplitude pulsators in our monitored sample is 100 percent (six out of six) with orbital periods ranging from 120 to 1800 days. Four objects out of these six have non-circular orbits [14].

The complete binary statistics will be published elsewhere and in Fig. 2 we show the period-eccentricity diagram of the sample obtained till now. Our obtained orbits are complemented with data taken from the literature ([14] and references therein). Since many of the objects do also show pulsational cycle-to-cycle variability it is needless to say that the detection of binary motion in the pulsating objects is a non-trivial observational challenge.

Given these preliminary results, we propose that the whole sample of post-AGB stars with a near-IR excess are likely binaries. Since the orbits detected so far are too small

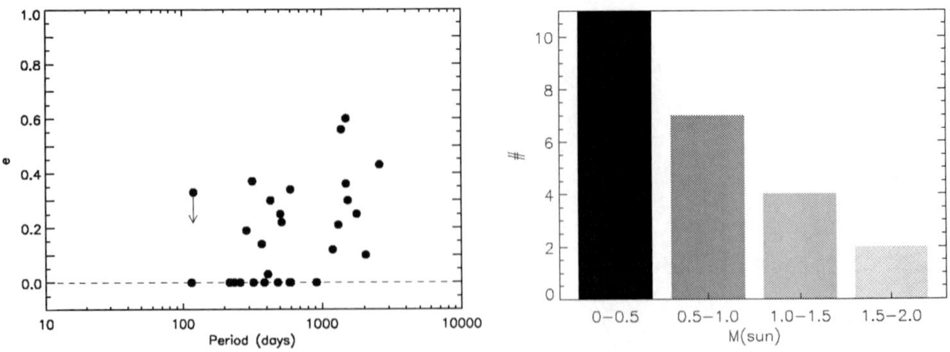

FIGURE 2. Left: the eccentricity versus orbital period diagram of all post-AGB binaries with known orbital elements so far. Data is coming from the literature and from our own radial velocity monitoring project [14]. Note the rather high eccentricities for relatively short orbits. Right: the distribution of minimal masses for the companion stars, assuming an inclination of 90 degrees and a mass of the post-AGB primary of 0.6 $M_\odot$.

to accommodate a cool AGB star, the objects must have been subject to severe binary interaction when the primary was a cool supergiant. The (albeit uncertain) luminosities and the measured effective temperatures indicate that the objects are now well within their Roche potential.

The inclination of the orbits are difficult to constrain from the SED alone. The distribution of mass-functions, which is an observable once the orbital parameters are known, do show that the minimal mass for the companion range from well below solar (e.g.[15]) up to about 2 $M_\odot$ (Fig. 2). In none of the multi-wavelength observations obtained till now, we have found evidence for a contribution of the companion. Moreover, in none of the systems, symbiotic activity was found either. The companions in most, if not all, systems are likely un-evolved main sequence stars.

4. NEED FOR INTERFEROMETRY

The near-IR excess was interpreted as indicating the presence of gravitationally-bound circumstellar material in the system ([3] and references therein). A simple back-of-the-envelope optically thin calculation, assuming a pure black-body approach to the dust grain radiative cross section and a grain in thermal equilibrium with the radiation field, yields a simple relation between the distance of the grain from the star and the grain temperature which reads:

$$T_{dust} = T_* \sqrt{\frac{R}{2d}} \tag{1}$$

With T_{dust} and T_* the dust grain temperature and the effective temperature of the star respectively; R the radius of the star and d the distance between the individual dust grain and the central star. As the dust excess of the SED starts at K and often even at H (see e.g. [7]), the distance of the inner circumstellar dust shell must be $\leq$ 10 A.U. from the central object, assuming the star is a supergiant with $L_* \geq 3000 L_\odot$. As the distances of the sample stars are typically one to several kpc, we do need infrared interferometric techniques to resolve the objects.

5. N-BAND INTERFEROMETRY WITH MIDI

We used the pupil-plane combiner MIDI (`http://www.eso.org/instruments/midi`) at ESO to obtain spectrally dispersed fringes in the N-band. In most cases a prism was used as dispersive element yielding a spectral resolution of about 30. We used baselines with the fixed 8m UT telescopes as well as the movable 2m AT telescopes, spanning baseline lengths from 15 to 128m. MIDI is a two-beam combiner, hence the number of measurements we could achieve on different baselines is limited, since a calibrated visibility on one baseline takes a - strongly overhead dominated - 1 hour of telescope time. For every object we obtained typically two to three different measurements with different lengths and projected angles.

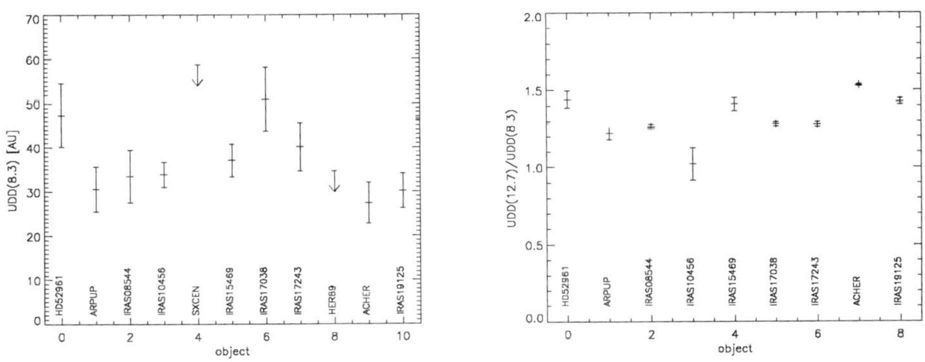

FIGURE 3. Left: the diameters at 8.3 μm of the sample of objects resolved with the MIDI interferometer. The errorbars are dominated by uncertainties on the distance estimates. Two objects (SX Cen and 89 Her) were not resolved with our set-up. Right: the ratio of the sizes of the emission at 12.7 over 8.3 μm. Note the very similar value for all sources indicating that interferometric wise, the objects are very homogeneous: the temperature structure of the dusty discs must be very similar. Figures from the PhD thesis of Pieter Deroo.

The main conclusion of our interferometric effort is that we resolved indeed the very compact N-band emission regions around almost all program stars (Fig. 3). The resolved sizes are not compatible with any outflow model and we can conclude that we do resolve the circumbinary discs.

Moreover, the discs around the sample stars have homogeneous interferometric characteristics, indicating that a similar physical model must apply to quantify the radiative transfer through the discs. Deroo et al. [16] show that on the basis of combined interferometric data in the K and N-band, a physically consistent model to describe the circumstellar disc is found assuming a passive, irradiated disc [17, 18, 19]. The disc structure is governed by the equations of hydrostatic equilibrium. The dust-free inner hole is confined by dust sublimation. At the dust sublimation radius, the dust in the disc receives head-on radiation of the central star and puffs up. Depending on the optical thickness of this inner part, the structure of the disc will adjust accordingly. We refer to the PhD thesis of Pieter Deroo (`http://www.ster.kuleuven.be/pub/index_en.html`) and to the publications in preparation, for a more detailed discussion. In Fig. 4 we show the best, self-consistent model at 2 and 10 μm respectively. The model parameters and its projection onto the sky, are constrained by the SED energetics, as well as by the interferometric observables (K-band visibilities on three baselines, the non-zero closure phase and its slope over the K-band, and two N-band base-line measurements).

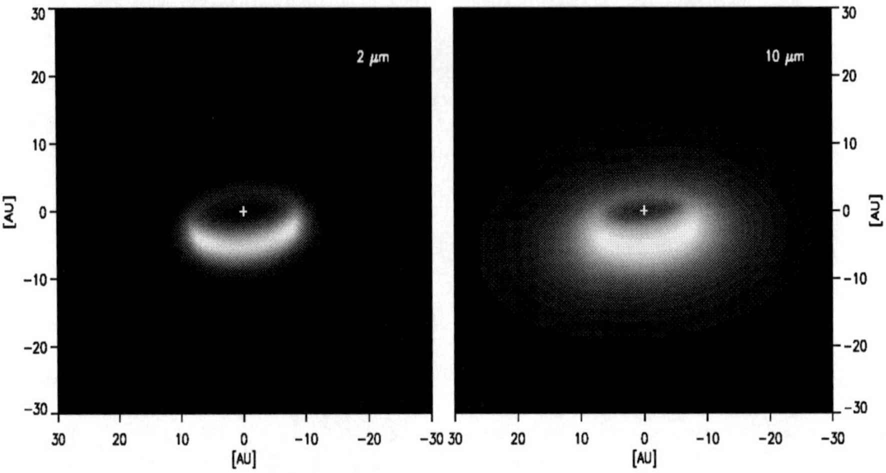

FIGURE 4. Image reconstructed from our self-consistent disc model which is constrained by both the SED and the interferometric observables in the K and N-band. The left panel gives the view at 2 μm, the right at 10 μm. The position of the central object is indicated with a +-sign. The non-resolved photospheric contribution of the central object has been omitted. The images are in gray scale with 25 equally spaced contours. The brightest region of the inner wall is on the far sight. North is up, east to the left. Figure from [16].

6. CONCLUSIONS AND OUTLOOK

We conclude that our interferometric experiments show that (part of) the circumstellar material around our specific sample of post-AGB objects, is trapped in a stable, compact but resolved circumstellar structure. Our best model, which is constrained by the SED and by the interferometric observables assumes that the dust is trapped in a stable but passive disc. Our radial velocity monitoring indicate that this is likely only possible in binary stars. The circumstellar discs are circumbinary.

A nice illustration of the versatile impact of the disc on the central star is that in many objects in our large sample display a photopheric chemical pattern similar to the gas phase abundance pattern in the ISM: the refractory elements like Fe, Ca, Si and the s-process elements are underabundant with respect to more volatile elements like Zn and S (Fig. 1, e.g. [3] and references therein). Although the depletion process is far from being understood, (see also Reyniers et al. these proceedings,), the presence of a circumstellar disc is likely a prerequisite for the process to occur. For pulsating stars with large amplitude variabilities and associated photospheric shocks, a chemical analysis based on static atmospheres must be interpreted with caution [20, 8].

The grain processing is very different in these discs than in normal AGB outflows and we refer to the contribution of Gielen et al., these proceedings as well as to [21] for a more detailed discussion.

The number of post-AGB stars with a near-IR excess and hence likely a circumbinary disc is significant. This was illustrated again by the first release of the SAGE project [22] which is the Spitzer survey of the LMC. The IR-data show that the depleted post-AGB stars in the LMC [23] all turned out to have SEDs similar to the Galactic disc sources (Reyniers, these proceedings). A systematic study of the disc sources of the LMC with their known distances and hence luminosities will be highly rewarding to probe the evolution of the disc under the changing radiation field of fast evolving central star. Such a study will allow a much better probe to study the impact of the disc on the evolution of the system as well as to the impact of the evolution of the central star on the structure of the circumbinary disc.

The final evolution of low to intermediate mass binaries seems to be accompanied by the formation of a circumstellar disc. This disc is stable. Any selection of post-AGB stars based on infrared fluxes will be biased towards the detection of these objects: during their post mass-loss evolution, they are much longer infrared luminous then outflow sources.

The impact of this circumbinary disc on the properties and evolution of the binary can hardly be overestimated.

ACKNOWLEDGMENTS

We thank the whole staff of the Institute in Leuven for their contributions to the radial velocity monitoring program.

REFERENCES

1. T. Blöcker, *A&A* **299**, 755 (1995)
2. R. Szczerba, N. Siódmiak, G. Stasińska, and J. Borkowski, *A&A, arXiv:astro-ph/0703717* **469**, 799 (2007)
3. H. Van Winckel, *ARAA* **41**, 391 (2003)
4. B. J. Hrivnak, *Proto-Planetary Nebulae (invited review)*, in *Planetary Nebulae: Their Evolution and Role in the Universe*, edited by S. Kwok, M. Dopita, and R. Sutherland, **209** of *IAU Symposium*, 113 (2003)
5. P. García-Lario, *Properties of Post-AGB Stars*, in *Planetary Nebulae in our Galaxy and Beyond*, edited by M. J. Barlow, and R. H. Méndez, **234** of *IAU Symposium*, 63 (2006)
6. H. Van Winckel, C. Waelkens, and L. B. F. M. Waters, *A&A* **293**, L25 (1995)
7. S. de Ruyter, H. van Winckel, T. Maas, T. Lloyd Evans, L. B. F. M. Waters, and H. Dejonghe, *A&A* **448**, 641 (2006)
8. S. Giridhar, D. L. Lambert, B. E. Reddy, G. Gonzalez, and D. Yong, *ApJ* **627**, 432 (2005)

9. K. Lodders and B. Fegley, *The planetary scientist's companion / Katharina Lodders, Bruce Fegley.* New York : Oxford University Press, 1998. QB601, L84 (1998)
10. G. Raskin, H. Van Winckel, and H. Lehmann, *Design of HERMES: a high-resolution fiber-fed spectrograph for the Mercator Telescope,* in *Ground-based and Airborne Instrumentation for Astronomy. Edited by McLean, Ian S.; Iye, Masanori,* **6269** of *Presented at the Society of Photo-Optical Instrumentation Engineers (SPIE)* (2006)
11. T. Maas, H. Van Winckel, and C. Waelkens, *A&A* **386**, 504 (2002)
12. T. Maas, H. Van Winckel, T. Lloyd Evans, L.-Å. Nyman, D. Kilkenny, P. Martinez, F. Marang, and F. van Wyk, *A&A* **405**, 271 (2003)
13. T. Maas, H. Van Winckel, and T. Lloyd Evans, *A&A* **429**, 297 (2005)
14. H. Van Winckel, *Baltic Astronomy, arXiv:astro-ph/0702245* **16**, 112 (2007)
15. V. Bujarrabal, H. van Winckel, R. Neri, J. Alcolea, A. Castro-Carrizo, and P. Deroo, *A&A, arXiv:astro-ph/0703718* **468**, L45 (2007)
16. P. Deroo, B. Acke, T. Verhoelst, C. Dominik, E. Tatulli, and H. van Winckel, *A&A* **474**, L45 (2007)
17. C.P. Dullemond, C. Dominik, and A. Natta, *ApJ* **560**, 957 (2001)
18. C. P. Dullemond and C. Dominik, *A&A, arXiv:astro-ph/0401495* **417**, 159 (2004)
19. E. Vidal and M. Baes, *Baltic Astronomy* **16**, 101 (2007)
20. T. Maas, S. Giridhar, and D. L. Lambert, *ApJ, arXiv:0706.2029* **666**, 378 (2007)
21. C. Gielen, H. van Winckel, L. B. F. M. Waters, M. Min, and C. Dominik, *A&A, arXiv:0709.3197* **475**, 629 (2007)
22. M. Meixner, K. D. Gordon, R. Indebetouw, J. L. Hora, B. Whitney, R. Blum, W. Reach, J.-P. Bernard, M. Meade, B. Babler, C. W. Engelbracht, B.-Q. For, K. Misselt, U. Vijh, C. Leitherer, M. Cohen, E. B. Churchwell, F. Boulanger, J. A. Frogel, Y. Fukui, J. Gallagher, V. Gorjian, J. Harris, D. Kelly, A. Kawamura, S. Kim, W. B. Latter, S. Madden, C. Markwick-Kemper, A. Mizuno, N. Mizuno, J. Mould, A. Nota, M. S. Oey, K. Olsen, T. Onishi, R. Paladini, N. Panagia, P. Perez-Gonzalez, H. Shibai, S. Sato, L. Smith, L. Staveley-Smith, A. G. G. M. Tielens, T. Ueta, S. V. Dyk, K. Volk, M. Werner, and D. Zaritsky, *AJ, arXiv:astro-ph/0606356* **132**, 2268 (2006)
23. M. Reyniers and H. van Winckel, *A&A, arXiv:astro-ph/0701327* **463**, L1(2007)

Dust-Grain Processing in Circumbinary Discs Around Binary Post-AGB Stars

Clio Gielen*, Hans Van Winckel*, Laurentius B.F.M. Waters*,†, Michiel Min† and Carsten Dominik†

*Instituut voor Sterrenkunde, Katholieke Universiteit Leuven, Celestijnenlaan 200D, 3001 Leuven, Belgium
†Sterrenkundig Instituut Anton Pannekoek, Universiteit Amsterdam, Kruislaan 403, 1098 Amsterdam, The Netherlands

Abstract. The number of binary post-AGB stars known in the galaxy is significant, yet their evolutionary status is far from understood. The orbital elements from our radial velocity monitoring program indicate that these objects must have undergone severe binary interaction, when the primary was at giant dimension. Evidence is growing that all these evolved binaries are surrounded by circumbinary compact Keplerian dusty discs. We present the first results of our SED-modelling and detailed study of the mineralogy using the rich Spitzer-IRS and TIMMI2 spectra. In all sampled stars, the dust is oxygen rich with a variable but, on average, high degree of crystallinity. Features of the crystalline olivine and pyroxene (forsterite and enstatite) dominate the spectra. Both cool and hot crystalline dust grains must be present to produce the observed dust emission features. Our 2D SED-modelling shows that the presence of a component of large grains is needed to account for the 850 micron flux. This, together with the high crystallinity, is evidence for efficient dust processing in these circumstellar discs. Since the binaries span a significant range in initial metallicity, orbital separation and mass functions we conclude that the circumstellar dust around a significant population of binaries is very different from dust formed and processed around single stars.

Keywords: Stars: AGB, Post-AGB - Stars: Evolution - Stars: Binaries - Stars: Circumstellar Matter
PACS: 95.85.Gn, 97.10.Fy, 97.30.Kn, 97.60.-s, 97.80.Fk

1. INTRODUCTION

Post-AGB stars are low- and intermediate-mass stars that evolve rapidly from the Asymptotic Giant Branch (AGB) towards the Planetary Nebula (PNe) phase, before cooling down as white dwarves. The processes governing the evolution from slowly expanding spherically symmetric AGB outflows and the observed large variety in proto-PNe and PNe geometries and kinematics are still badly understood. There is growing evidence that binary interaction in these evolved stars could be responsible for a large fraction of the observed asymmetric nebular morphologies (e.g. [1]).

To study late stellar evolution in binary systems, optically bright, less obscured post AGB stars are ideal candidates and in recent years it became clear that binarity is indeed very common. On basis of distinct characteristics of the SED, a more systematic search for binary post-AGB stars was launched. These binaries all show a broad IR-excess, pointing to the presence of both cool and hot dust, around the system. The original selection of 51 objects [2], has increased now to about 60 and we started an extensive multiwavelength study, including radial velocity monitoring, high spectral resolution

CP1001, 9th Torino Workshop on Evolution and Nucleosynthesis in AGB Stars and the 2nd Perugia Workshop on Nuclear Astrophysics, edited by R. Guandalini, S. Palmerini, and M. Busso

optical and infrared studies, submillimetre bolometric observations and high spatial resolution interferometric studies with the VLTI.

Our radial velocity monitoring campaigns are still ongoing but we can safely conclude that we indeed found the suspected high binary rate [3, 4]. The objects for which we do not have orbital parameters yet have pulsational amplitudes which are too large for easy detection of binary motion. The distribution of orbits and eccentricities (Fig. 1a) is surprising, since theory predicts efficient circularization by tidal forces on very short timescales [5]. The orbits are now not in contact, but they are too small to have accommodated an AGB star. The global picture that emerged is therefore that of a binary star evolving in a system which is too short to accommodate a full grown AGB star. During a badly understood phase of strong interaction, a circumbinary disc was formed, but the binary system did not suffer dramatic spiral-in. The dust must be circumbinary since all the determined orbits are within the sublimation radius of the dust.

In this contribution we focus on the infrared diagnostics. Infrared spectra are ideal tracers of the dust in these circumbinary discs. The chemistry can be traced by resonances of the dust species and the feature profiles can be used to constrain the chemico-physical conditions of the dust grains. The temperature distribution of the dust species throughout the disc is probed by the ratios of the strengths of the features of the same species at different wavelengths.

2. OBSERVATIONS

Spitzer-IRS high- and low-resolution spectra were obtained for 21 sample stars, ranging from 10 μm to 36 μm, with spectral resolutions of respectively $R \sim 600$ and $R \sim 100$. For some stars we obtained additional ground-based N-band infrared spectra with the Thermal Infrared Multi Mode Instrument 2 (TIMMI2), mounted on the 3.6 m telescope at the ESO La Silla Observatory. At submillimetre wavelength there are SCUBA 850 μm measurements available for a few stars only.

Orbital parameters are determined using radial velocity measurements obtained with the CORALIE spectrograph attached to the 1.2 m Swiss Euler Telescope. So far, orbital parameters for 16/21 sample stars are obtained (Fig. 1a). For the stars with small pulsational amplitudes, the binary rate is 100%.

3. MINERALOGY

3.1. General

In all objects the dust is oxygen rich and highly crystalline (Fig. 2), with strong emission features which we can identify as features of the Mg-rich end members of crystalline olivine and pyroxene, namely forsterite (Mg_2SiO_4) and enstatite ($MgSiO_3$). In none of the spectra there is evidence for a carbon-rich component. Two objects do show CO_2 gas emission lines. We find a ratio $^{12}C/^{13}C$ smaller than 10, meaning that ^{12}C is not enriched by the third dredge up during the previous AGB evolution.

In nearly all the spectra there appears to be a shift from the amorphous 18 μm feature towards the right, when comparing with synthetic spectra of standard amorphous olivine and pyroxene mass absorption coefficients. This points to the dominance of Mg-rich amorphous dust, also in the amorphous component. The lack of iron detected in the silicates is surprising since we expect the refractory iron to be present in the dust. Photospheric depletion in iron, which we detect in a sample of our stars [6], can be understood when the iron is indeed locked up in the circumstellar dust grains and not in the gas-phase [7] (see also Reyniers in these proceedings). If both the crystalline and amorphous silicates are devoid of iron, this could mean that iron is stored in metallic iron or iron-oxide [8].

One of the most remarkable features in the Spitzer spectra is the amazing diversity in spectral appearance (see Fig 2 and [9]). For the moment it is not clear if and how this diversity relates to specific structural elements of the disc, the star and/or the orbit or whether we witness directly an evolutionary change between different sources.

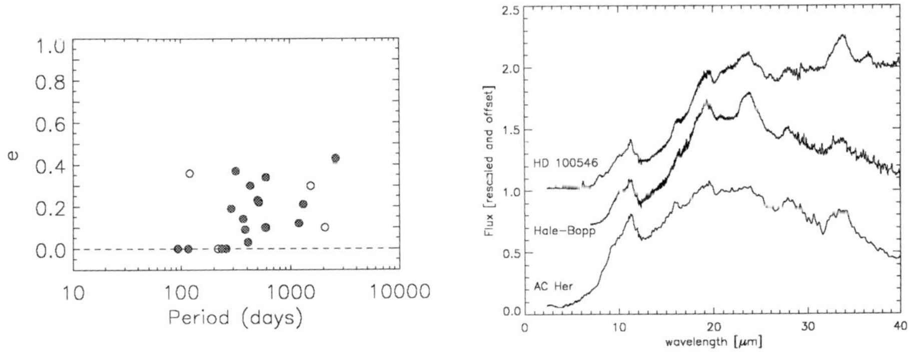

FIGURE 1. Left: The eccentricity-period diagram of the binaries determined on the basis of our radial velocity monitoring program. Empty symbols indicate an upper limit of the eccentricity. Right: Comparison between the binary post-AGB star AC Her, the young stellar object HD 100546 and the comet Hale-Bopp. The dust emission features in the 3 sources are very similar.

Despite the very different evolutionary history and timescales involved, the mineralogy of some of these binary post-AGB disc sources is extremely similar to what is found in some young stellar objects (YSO) such as HD100546 [10, 11] and solar-system comets such as Hale-Bopp (e.g. [12, 13])(see Fig 1b). In YSO as well as in comets, the dust processing is thought to be the tracer for the process of disc clearing and planet building in the proto-planetary disc. Similar processes may well be going on in the discs around the evolved objects.

3.2. Spectral Syntheses

We fit the observed dust emission features of our sample stars with synthetic spectra of amorphous and crystalline olivine and pyroxene [9]. We have access to a large sample of mass absorption coefficients of various dust shapes and sizes. Since different dust shapes result in different emission profiles, as a first step, we use the observed emission profiles to determine the best adopted dust opacity descriptions. The best fit is obtained when using Mg-rich amorphous silicates and small ($< 0.1\,\mu$m) and big ($2.0\,\mu$m) forsterite and enstatite particles in GRF (Gaussian Random Fields, [14]) approximation. As a next step we perform a full spectral fitting where we determine the best underlying dust continuum and different dust fractions (Fig. 2). We allow three dust temperatures to reproduce the observed emission features along the full spectral range. We find that both hot and cool amorphous and crystalline dust must be present.

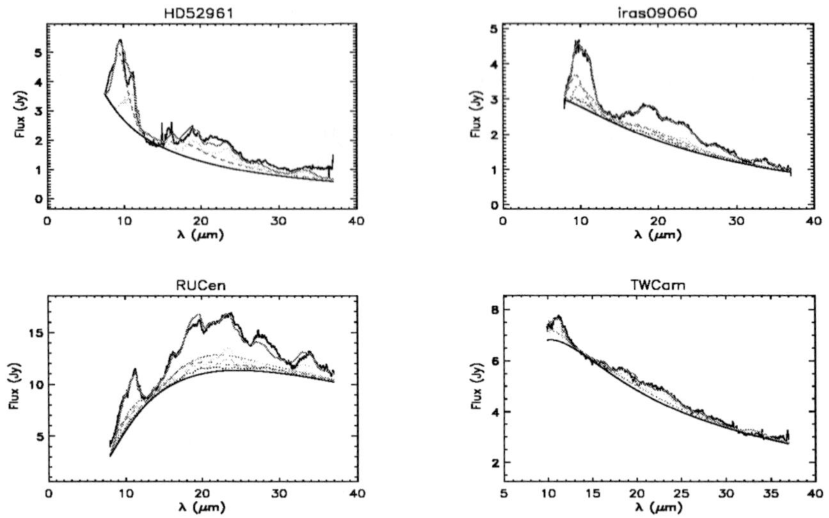

FIGURE 2. High-resolution Spitzer-IRS spectra, illustrating the rich mineralogy. Note the wide variety of spectral signatures but with a strong crystalline component in almost all objects. The dust emission features can be identified as due to enstatite and forsterite grains. Overplotted the result of our full spectral fitting which includes a continuum determination and allows different fractions of amorphous and crystalline silicate dust. Three different dust temperatures are allowed in order to reproduce the observed emission features in the full spectral range.

4. SED MODELLING

A study of the spectral energy distributions (SEDs) of our sample stars shows that the SEDs of all sample stars are very similar. Dust excess starts near sublimation temperature (≈ 1500 K), irrespective of the parameters of the central star [2]. Our first attempt to model the SEDs using a spherically symmetric model failed, since we could not model the SED within acceptable evolutionary timescales.

As a next step we performed a SED-fitting using a 2D radiative transfer code assuming a passive disc model [15]. The structure (scale height) of the disc is computed self-consistently using the gas pressure and assuming the structure is determined by hydrostatic equilibrium. A significant fraction of large grains ($850\,\mu$m) is needed to account for the high submillimetre flux. Since dust settling times for such large grains are short compared to disc lifetime scales, the picture that emerges is that of an inhomogeneous disc of small grains and a cooler midplane of large grains. The small grains will dominate the near- and mid-infrared part of the SED, while the large grains will be the main contributor to the far-infrared part of the SED.

When modelling the near- and mid-infrared part of the SED of one our sample stars, RU Cen, we find that the feature-to-continuum ratio of the silicate features is too strong in comparison with the ratio observed in the infrared spectra [16]. Including an extra opacity source, like large grains and/or metallic iron, is needed to reduce the strength of the features (Fig. 3).

For a large fraction of our stars, the inner rim of the disc is puffed up due to gas pressure which causes a significant fraction of the luminosity to be converted to infrared wavelengths. Our sample stars show a range in values for the energy ratio $L_{IR}/L_{\odot}$ ranging from 0.1 to 3. Some stars, like RU Cen and AC Her, show an infrared excess starting at longer wavelengths and they do not show this strong puffing up of the inner rim. Because of the strong flaring of the disc, they do however, still convert about 30% of light emitted by the central star to IR-wavelengths [16].

The SED modelling also gives possible values for the inclination of the system. With these values a minimal mass for the companion star can be estimated, using the mass function and a typical value for the primary of $M_1 = 0.6 - 1.0\,M_{\odot}$. For RU Cen, with a mass function of $f(M) = 0.83\,M_{\odot}$, this yields a minimal mass for the companion star of $M_2 = 1.8 - 2.2\,M_{\odot}$. This means that the companion is likely an unevolved main sequence star, but with a considerable mass.

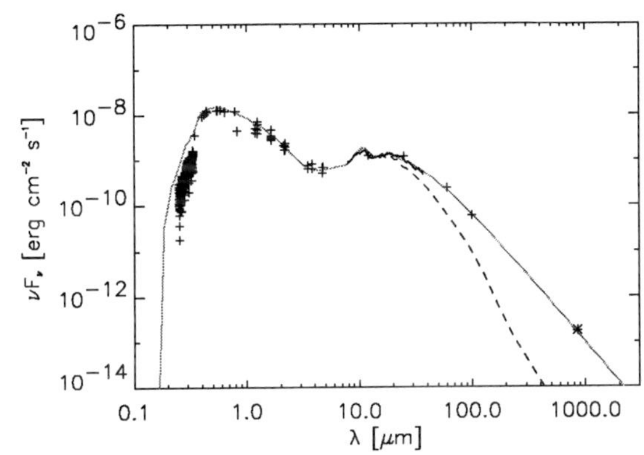

FIGURE 3. SED disc modelling of RU Cen. The dashed line represents the homogeneous disc model consisting of grains between 0.1 μm and 20 μm. The full line gives the disc model with an added blackbody to represent the cool midplane. Crosses represent photometric data and the observed Spitzer spectrum is also plotted.

5. CONCLUSIONS

Our spectral syntheses and SED modelling give conclusive evidence for the presence of stable Keplerian discs around all our sample stars. Although there is a wide variety in observed infrared spectra, all objects are oxygen-rich and show evidence of strong processing of the dust grains. Nearly all objects show a high degree of crystallinity, with both hot and cool dust present in the disc. The observed crystalline emission features are due to small irregular grains, while the submillimetre flux points to the presence of large grains. The presence of cool crystalline grains shows that either radial mixing from the hot inner boundary to far out in the disc must have occurred or that the formation process and thermal history of these grains is quite different in discs than in outflows.

The SED modelling is still a degenerate problem so interferometric data is necessary to constrain further the disc geometry [17]. The silicates dominated mineralogy and mass estimates for the companion show that these stars did not evolve on standard evolutionary tracks. The chemical evolution has been shortcut by binary interaction. It is likely that the formation of the circumstellar disc in these objects is closely related to the binary interaction.

The physical processes in the hydrostatic discs of these evolved binaries are very similar to the ones governing in protoplanetary discs around young stellar objects. We are in the process of expanding our study to a wider sample of evolved binaries, not only in the Galaxy but also in the LMC. This will enable us to build up a broader view of the chemico-physical dust processes of the grains around rapidly evolving stars.

ACKNOWLEDGMENTS

The authors want to acknowledge: the Geneva Observatory and its staff for the generous time allocation on the Swiss Euler telescope; the 1.2 m Mercator staff as well as the observers from the Instituut voor Sterrenkunde who contributed to the monitoring observations using both the Euler and Mercator telescopes. CG acknowledges support of the Fund for Scientific Research of Flanders (FWO) under the grant G.0178.02. and G.0470.07. We also thank Fred Lahuis for his assistance with the Spitzer data reduction.

REFERENCES

1. M. Moe and O. De Marco, *ApJ* **650**, 916 (2006), arXiv:astro-ph/0606354
2. S. De Ruyter, H. van Winckel, T. Maas, T. Lloyd Evans, L.B.F.M. Waters, and H. Dejonghe, *A&A* **448**, 641 (2006), astro-ph/0601578
3. H. van Winckel, *ARA&A* **41**, 391 (2003)
4. H. van Winckel, *Baltic Astronomy* **16**, 112 (2007), arXiv:astro-ph/0702245
5. O.R. Pols, A.I. Karakas, J.C. Lattanzio, and C.A. Tout, "Can Standard Evolution Models Explain the Properties of Barium Stars?," in *Astronomical Society of the Pacific Conference Series*, edited by R.L.M. Corradi, J. Mikolajewska, and T.J. Mahoney, 2003, vol. 303 of *Astronomical Society of the Pacific Conference Series*, p. 290
6. T. Maas, H. Van Winckel, and T. Lloyd Evans, *A&A* **429**, 297 (2005), astro-ph/0501469
7. L.B.F.M. Waters, N.R. Trams, and C. Waelkens, *A&A* **262**, L37 (1992)
8. U.J. Sofia, K.D. Gordon, G.C. Clayton, K. Misselt, M.J. Wolff, N.L.J. Cox, and P. Ehrenfreund, *ApJ* **636**, 753 (2006)
9. C. Gielen and H. van Winckel (in prep.) (2008)
10. K. Malfait, C. Waelkens, L.B.F.M. Waters, B. Vandenbussche, E. Huygen, and M.S. de Graauw, *A&A* **332**, L25 (1998)
11. C.M. Lisse, K.E. Kraemer, J.A. Nuth, A. Li, and D. Joswiak, *Icarus* **187**, 69 (2007)

12. J. Bouwman, A. de Koter, C. Dominik, and L.B.F.M. Waters, *A&A* **401**, 577 (2003), astro-ph/0301254
13. M. Min, J.W. Hovenier, A. de Koter, L.B.F.M. Waters, and C. Dominik, *Icarus* **179**, 158 (2005), astro-ph/0505603
14. M. Min, L.B.F.M. Waters, A. de Koter, J.W. Hovenier, L.P. Keller, and F. Markwick-Kemper, *A&A* **462**, 667 (2007), astro-ph/0611329
15. C.P. Dullemond and C. Dominik, *A&A* **417**, 159 (2004), astro-ph/0401495
16. C. Gielen, H. van Winckel, L.B.F.M. Waters, M. Min, and C. Dominik, *A&A* **475**, 629 (2007), arXiv:0709.3197
17. P. Deroo, A. Acke, T. Verhoelst, C. Dominik, E. Tatulli, and H. Van Winckel, *A&A* **468** (2007)

LIST OF PARTICIPANTS:

Abia Carlos
Departamento de Física Teórica y del Cosmos, Universidad de Granada,
Granada, Spain
cabia@ugr.es

Arnett David
Steward Observatory, University of Arizona,
Tucson, USA
darnett@as.arizona.edu

Battiston Roberto
INFN sez. di Perugia and Dipartimento di Fisica, Università di Perugia,
Perugia, Italy
r.battiston@tiscali.it

Bisterzo Sara
Dipartimento di Fisica Generale, Università di Torino,
Torino, Italy
bisterzo@ph.unito.it

Busso Maurizio
Dipartimento di Fisica, Università di Perugia,
Perugia, Italy
busso@fisica.unipg.it

Cescutti Gabriele
Dipartimento di Astronomia, Università di Trieste,
Trieste, Italy
cescutti@oats.inaf.it

Chieffi Alessandro
INAF-IASF Roma
Roma, Italy
Alessandro.Chieffi@iasf-roma.inaf.it

Ciprini Stefano
Dipartimento di Fisica, Università di Perugia,
Perugia, Italy
stefano.ciprini@pg.infn.it

Collon Philippe A.
University of Notre Dame,
Notre Dame, USA
pcollon@nd.edu

Cristallo Sergio
INAF Osservatorio Astronomico di Collurania,
Teramo, Italy
cristallo@oa-teramo.inaf.it

D'Antona Francesca
INAF Osservatorio Astronomico di Roma,
Roma, Italy
dantona@oa-roma.inaf.it

Davis Andrew
University of Chicago,
Chicago, USA
a-davis@uchicago.edu

Domìnguez Inma
Departamento de Física Teórica y del Cosmos, Universidad de Granada,
Granada, Spain
inma@ugr.es

Gallino Roberto
Dipartimento di Fisica Generale, Università di Torino,
Torino, Italy
gallino@ph.unito.it

Gielen Clio
Instituut voor Sterrenkunde, KU Leuven,
Leuven, Belgium
clio.gielen@ster.kuleuven.be

Gil-Pons Pilar
Universitat Politecnica de Catalunya,
Barcelona, Spain
pilar@fa.upc.edu

Goerres Joachim
University of Notre Dame,
Notre Dame, USA
jgoerres@nd.edu

Guandalini Roald
Dipartimento di Fisica, Università di Perugia,
Perugia, Italy
guandalini@fisica.unipg.it

Gutierrez Jordi
Universitat Politecnica de Catalunya,
Barcelona, Spain
jordi@fa.upc.edu

Hammache Fairouz
IPN Orsay,
Orsay, France
hammache@ipno.in2p3.fr

Hoppe Peter
Universität Mainz and Max Planck Institut für Chemie,
Mainz, Germany
hoppe@mpch-mainz.mpg.de

Hron Josef
Department of Astronomy, University of Vienna,
Vienna, Austria
hron@astro.univie.ac.at

Husti Laura
Dipartimento di Fisica Generale, Università di Torino,
Torino, Italy
husti@ph.unito.it

Imbriani Gianluca
Università' Federico II di Napoli and INFN sezione di Napoli
Napoli, Italy
gianluca.imbriani@na.infn.it

Ivans Inese I.
Princeton University Observatory, Princeton,
Princeton NJ, USA
iii@ociw.edu

Izzard Robert
University of Utrecht,
Utrecht, Netherlands
r.g.izzard@phys.uu.nl

Johnson Torrence
Jet Propulsion Laboratory, California Institute of Technology
Pasadena, USA
torrence.v.johnson@jpl.nasa.gov

Käppeler Franz
Forschungs Zentrum Karlsruhe, Institut für Kernphysik,
Karlsruhe, Germany
franz.kaeppeler@ik.fzk.de

Kratz Karl Ludwig
Max Planck Institut für Chemie, Otto Hahn Institut,
Mainz, Germany
kl.kratz@uni-mainz.de

La Cognata Marco
Università di Catania and INFN - Laboratori Nazionali del Sud,
Catania, Italy
lacognata@lns.infn.it

Lamia Livio
INFN - Laboratori Nazionali del Sud,
Catania, Italy
llamia@lns.infn.it

Lattanzio John
Centre for Stellar and Planetary Astrophysics, School of Mathematical Sciences,
Monash University, Australia
John.Lattanzio@sci.monash.edu.au

Lau Herbert
Institute of Astronomy,
Cambridge, UK
hbl21@cam.ac.uk

Lebzelter Thomas
Department of Astronomy, University of Vienna,
Vienna, Austria
lebzelter@astro.univie.ac.at

Lederer Micheal T.
Department of Astronomy, University of Vienna,
Vienna, Austria
lederer@astro.univie.ac.at

Lugaro Maria
Sterrenkundig Instituut, University of Utrecht,
Utrecht , Netherlands
M.Lugaro@phys.uu.nl

Mantovani Giancarlo
Dipartimento di Fisica, Università di Perugia,
Perugia, Italy
mantovani@fisica.unipg.it

Marengo Massimo
Harvard-Smithsonian, Center for Astrophysics,
Cambridge, USA
mmarengo@cfa.harvard.edu

Marigo Paola
Dipartimento di Astronomia, Università di Padova,
Padova, Italy
paola.marigo@unipd.it

Marrone Stefano
INFN Sezione di Bari and Università di Bari,
Bari, Italy
Stefano.Marrone@ba.infn.it

Nordhaus Jason
Dept. of Physics and Astronomy, University of Rochester,
Rochester, USA
nordhaus@pas.rochester.edu

Palmerini Sara
Dipartimento di Fisica, Università di Perugia,
Perugia, Italy
palmerini@fisica.unipg.it

Pignatari Marco
Astrophysics Group,
Keele University, UK
pignatar@to.infn.it

Primas Francesca
European Southern Observatory,
Garching, Germany
fprimas@eso.org

Pumo Maria Letizia
Dip.S.F.A. Università di Palermo and INAF Osservatorio Astronomico di Roma, Roma, Italy
mlpumo@astrct.oact.inaf.it

Reyniers Maarten
Instituut voor Sterrenkunde, K.U.Leuven,
Leuven, Belgium
maarten@ster.kuleuven.be

Roederer Ian U.
University of Texas,
Austin, USA
iur@astro.as.utexas.edu

Straniero Oscar
INAF Osservatorio Astronomico di Collurania,
Teramo, Italy
straniero@oa-teramo.inaf.it

Tosti Gino
Dipartimento di Fisica, Università di Perugia,
Perugia, Italy
gino.tosti@pg.infn.it

Tout Christopher A.
Institute of Astronomy,
Cambridge,UK
cat@ast.cam.ac.uk

Uttenthaler Stefan
Department of Astronomy, University of Vienna,
Vienna, Austria
suttenth@eso.org

van Aarle Els
Instituut voor Sterrenkunde, KU Leuven,
Leuven, Belgium
Els.vanAarle@student.kuleuven.be

van Loon Jacco Th.
Astrophysics Group, Lennard-Jones Laboratories,
Keele University, Staffordshire, UK
jacco@astro.keele.ac.uk

van Raai Mark
Sterrekundig Instituut, University of Utrecht,
Utrecht, Netherlands
m.a.vanraai@phys.uu.nl

Van Winckel Hans
Instituut voor Sterrenkunde, KU Leuven,
Leuven, Belgium
Hans.VanWinckel@ster.kuleuven.be

Ventura Paolo
INAF Osservatorio Astronomico di Roma,
Roma, Italy
ventura@oa-roma.inaf.it

Wasserburg Gerald J.
Divisionof Geological and Planetary Sciences, California Institute of Technology,
Pasadena, USA
gjw@gps.caltech.edu

Wood Peter
Research School for Astronomy and Astrophysics, Australian National University,
Canberra, Australia
wood@mso.anu.edu.au

Worley Clare
Department of Physics and Astronomy, University of Canterbury,
Christchurch, New Zealand
ccw29@student.canterbury.ac.nz

Wylie Liz
RSAA, Mt Stromlo Observatory, Australian National University,
Canberra, Australia
ewylie@mso.anu.edu.au

Yong David
RSAA, Mt Stromlo Observatory, Australian National University,
Canberra, Australia
yong@mso.anu.edu.au

Zinner Ernst
Washington University,
St.Louis, USA
ekz@wustl.edu

HOTEL

HOTEL

Author Index

G

H

I

J

K

L

M